WITHDRAWN

Introduction to Bacteria
and Their Ecobiology

Introduction to Bacteria and Their Ecobiology

R. N. Doetsch and T. M. Cook

University Park Press

Baltimore · London · Tokyo

UNIVERSITY PARK PRESS
International Publishers in Science and Medicine
Chamber of Commerce Building
Baltimore, Maryland 21202

Copyright © 1973
By University Park Press

Printed in the United States of America
Second printing - July, 1974

Library of Congress Cataloging in Publication Data

Doetsch, Raymond Nicholas, 1920–
  Introduction to bacteria and their ecobiology.

  1. Bacteria. I. Cook, Thomas Melbourne, 1931–
joint author. II. Title. [DNLM: 1. Bacteria
Ecology. QW 4 D653i 1972}
QR41.2.D63   589.9   72-4523
ISBN 0-8391-0736-6

Contents

Preface

The purpose of this book is to illustrate a selection of biological properties of bacteria that reveal them as important living beings. We have primarily addressed readers who have had some previous education in the natural sciences, and we have assumed a modest understanding of elementary chemical and biological principles. Our aim is to provide a brief survey of bacterial forms and structures, placing special emphasis on the activities of bacteria in their environment and some important interrelations within it. *Bacterial ecobiology* is the study of those aspects of bacteria that influence, and are influenced by, environmental phenomena.

Some material traditionally covered in standard texts—such as medical bacteriology and immunology, applied bacteriology, and bacterial classification—will not be found here, because it is our opinion that these are peripheral to the idea of ecobiology and because numerous excellent treatments of this material are readily available. There is also no formal presentation of bacterial genetics or of molecular biology *per se* in this book. However, mention of phenomena involved in these subjects is made where considered appropriate.

Bacteria are especially worthy of study because they are ubiquitous and they have extensive powers of adaptation. They also show a remarkably wide range of form and function, and have a profound influence on the economy of nature. We emphasize the importance of understanding bacteria as whole, functioning organisms in their natural habitats, because such an appreciation will be critical in the efforts to meet the challenge of environmental crisis during the next decade. We make no claim of originality for this concept; we simply reassert a fundamental biological point of view, which derives from the great traditions developed by such masters as Beijerinck, Winogradsky, Kluyver, and van Niel, but which is too little in evidence among contemporary works on bacteriology. While there are books treating bacteria as organisms "in their own right," there is now justification for emphasizing this aspect of bacteria in modern terms.

The bacterial world is vast, both quantitatively and qualitatively, but there is a tendency to equate the properties of all bacteria to the "white rat" of bacteriology, *Escherichia coli.* We will try to dispel this myth and show that the bacterial cosmos is rich in diversity, and that familiarity with the broad range of bacterial life is essential to unraveling the complexities of ecological processes.

This book presents some aspects of the natural history of bacteria, but can by no means be considered a definitive treatise or an exhaustive compendium. There has been no attempt at inclusiveness in our presentation. Rather, we have selected features of bacteria which illustrate characteristic modes of existence, and have discussed how these features account for the persistence of bacteria in a given environment. We have not encumbered the book with an extensive documentation of references; however, we have listed recent representative treatises, reviews, and important papers at the end of each chapter. We have relied, of course, on this material for most of the factual statements presented in the text. The book is intended to serve, in part, as a commentary on, and a critique of, certain aspects of bacteriological science, rather than merely as a descriptive exposition. We believe that only by reference to the antecedents of current laboratory techniques for the study of bacteria, and the assumptions implicit in these techniques, can one hope to apply them intelligently and to make improvements on them.

Although we cannot note individually the names of the many colleagues who generously contributed materially toward the completion of this work, it is a pleasure to acknowledge our indebtedness, and we hereby express our appreciation to all who have helped us. We especially acknowledge the assistance of H-D. Babenzein, G. J. Hageage, R. L. Harold, T. Y. Kingma- Boltjes, N. Pfennig, M. J. Thornley, Z. Vaituzis, and R. B. Wildman. Our thanks go to Rita Colwell and A. L. Demain for reading and criticizing portions of the manuscript, and to our editors for invaluable assistance in the final preparation of the book. We alone take responsibility for such errors of fact or interpretation as may remain. Also, we thank Mrs. Betty Swope and Mrs. Arlene Cook for their indefatigable efforts in typing the manuscript through its various versions.

College Park, Maryland

R. N. Doetsch
T. M. Cook

one | Bacteria in Nature

The relationships of individuals and populations to their environments are of both theoretical and practical importance. It is through the investigation of these relationships that normal ("natural") modes of existence may be apprehended. It is unfortunate that the ecological framework for bacteriology is of primitive construction, and that there exists no critical body of observations from which one can make general and unifying statements.

One might hope that a natural history of bacteria would cover details similar to those in treatises on reptiles or mammals. Thus one would hope to learn how the structures of bacteria and their functions enable them to survive and proliferate in various environments. Knowledge of the natural relationships among bacteria would provide man with additional means of coping with current environmental disruptions. However, the life of bacteria as maintained in the laboratory is not equivalent to the life of bacteria in nature. If one hopes to be able to direct bacterial processes, one must know what these processes are in nature. The problem is that the means of obtaining this information are not available.

The natural history of bacteria includes a core of phenomena, among which are: locomotion and locomotor behavior; developmental processes; modes of reproduction; aging and death; dormancy; habitats and their influence on form and function; interactions with other organisms, plants, and animals; and effects of bacterial activities on the environment. Detailed knowledge about many of these phenomena is scant in comparison with what is known about higher organisms, and fresh and innovative approaches are needed.

Many of the above-mentioned functions have been studied and described as characteristics of whole organisms and whole populations, and they cannot be investigated on the subcellular level of organization. Inevitably the technical methods used in dealing with subjects as intractable as individual living bacteria are crude when contrasted with the sophisticated biophysical and biochemical approaches used in dealing with molecules derived from fragmented, dead organisms. The "complexity," "inhomogeneity," and "individuality" of living organisms, to use Elsasser's terms, stand in sharp contrast to the overwhelming uniformity of the atoms or molecules studied in physics and chemistry. An analysis of the function of an organism and a *complete* account of its molecular basis are limited, because an intact organism retains a degree of autonomy while these phenomena are being considered. Natural history studies are carried out at the highest level of organismal complexity and deal with phenomena that are expressions of organizational hierarchies.

The use of the words "natural history" perhaps requires an explanation. Today there is a tendency to consider the term antiquarian and definitely out of date, but we hold that there is a pressing need for biologists, "molecular" as well as "nonmolecular," to become familiar with a *whole* organism and its natural relationships with its living and nonliving environment. The connection between these aspects of bacteria and specific properties of bacteria poses a fundamental biological problem which awaits investigation by imaginative minds. We view bacteria as organisms worthy of study for their intrinsic merits, rather than because they are "model cells," "bags of enzymes," "convenient genetic material," or "commercially exploitable reagents." Thus, by the term "natural history" we imply that, although the examination of bacteria as subjects in experiments leading to interpretation of some of their properties in molecular terms is a fruitful and complementary scientific pursuit, there exist many facets of bacterial activities *not* amenable to this approach. The study of bacteria at nonmolecular levels, particularly at the level of a functioning organism, and in relation to its environment, that is, its "natural history" or ecobiology, is a field that is only now coming into full view for the bacteriologist. This gives rise to the recognition of the unique historical development of bacteriology, in which much more information has been accumulated about bacterial activities in chemical and medical terms than about fundamental biological properties and relationships of the bacteria themselves.

1.1 | The Origin of Bacteria

Nothing is known about the origin of bacteria, but because of their relative structural simplicity it seems likely that they preceded the eukaryotic forms (organisms with a "true" nucleus) in the course of evolution. This notion is strengthened by the knowledge of the amino acid sequences of the respiratory enzyme, cytochrome c, which indicates that bacteria must have branched off the "evolutionary tree" prior to fungi and other eukaryotes.* Bacteria are, therefore, representatives of an ancient and obviously successful form of life. Their persistence and almost universal distribution are indicative of great adaptability, but in no way prove that modern bacteria are the proto-organisms from which others developed. We have no way of knowing to what extent bacteria themselves have evolved from some ancestral life form.

Reconstruction of the evolutionary development of a particular species rests on evidence from several sources: comparative anatomy, paleontology, and embryology. Bacteria are too simply organized for elaborate anatomical analysis; consequently they are usually characterized in physiological and chemical terms. Most bacteria lack structural components suitable for fossilization, and, in fact, their fossil record is extremely scanty. Finally, bacteria do not show any process akin to embryological development, which would permit recognition of phylogenetic relationships and evolutionary origins.

---

*The amino acid sequences of cytochrome c from various organisms are generally similar. Specific individual differences can be interpreted as reflections of discrete amino acid replacements during the course of evolution. Analysis permits reconstruction of the relative order of such events.

Fossils are remains, traces, or impressions of living organisms from past geological ages. It is questionable whether bacterial fossils exist. It would appear that bacteria, lacking bony or calcareous structures, would not be likely to leave fossil remains, but results of recent investigations suggest that rocks as old as 3×10^9 years (Precambrian) contain fossils of filamentous forms, some of which might be bacteria. Chert from the Gunflint Formation in Ontario, Canada, the age of which is estimated at 2×10^9 years, also shows structures which can be interpreted as fossil bacteria.

The more recent Green River Formation of western Colorado, eastern Utah, and southern Wyoming is estimated to be between 5 and 6×10^7 years old, and prokaryotic-type microfossils have been described in carbonate oil-shale specimens from this source. Samples were treated with hydrofluoric acid and nitric acid and softened with phenol; they were then washed in water, dehydrated, and embedded in paraffin, then sectioned with a microtome. The treated material showed two kinds of fossils, one consisting of filaments 1–5 μm in diameter, and the other of rod-like organisms 1 μm in diameter and 20 μm in length.

If bacteria are, indeed, the most primitive examples of life existing on earth, the time of their first appearance is very important. It is possible, of course, that forms structurally simpler than filaments have left absolutely no trace of their existence, and also that some putative fossils may be artifacts. The fossil record is fragmentary at best, and such microfossils as have been observed cannot be adequately evaluated or compared with modern bacteria because the available physiological (functional) tests can be applied only to the living organism. The relative antiquity of bacterial life suggested by such fossils is also supported by evidence of another sort. Biological activity associated with modern bacteria, namely dissimilatory sulfate reduction, was taking place as long as 2–3×10^9 years ago (Section 3.17 and Section 6.11).

1.2 | The Uncertainty of Knowledge Concerning Bacteria

The development of knowledge about the biology of bacteria at the organismic level is hampered by their small size. Consequently, the measurements of intracellular pH, membrane potentials, redox potentials, and respiratory rates of individuals are beset by technical difficulties. It is impossible to state when a single bacterium is dead because the definition of bacterial death is not precise. Physical manipulations of individual bacteria are tedious at best, and it is not feasible to perform microsurgical operations on them of the sort used for enucleation of *Amoeba* or *Acetabularia*. A single bacterium may be isolated by the use of a micromanipulator. However, further experimentation is limited to observation of sporulation, germination, locomotion, multiplication, etc. The organism cannot easily be followed in a natural population unless it is morphologically unique, or unless it occurs in a habitat that limits the number of similar organisms. Methods employing Cartesian divers or micro-electrodes, which have been devised for single protozoan or mammalian cells, are still too gross or insensitive for an individual bacterium. Radioautographic techniques have been employed to follow metabolic reactions; for example, the incorporation of tritium-labeled thymidine during deoxyribonucleic acid synthesis. Although radioautography is useful in some cases, the process kills the bacterium and hence precludes a series of measurements on one living individual.

Knowledge of bacteria is derived mainly from studies of populations of presumably similar individuals. Although some investigations of structure can be carried out on an individual bacterium, elucidation of function requires a large number of organisms. This is not always evident in the literature because bacteriologists frequently express themselves as if they had discovered characteristics from the examination of individuals when, in fact, they have inferred them from populations. One may find a statement, for example, specifying the average weight of a bacterium. How does this differ from the specification of the weight of an average white rat? In the latter case measurements would be made on a large number of individuals, and the average (mean) would be calculated, with an indication of the extent of variation from the mean value. The small size of bacteria precludes individual weighings and, therefore, their average weight is determined indirectly. First, the number of bacteria present in a population is estimated by one of several methods, and the weight of the total population is ascertained. Assuming that bacteria are of uniform volume, one divides the weight by the estimated number of bacteria, and concludes that this is the weight of an "average" cell. One assumes that the weighing process could be done with a high degree of accuracy, but since an average organism weighs many times less than can be measured by the most precise modern balance, there is considerable error caused by very minute deviations from the actual weight. An average value may be useful for some purposes, but this should not obscure the fact that the actual weight of any particular bacterium of the population is unknown. Indeed, this may be cast in the form of a "Bacterial Uncertainty Principle"; although it is possible to determine the general properties of a bacterial population with precision, it is not often possible to quantitatively specify individual characteristics.

The tacit assumption of uniformity is difficult to maintain. In the weighing procedure described, one cannot obtain an indication of individual variability, but it is just such variability that is inherent in living beings. It is a biological principle of great generality that each organism is a historically unique individual, unlike any other. Atoms and molecules are ahistorical. If one argues that they have a history, it can still be ignored in most physical studies. The concept of individuality is derived from studies of more complexly organized multicellular organisms in which there is a genetic inheritance from both parents, a result of sexual reproduction. Greater uniformity might be expected in bacteria since they generally multiply by asexual processes and are usually haploid in chromosome number. In the absence of mutations all progeny arising from a single bacterium by repeated divisions ought to have the same genetic endowment. However, a simple microscopic examination of the progeny of a given bacterium reveals obvious variations in gross dimensions of individuals, and this probably applies to their internal mechanisms as well.

1.3 │ Some Methods for Studying Bacteria in Natural Habitats

In the further development of precise methods for studying bacteria in their natural habitats, current techniques will probably have to be modified and improved. There exist some methods of unrecognized merit which, although of long standing, have been used only occasionally. The following brief descriptions are illustrative, but not all-inclusive, of useful methods for such investigations:

1. The *immersion technique* consists of burying or immersing clean, sterile glass microscope slides in soil or water for varying periods of time, after which they are carefully removed, fixed, stained, and

examined under a microscope. Bacteria capable of attaching themselves to glass surfaces can be seen and their structural features can be delineated. Many of the kinds of bacteria that can be observed by this method are not amenable to cultivation on laboratory-concocted media. Unfortunately, glass microscope slides are not always easy to use without disturbing the natural environment, and it is extremely difficult to isolate from such surfaces the interesting or unusual organisms observed. Successful use of flat glass microcapillaries has been reported. They are inserted into the environments under study and colonization by various bacteria may take place directly in them. They may be removed periodically and their contents examined with a microscope, and, if desired, inoculated into culture media.

2. *Radioautographic methods* may be used to determine which substrates are utilized by various morphologically distinguishable organisms in natural environments. Radioactively labeled compounds are added to samples containing the microflora under investigation. These are incubated in containers placed directly in the environment from which they were removed, such as the sea, or the rumen. Portions of the mixtures are removed periodically, fixed, and developed using microscopic-radioautographic techniques.

3. It is often desirable to *estimate the biochemical activity* of normal flora under natural conditions. This may be accomplished by adding a radioactively labeled substrate to a sample taken from the ecosystem under study and by measuring the rate of its disappearance (turnover). The distribution and specific activity in the organisms and metabolic end products may be determined. In soil perfusion methods, for example, the conversion rate is determined by continuously recirculating a solution of the compound over the soil sample. Effects of imposed physicochemical conditions on such rates may be observed. This approach also can be used with nonradioactive substances. The classical work of Schloesing and Müntz (1877) on soil nitrification was performed using a simple version of this technique.

4. *Measurement of rates of oxygen uptake* may be made by using an oxygen electrode or manometric apparatus. Known amounts of substrate are added directly to samples of a given environment, such as river water, or to a mixed flora of organisms separated from such environments by differential centrifugation. Respiratory rates and the chemical identity of end products may be determined using this technique, and hence one obtains information on the overall biochemical activities of the interacting microflora.

Radiorespirometry provides another way of assessing biodegradative activities of the microflora of natural habitats. This procedure is based on the measurement of radioactive carbon dioxide evolved in the oxidation of a ^{14}C-labeled substrate. The great sensitivity of measuring devices for radioactivity permits use of small amounts of material, which is important in studies with potentially inhibitory compounds such as pesticides, phenolic compounds, and hydrocarbons. The radioactive compound is mixed with the sample of soil or water to be evaluated and placed in a closed vessel with a sidearm containing alkali to absorb evolved carbon dioxide. Breakdown of the substance into $^{14}CO_2$ by bacterial respiratory processes is followed by periodic sampling of the alkali for radioactivity.

5. If pure cultures of bacteria, capable of reisolation because of *distinctive genetic markers,* are added to natural ecosystems, their fate after various periods of time may be determined, and their overall effect on the activities of the indigenous microflora may be evaluated.

All of these methods have inherent limitations, and their underlying assumptions must be kept clearly in mind when applying them to specific situations and when assessing the results obtained from their use.

1.4 | Functions of Bacteria in Nature

Bacteria are particularly significant to, and occupy a key position in, the economy of nature. They are the great scavengers of the biotic world. As such they contribute to, or are responsible for, many natural phenomena. One may note especially their roles as decomposers and rearrangers of materials. They are essential participants in the biological cycles of the elements, and they are the effectors of large-scale chemical changes which result in alterations of the environment.

The degradative activities of bacteria are very important to transformations of organic materials. In both soil and water they decompose a wide variety of substances, many of which can not be broken down by any other organisms. A single bacterial species may be capable of metabolizing as many as a hundred or more different organic compounds. While there are limits to the abilities of any given kind, as a whole bacteria are able to degrade nearly all naturally occurring and many synthetic compounds, including formaldehyde, phenol, ethanol, benzene, naphthalene, kerosene, paraffin, methane, testosterone, nicotine, chitin, rubber, and various herbicides and insecticides.

Chemical changes brought about by bacteria are the result of their life processes. These changes are not simply chance occurrences, and, in general, they are connected with either bacterial energy metabolism or biosynthesis. It may be assumed that there is a reason for every activity, because nature is too parsimonious to allow for random or purposeless functions. Natural selection tends to conserve only those processes which contribute to the organism in a positive way, but this does not mean that every chemical reaction observed is actually of physiological significance. Some may be simply epiphenomena, in which the reaction observed is not the one which serves a function. The specificity of many enzymes is sufficiently broad to accommodate nonphysiological substrates; for example, it is probable that the reduction of nitro groups ($-NO_2$) to the corresponding amines ($-NH_2$) serves a vital function in an organism, even though it is doubted that the conversion of nitrobenzene to aniline by certain bacteria is physiologically significant.

Individual reactions and reaction sequences are small segments or units of larger cycles in nature and ought not be viewed in isolation. The chemical activities of bacteria make substantial and necessary contributions to the cycles of energy and of the elements on which continuance of life on earth depends. In regard to the elements, the earth is essentially a closed system with a finite amount of material. Biologically important elements, such as carbon, nitrogen, phosphorous, and sulfur, are converted from one chemical compound to another by the activities of living organisms, and the overall result of these changes is the recycling of elements through the biosphere. From an anthropocentric point of view some bacterial activities may appear beneficial while others seem deleterious. The formation of humus in the soil from plant and animal material would seem desirable, whereas the loss of nitrogen from the soil by the actions of denitrifying bacteria would appear undesirable. In the larger sense all of these processes are essential to the continued operation of the cycles of matter and, therefore, are desirable. To complete the cycles it is essential that material should not accumulate at any one stage, and it is in this regard that the dissimilation by bacteria assumes greatest significance. These essential processes of decomposition proceed rapidly and relentlessly, largely unperceived and unappreciated by man. Only occasionally, perhaps when a pesticide or pollutant proves recalcitrant, is the crucial nature of the biodegradation of materials brought to our attention. Often these bacterial reactions take place on a large scale and produce massive biogeochemical changes. Among such phenomena are the formation, deposition, solubilization, and removal of certain minerals such as chalk, elemental sulfur, and metallic sulfide ores. Although their exact role remains uncertain, bacteria may be involved in the formation of deposits of petroleum, coal, and peat.

The components of an environment, both chemical and biological, are interdependent and in a state of dynamic equilibrium. To disturb one component is to disturb them all. One of the important roles of bacteria in nature is the maintenance of a steady state. Bacterial activities influence the physical environment and the biological community, and are in turn influenced by both. The steady state which results from these opposing effects may appear more stable than it really is. Bacteria, while versatile, have limits to their capabilities and to the rates at which they can transform materials. If nondegradable substances are introduced into a system, or if degradable materials are added at too rapid a rate, the equilibrium may be upset.

One of the important consequences of bacterial activities is exertion of considerable influence on environments by participation in, and alterations of, food chains. In respect to alterations in environments and food chains, chemical activities of bacteria may be viewed in several ways. One important action of bacteria is to solubilize insoluble biopolymers such as certain proteins, cellulose, and chitin. Related to this, and at times a direct result of such solubilizations, is the conversion by bacteria of "nonusable" substances to more easily metabolized compounds. An example of this is the conversion of petroleum hydrocarbons to bacterial proteins.

As a result of their metabolism, bacteria may produce toxic materials (acids, alcohols or other solvents, antibiotics) which inhibit the growth of other organisms, or they may produce stimulatory factors (vitamins, amino acids, nucleotides) which promote the growth of other organisms. One simple consequence of bacterial activities is the increase in the competition for nutrients. Because of their rapid rate of growth and metabolism, bacteria may withdraw sufficient amounts of available nutrients from an environment to inhibit development of other organisms. This can be observed when large amounts of straw or sawdust are added to the soil. The rapid development of cellulose-degrading bacteria causes a depletion of available nitrogen, and as a result, plant growth is temporarily retarded. Another bacterial process which has important effects on the availability of nutrients is the removal of organic material from the environment by converting it to carbon dioxide and other inorganic compounds. The general term for this is *mineralization.* In respect to organic compounds this process usually requires oxygen, and depends on the oxidation of such compounds to carbon dioxide via aerobic respiration. Mineralization is one of the primary objects in sewage treatment processes, since it is desired to remove organic matter which would permit growth of undesirable organisms in rivers, thereby depleting the oxygen supply for fish and aquatic life.

The effects of bacterial activities are by no means confined to chemical changes in the nonliving environment. Bacteria may also produce alterations in the biotic community of which they are a part. In this regard, one may note interactions among the bacteria themselves, between bacteria and other microbes, and between bacteria and multicellular organisms.

Bacteria influence each other not only in all the ways already mentioned such as by competition for available nutrients, production of growth factors and inhibitory metabolites, and mobilization of nutrients, but also by the direct attack of one organism on another. Some bacteria, notably the myxobacteria, produce exocellular enzymes which dissolve or lyse other bacteria. Another way in which bacteria interact with other indigenous populations is by serving directly as nutrients for predatory protozoans. It is known that some protozoa ingest bacteria as a source of food; cells of a single bacterial species have been shown to provide all the required nutrients for the growth of the intestinal parasite *Entamoeba histolytica* in artifical culture. Bacteria serve as prey in nature, and it has been observed that in the soil an inverse relationship exists between the numbers of protozoans present and the bacterial population.

Intimate associations between bacteria and more complex organisms often involve residence of the bacteria within the larger organism, in some cases actually inside the tissue cells. These interactions may take the form of ecto- and endosymbioses and may have deleterious, neutral, or beneficial results.

Returning now to an examination of life processes subsumed under our view of what constitutes the natural history of bacteria, what does one find? The underlying mechanisms of flagellar and gliding motility are unknown, although significant advances have been made, particularly with regard to the former. If bacterial locomotor activity is a process of regulation at the cellular level, motility is possibly an expression of "behavior" aimed at maintaining an organism in a physiologically optimal area. The process by which the energy of environmental signals is transduced into flagellar motion is an intriguing biological problem, and the so-called tactic responses of bacteria are excellent phenomena for studying "bacterial ethology."

Bacteria are osmotrophic (taking food in soluble form); none is known that is phagotrophic (taking food in solid form). There are, however, interesting problems involving materials utilized by certain bacteria. In the case of organisms capable of utilizing insoluble substances, sulfur or hydrocarbons, for example, the question arises as to how the bacteria "know" that these are "out there." Even if bacteria fortuitously adhere to insoluble solid particles or dispersed liquid droplets, what mechanism allows them to degrade the particles to a size and in a form in which they can pass through the cell wall and cytoplasmic membrane? Bacterial populations characteristic of environments in which insoluble polymers are degraded consist of one group of bacteria that is able to carry out the initial breakdown, and others that proliferate on polymer fragments. Studies of interactions and population dynamics among such groups should prove most interesting.

Some bacteria develop resistant forms which are able to remain dormant for long periods in hostile environments. The biological significance of bacterial endospores is still being debated, even though research has been vigorously carried out on every conceivable aspect of them. The concept of dormancy has a broad biological base. The natural history of bacteria involves endospores in terms of protective or timing devices, as parts of a life cycle, or as dissemination mechanisms. There has been discussion of the possibility of the transfer of endospores through outer space but so far there is no evidence for this.

Eubacteria multiply by transverse fission; noneubacterial forms may reproduce in various ways. Among these are unequal division by budding, division by fragmentation, and, especially in aquatic forms, the development of a motile "swarmer" from a sessile form. The microcysts (myxospores) of the myxobacteria and the conidiospores of actinomycetes are disseminative as well as reproductive devices. The processes involved are excellent models for study by developmental biologists. Interactions between the animate and inanimate elements of a microenvironment that trigger production of these forms are unknown in most cases.

We return to the theme of this book, namely, that the ecobiology of bacteria includes their activities and the interrelationships between them and their environment, and that it must investigate morphological, physiological, and ecological differences found between different groups, and account for these differences in functional and structural terms. The subsequent chapters will discuss these topics in greater detail.

References

Alexander, M., *Microbial Ecology*, Wiley, New York, 1971.

Bott, T. L., and T. D. Brock, Growth and Metabolism of Periphytic Bacteria, *Limnol. Oceanogr.* **15,** 333 (1970).

Brock, T. D., *Principles of Microbial Ecology,* Prentice-Hall, Englewood Cliffs, N.J., 1966.

Brock, T. D., Microbial Growth Rates in Nature, *Bacteriol. Rev.* **35,** 39 (1971).

Casida, L. E., Observation of Microorganisms in Soil and Other Natural Habitats, *Appl. Microbiol.* **18,** 1065 (1969).

Cleaver, J. E., *Thymidine Metabolism and Cell Kinetics,* North Holland Publishing Co., Amsterdam, 1967.

Collins, V. G., and C. Kipling, The Enumeration of Waterborne Bacteria by a New Direct Count, *J. Appl. Bacteriol.* **20,** 257 (1957).

Elsasser, W. M., Atoms and Organism Theory, in S. Devons (ed.), *Biology and the Physical Sciences,* Columbia University Press, New York, 1969.

El-Shazly, K., and R. E. Hungate, Fermentation Capacity as a Measure of New Growth of Rumen Microorganisms, *Appl. Microbiol.* **13,** 62 (1965).

Maynell, G. G., Use of Superinfecting Phage for Estimating the Division Rate of Lysogenic Bacteria in Infected Animals, *J. Gen. Microbiol.,* **21,** 421 (1959).

Perfil'ev, B. V., and D. R. Gabe, *Capillary Methods of Investigating Micro-Organisms,* Oliver and Boyd, Edinburgh, 1969.

Swain, F. M., Paleomicrobiology, *Ann. Rev. Microbiol,* **23,** 455 (1969).

two | Some General Structural Features of Bacteria

"What was formerly called natural history is the perennial foundation of the biological sciences. It has given rise to all the theoretical branches and will no doubt give rise to others in the future."

William Morton Wheeler (1934)

2.1 | Bacteria as Prokaryotic Forms

Bacteria are distinct living entities whose structural organization differs from that of all other microorganisms except blue-green algae. A brief review of their anatomical features will clearly reveal those characteristics by which they are identified. We will consider here mainly, but not exclusively, the *eubacteria* ("true bacteria"), which possess rigid walls, divide by transverse binary fission and, when motile, swim by means of flagella.

The organizational unit of bacteria and blue-green algae is the *prokaryotic cell*. (Prokaryotic is a word combined from the Greek, *pro* meaning before, and *karyon* meaning kernel, or in this case nucleus; thus the word implies an organism not showing a "classical" nuclear organization, and suggests a more primitive or prior condition, Fig. 2.1A and B.) Microorganisms such as the other algae (brown, red,

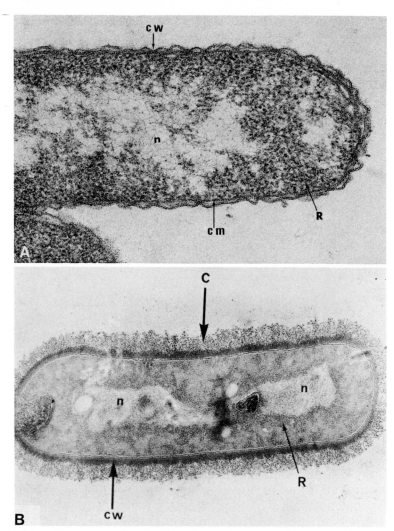

Fig. 2.1. *A,* Thin section of *Escherichia coli,* a prokaryotic eubacterium. The paucity of internal membranous organization is evident. Fine structural features are limited to the ribosome-studded cytoplasm (R), the nuclear region (n), the cytoplasmic membrane (cm), and wavy cell wall profile (cw). (From W. van Iterson, Symposium on the fine structure and replication of bacteria, II. Bacterial cytoplasm, *Bacteriol. Rev.* **29,** 229 (1965) Courtesy W. van Iterson and permission of the American Society for Microbiology.)

B, Thin section of *Bacillus megaterium.* In this prokaryotic eubacterium the nuclear regions (n), ribosomes (R), cell wall (cw), and capsule (C) are clearly discernable. Note especially the difference in appearance of the cell wall profile from that shown in Fig. 2.1*A.* (From D. J. Ellar, D. G. Lundgren, and R. A. Slepecky, Fine structure of *Bacillus megaterium* during synchronous growth, *J. Bacteriol.* **94,** 1189 (1967). Courtesy R. A. Slepecky and permission of the American Society for Microbiology.)

and green), fungi, and protozoa have *eukaryotic* cell forms, that is a "true" or classical nuclear organization* (Fig. 2.2). In eukaryotic organisms, in addition to a boundary membrane, there is a membrane around the nuclear material in its interdivisional state, and there are distinct membranes enclosing respiration *(mitochondria)* and photosynthesis *(chloroplasts)* organelles. The complex of smooth and ribosome-studded (Section 2.9) membranes known as the *endoplasmic reticulum* is seen only in eukaryotic forms. Their membranes contain sterols in higher concentrations than found in blue-green algae, for example, and nuclear division is generally accomplished by mitosis.†

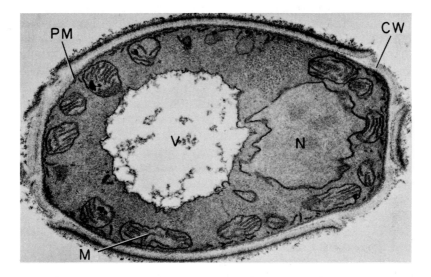

Fig. 2.2. Thin section of *Candida parapsilosis,* a eukaryotic organism. The cell wall (CW), cytoplasmic membrane (PM), membrane-bounded nucleus (N), vacuole (V), and mitochondria (M) are clearly shown. (From G. M. Kellerman, D. R. Biggs, and A. W. Linnane, Biogenesis of mitochondria, XI. A comparison of the effects of growth-limiting oxygen tension, intercalating agents, and antibiotics on the obligate aerobe *Candida parapsilosis, J. Cell Biol.* **42,** 378 (1969). Courtesy A. W. Linnane and permission of the Rockefeller University Press.)

---

*In eukaryotic cells there are two or more genetic systems each of which is contained and replicated in a separate intracellular structure, that is, the nucleus, mitochondria, and, if present, chloroplasts.

†Dinoflagellates, as exemplified by *Gyrodinium cohnii,* appear to combine features of prokaryotic and eukaryotic mechanisms in their mode of nuclear division.

In contrast, the prokaryotic unit of organization has a paucity of internal detail. There is no nuclear membrane, and the complement of nuclear material is apportioned among dividing cells by a process involving little, if any, change in gross appearance. There are no mitochondria in prokaryotic organisms, but the tubular, lamellar, and vesicular invaginations of the cytoplasmic membrane may serve a similar purpose. The photosynthetic apparatus, when present, also may be associated with elements derived from this membrane (Section 3.19). In general, prokaryotic organisms possess only a single boundary membrane and the functional elements are differentiated from it and are situated in a stationary cytoplasmic *milieu*.*

Although both prokaryotic and eukaryotic organisms swim through fluid environments by means of flagella, prokaryotic flagella are simple, noncontractile, helical proteinaceous tubes, whereas eukaryotic flagella have an internal structural complexity and are of larger cross-sections. The name "flagella" has been applied therefore to analogous rather than homologous structures.

Not all prokaryotic cells†are eubacteria. Later we will discuss the *Mycoplasma* (Section 3.9), organisms completely and permanently devoid of cell walls; the budding bacteria (Section 3.1), in which cell division is unequal; the myxobacteria (Section 3.4), whose motility is a slow and stately flagella-less gliding; and certain other noneubacterial forms.

There is a fundamental divergence between the prokaryotic organizational pattern and the eukaryotic form, but the latter, once established, did not undergo major alterations during the subsequent course of biological evolution. Prokaryotic organisms may be vestigial representatives of extinct proto-organisms, and they probably predate the development of eukaryotic forms. It has been postulated that the semi-autonomous mitochondria and chloroplasts of eukaryotes may be highly modified relics of original prokaryotic endosymbionts.

---

*Gas vesicles of some purple and green bacteria, as well as those of some extreme halophiles and the "chlorobium vesicles" of *Chlorobium* and *Pelodictyon* are enclosed by membranes structurally distinct from the cytoplasmic membrane (Section 3.19).

†It has been argued that bacteria are not, in fact, cells. If cells are defined as constituent parts of a multicellular living being, they do not exist independently. Bacteria are organisms capable of an independent existence, and their place in nature may be specified in ecological terms. In general, bacteria may be regarded as homologous with multicellular living things, not with their component cells. Long usage has sanctioned describing bacteria as cells and is employed here.

2.2 | Bacterial Shapes and Some Consequences of Their Being Very Small

There are three common bacterial shapes—spherical, cylindrical, and spiral (Fig. 2.3A). These remain fairly constant for a given organism, but aging or certain environmental conditions may result in the appearance of aberrant cells called *involution forms* (Fig. 4.6). Variations on the patterns also occur as, for example, in oval, lanceolate, spindle, or comma types. The mode of division and of cell attachment thereafter determines, in large measure, the overall appearance of aggregations comprised of these forms (Fig. 2.3A).

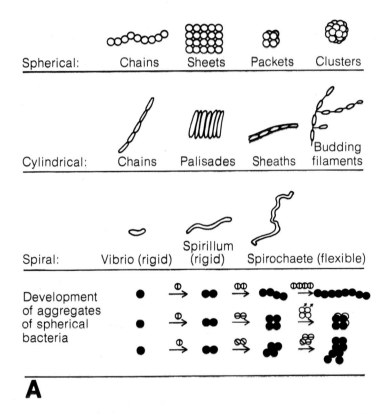

Fig. 2.3. *A,* Common shapes of bacteria and aggregations of individuals. Cocci, or spherical bacteria, show more aggregational variations because of the greater number of possible division planes.

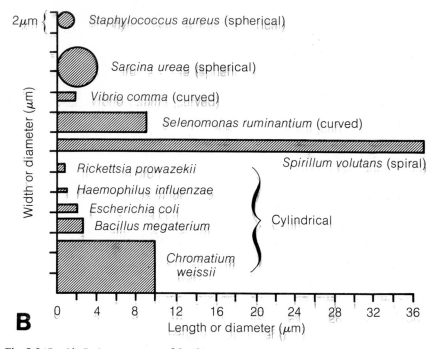

Fig. 2.3 *(Cont'd). B*, A comparison of the dimensional ranges of spherical, curved, cylindrical, and spiral bacteria.

In spherical bacteria, division in a single plane leads to formation of short or long chains (streptococci), whereas random planes of division result in clumps of spherical cells (staphylococci). Among the more interesting arrangements are quartet (tetrad) and octet groupings. Since two or three divisions are made in sequence, the plane of each being at a 90° angle to the previous one, there results a group of from four to eight spherical cells adhering together. Little is known about the control and mechanics involved in these modes of division. Often overlooked is the fact that the next division following the formation of the quartet or octet ought to lead to complete separation of these particular daughter cells so that the process may begin again. It is possible, of course, that quartets and octets of cells are unstable arrangements, so that the process is repeated from detached single cells, but in typical cultures the majority are observed in these aggregated forms. Individuals remain associated after division because of cell surface properties, the firmness of attachment depending largely upon enzymatic and mechanical factors, such as cell wall composition, mode of wall septation, and autoproduction of cell wall lytic enzymes.

In cylindrical and spiral types several aggregations are recognized, the most common of which is a chain of organisms similar to the streptococcal arrangement. Curved and spiral forms of bacteria are generally motile and possess rigid cell walls, but the spirochetes are helically shaped motile organisms that are flexible and have a very thin wall. Both short and long chains of organisms occur, depending to some extent upon whether the mode of division is by constriction (short chains) or by septation without constriction (long chains). Packets of cylindrical cells are observed in cultures of mycobacteria, among others. Presumably, these result from postfission movements (Section 2.13) in which organisms come to lie side by side like stacked logs. Bacteria in aggregations apparently retain their independence and individuality, that is, there is no division of labor among them. It may be, however, that prokaryotic organisms evolved modes of division and aggregational patterns that led to multicellular or *coenocytic* forms, the latter being organisms made up of multinucleate protoplasmic masses enclosed by a single cell wall.

Bacteria are described as "microscopic organisms," but this conveys little quantitative meaning and more definitive remarks concerning their dimensions must be made here. It is difficult to obtain anything other than representative or average values, not only because individual bacterial cells vary in dimensions, but also because the methods employed for measuring them yield only approximations. Microscopic examination and measurement of dried, negatively stained bacteria lead to dimensions somewhat too large because they may include the capsule or slime layer outside the cell wall (Section 2.6). Furthermore, the black, negatively charged colloids used (nigrosin or carbon) do not react with the bacterial surface because at physiological pH they are also negatively charged. When films of these colloids dry in negative stains, the similar charge on both the bacterium and the colloidal particle causes the edge of the dark film to dry at a small but significant distance from the actual cell boundary. This causes the visible profile of the cell to be significantly larger than the cell itself, even if no capsule is present. Low values result from measurements of positively stained organisms since shrinkage due to fixation processes may be as much as one-third of the apparent cell size and even bacterial cell wall stains involve shrinkage. Measurements on living bacteria are difficult to make, although phase microscopy is of some use. The electron microscope has not significantly improved upon measurements of bacteria determined by light microscopy. It is possible to base size estimates upon *volume* rather than length and width measurements, although errors in determining the volume of centrifuged packed cells occur in terms of variations in degree of packing, need for intercellular space corrections, and other uncertainties.

Bacteria range from 0.15 to 4.0 μm (a micrometer [μm] is equal to 10^{-6} meter) in width and from 0.2 to 50 μm in length. In cylindrical forms, length appears more variable than width (Fig. 2.3B). Cell volumes are from about 0.1 μm^3 to 5×10^3 μm^3 (there are 10^{12} μm^3 in 1 cm^3); most of them are between 1–5 μm^3. The consequences of very small or very large bacterial dimensions may be deduced from theoretical considerations of molecular and biophysical phenomena. What factors govern minimum cell dimensions? Free-living organisms possess a minimum number of enzymes, each in turn having a minimum volume, and all the essential enzymes, perhaps 10^3 in a bacterium the size of *Escherichia coli* (volume 0.5 μm^3), could be packaged in a small space, but rarely does an organism have a diameter *less* than 0.33 μm or a volume less than 10^{-2} μm^3. The large number of different kinds of bacteria having cell volumes of a few μm^3 suggests that this is optimal for them. Observed minimum cell volumes of free-living microorganisms are greater than those attributable solely to efficient protein packing, even including the volume occupied by nuclear material, hence there probably are other unknown restrictions governing minimal cell volume. Thus far, there are no known authentic examples of free-living "submicroscopic bacteria," that is, less than 0.15 μm in diameter, and it is doubtful that any exist.*

Organisms of the so-called pleuropneumonia group are among the smallest prokaryotic forms. They may develop from elementary bodies (100 nm in diameter; a nanometer, nm, is equal to 10^{-9} meter) into actively growing forms having diameters of from 200 to 250 nm. Their average volume is approximately 4×10^6 nm^3, and assuming the presence of 80% water, the total "dry volume" is about 10^6 nm^3, a space sufficient for packing about 1600 essential macromolecules.

---

*The amount of double-stranded DNA required is around 20 times the weight of the protein it specifies, and bacteria smaller than *Dialister pneumosintes* ($\sim 2.8 \times 10^{-14}$ g dry weight/cell) would not contain sufficient DNA to specify the necessary enzymes to allow an independent metabolism in them.

One of the largest prokaryotic organisms, *Thiospirillum jenese,* has a volume of about 5×10^3 μm^3, and a length up to 50 μm, but there are likely to be limitations imposed on maximum attainable size. These are in terms of volume of cytoplasm that may be efficiently operated with pure diffusion methods. This problem has been solved in various ways, such as by protoplasmic streaming, although this is not frequently observed in bacteria; by extensive membranous invagination, which increases both total external surface and contact with the interior of the cell; and by forming the cell into long tubes or threads where the surface-to-volume ratio remains constant. There is also the problem of the maximum distance that messenger ribonucleic acid (mRNA) may be expected to operate from the nuclear deoxyribonucleic acid (DNA), and some large organisms overcome this difficulty by developing multinuclear forms. It is clear that the maxima and minima of size observed in microscopic organisms dictate to some extent their morphological details; furthermore, molecular and biophysical considerations govern what these limits will be.

The specific gravity of bacteria is about 1.07, depending principally upon the degree of cell hydration, the amounts and kinds of reserve materials (Section 2.11), and lipid content. This figure is of interest since distribution of nonmotile organisms in a fluid medium is controlled in part by their density and that of their *milieu.* If gravitational forces alone were operative, nonmotile bacteria would eventually be concentrated at the bottom of their suspending medium; however, mixing effects of convection currents, turbulence, and inhomogeneities in medium viscosity prevent this to some degree.

If the specific gravity of common bacterial forms is around 1.07, and their average volume is 1 μm^3, then the "wet weight" of a *single* organism would be about 10^{-12} g or 1 picogram, and it follows that 10^{12} organisms would weigh 1 gram. The ratio of surface area (SA) to the volume (V) is also of great importance. Again using *E. coli* as an example, if the SA of a single cell is 3.0 μm^2, and the volume is 0.5 μm^3, the SA : V ratio is 6; then 10^{12} cells would have a *total* SA of 3 m^2. The overall metabolic rate per unit organism is generally inversely proportional to its size (Section 4.2), and organisms of bacterial dimensions have high rates of substrate conversion, that is, two or three times their weight per hour. Furthermore, this metabolic rate is coupled with short generation times or high division rates, and taken together, these two characteristics constitute the *sine qua non* of bacterial existence.

The study of intact, living organisms—a necessary requisite for investigations of problems in which the phenomenon is not shown at any other (lower) organizational level—has not been approached with the same profusion of biophysical and biochemical techniques as those found in the study of dead organisms and their fragments. Improvements in microscopy, especially development of phase, fluorescent, and interference methods, have been useful additions to available cytological techniques.

Fine structure information on bacteria derived from electron microscopic studies is impressive and illuminating, although in some instances it is simply confirmation of older, light microscope work. Increased resolution and clarity of detail have been critical factors in developing a more precise understanding of structure-function relationships in prokaryotic organisms. There remains a *caveat*, namely, that caution must be exercised in the interpretation of electron micrographs of fixed, embedded, sectioned, and stained bacteria. However, the present overall outlines of anatomical details of prokaryotic organisms are generally agreed upon.

Bacteria possess various structures based on a common plan; among these are capsules, pili, flagella, cell walls, membranes, and endospores. These features impart characteristics to the organisms which influence the range and degree of their interaction with the environment. In the following sections brief accounts of bacterial structures and their probable functions are given. It is recognized that a great amount of important work has been done on ribosomes, nucleic acid composition, cell wall biosynthesis, and the like, but while the value of this information is appreciated, it lies outside the main purpose of this book and will not be treated in detail.

2.3 | Capsules

Immediately outside the cell wall of many bacteria there occur amorphous and diffuse excretions of polymeric materials known as the *slime layer*. If these substances are arranged around the organism with some degree of relation to the cell wall they are termed *capsules* (Fig. 2.4A and B), and they are the only structural elements external to the cell wall that may be seen in their natural state with the light microscope. "Microcapsules" are 200 nm or less in thickness and they are below the resolving power of the light microscope. Slime layers

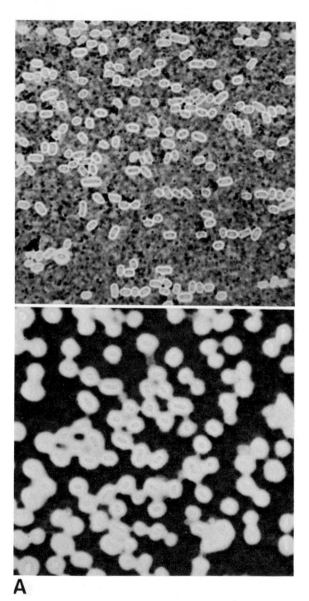

A

Fig. 2.4. *A,* Small (top) and medium-sized (bottom) capsules as revealed in wet films of *Aerobacter aerogenes* with India ink. (From J. P. Duguid, The demonstration of bacterial capsules and slime, *J. Pathol. Bacteriol.* **63,** 673 (1951). Courtesy J. P. Duguid and permission of Oliver and Boyd, Publishers.)

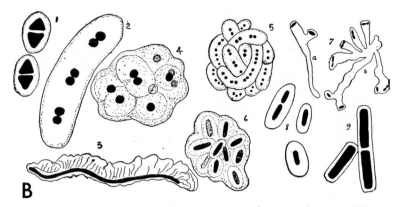

Fig. 2.4 *(Cont'd)*. *B,* Capsular formations seen around various bacteria. (1) *Diplococcus pneumoniae,* (2, 4, 5) *Leuconostoc mesenteroides,* (3) *Lactobacillus vermiformis,* (6) rods embedded in a zoogloeal mass, (7) *Bacterium pediculatum,* (8) rods in slimy milk, (9) *Bacillus anthracis.* (From R. E. Buchanan and E. D. Buchanan, *Bacteriology,* 5th ed., Macmillan, New York, 1951. Courtesy R. E. Buchanan and permission of the Macmillan Company.)

usually are less viscous, more water soluble, and less structure oriented than capsules, the latter tending to remain attached more intimately to the organism. Large quantities of capsular material may be formed per cell under specific conditions, for example, *Leuconostoc mesenteroides* synthesizes a capsular dextran from sucrose, but not from glucose or fructose, and *Bacillus anthracis* requires carbon dioxide to form its capsule. Capsulated bacteria give rise to colonies, termed smooth (*S*), or mucoid (*M*), whose surface texture and viscosity reflect the nature of the organisms comprising it. Colonies composed of capsulated bacteria, when touched with a probe, may be drawn into threads several centimeters long (Section 2.13). Capsules are not essential to the continued life of an organism, and they may be removed without causing death of the cell.

The exterior location and ease of removal with neutral aqueous solvents have facilitated purification and chemical analysis of capsular materials. In general, these have been found to be polymers of glucose, fructose, or sucrose, of molecular weights up to 10^6. *Bacillus subtilis* and *B. anthracis* produce capsules of poly-D-glutamic acid (glutamyl polypeptide), and some streptococci form nitrogen-containing polymers such as hyaluronic acid (N-acetyl-glucosamine and glucuronic acid polymer). *B. megaterium* produces a polysaccharide-polypeptide capsule, but mixed types of this kind are rare.

The external location of capsules is interesting in that while excreted they are products of biosynthetic rather than biodegradative reactions. It seems possible that polymer subunits are assembled into the capsular structure at sites on or external to the cytoplasmic membrane. There are cases in which the capsule is oriented in a definite pattern around the organism (Fig. 2.5) and this suggests that there are, in fact, specific loci at which capsular material is synthesized.

Capsules may be related to virulence or the disease-producing ability of pathogenic bacteria, and usually if the capsulated form is virulent, loss of capsule-forming ability is correlated with nonvirulence. Freshly isolated cultures are often composed of capsulated organisms. This suggests that capsule forming ability has survival value in the host, or that nutritional conditions in the latter are such as to lead to capsule formation. A viscous or slimy character is imparted to fluids in which large numbers of organisms synthesizing capsules are growing. Viscous environments have interesting properties, such as a high degree of

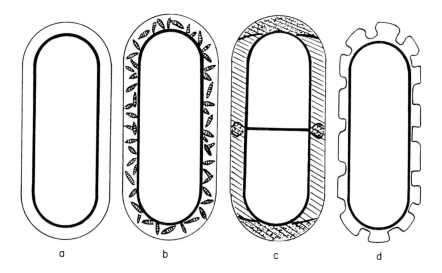

a   b   c   d

Fig. 2.5. Capsular orientation and structural patterns shown in various bacteria. *A*, Continuous layer; *B*, banded fibrils in *Escherichia coli* Lisbonne; *C*, complex capsule with localized patches of polysaccharide (poles and cross-wall) and polypeptide in *Bacillus megaterium* M; *D*, discontinuous layer in *Bacillus megaterium*. (From M. R. J. Salton, Surface layers of the bacterial cell, I. C. Gunsalus and R. Y. Stanier (eds.), *The Bacteria*, Vol. I, Ch. 3, Academic Press, New York, 1960. Courtesy M. R. J. Salton and permission of Academic Press.)

resistance to diffusion of materials, susceptibility to sudden reductions of viscosity because of enzymatic hydrolysis, and ability to trap and accumulate particulate elements.

Capsules may have various functions in nature, but these often have been proposed without proof being forwarded. It is possible, for example, that bacteriophage adsorption is blunted by the presence of a capsule; they may serve as "traps" for aiding in uptake of various essential ions; they may influence the electrokinetic characteristics of the cell and stability of suspensions; they are hydrophilic and thus could prevent death by desiccation under adverse environmental conditions; they may protect an organism from phagocytosis, either by host macrophages or by amebae, if in the soil; or, they may be useful in allowing organisms to adhere to surfaces. Most bacteria seem unable to hydrolyze their own capsular polymers if purified and supplied exogenously, and utilization of capsules as a reserve energy substance seems precluded. Heterologous organisms, however, may produce capsule-degrading enzymes; for example, *Bacillus palustris*, a soil organism, can hydrolyze capsular polysaccharides of pneumococcus types III and VIII when these are supplied as sole carbon sources.

2.4 | Pili

Located exterior to the wall of many gram-negative bacteria (Section 2.6) are filaments called *pili*, or *fimbriae* by British workers, from the Latin *pilus*—hair; or from the Latin *fimbria*—fiber or thread (Fig. 2.6). These are long, stiff-appearing, straight proteinaceous hollow tubes ranging from 3 to 25 nm in diameter and from 0.2 to 20 μm in length. The type I pili (there are numerous types, Table 2.1) are composed of a protein termed *pilin*, of a molecular weight around 16,600. Type I pili have an axial hole 2–2.5 nm in diameter around which pilin subunits are wound. There may occur as few as from one to four of those designated as F pili to many hundreds per cell of those designated as morphological types I through IV. Pili react with viruses, red blood cells, other bacteria, and antibodies, and nonspecifically with hydrophobic surfaces; nonetheless their *precise* function is unknown. As in the case of capsules, pili are not needed for continued cell life, although as with capsules, bacteria capable of producing pili, when freshly isolated from natural sources, are usually "piliated." Bacteria may possess two or more types of pili simultaneously on the same cell.

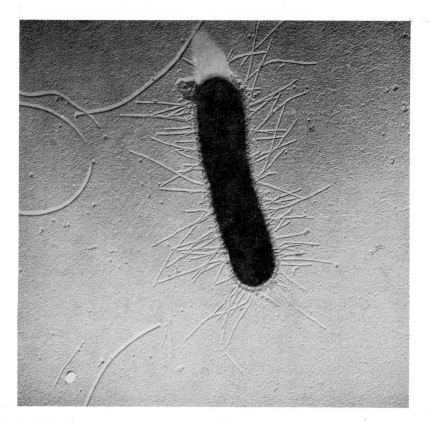

Fig. 2.6. Pili of *Escherichia coli*. A few detached flagella are also shown. There are about 100 pili on this bacterium. (Courtesy G. J. Hageage.)

Table 2.1.   Morphological Characteristics of Pili*

Type	Average width (nm)	Average length (μm)	Average pili per cell	Location on cell
I	7.0	0.5–2.0	150	Peripilous
II	4.8	0.5–2.0	150	Peripilous
III	3.5	2.0–6.0	350	Peripilous
IV	7.0	1.0–2.0	150	Peripilous
V	25.0	1.0–2.0	1	Not specific
VI	5.0	2.0–10.0	3	Polar
VII	15.0	0.2–0.5	15	Polar
F†	8.5	1.0–20.0	2	Not specific

*Compiled from various sources.

†F pili are associated with bacteria possessing the fertility plasmid, F, which confers "maleness" or ability to donate the chromosome.

Pilus attachment is evidently weak, or else the pili are very fragile, since electron microscopic examination of laboratory cultures reveals many unattached pili. In obligately aerobic bacteria growing under conditions of limited oxygen supply, type I piliated forms outgrow nonpiliated bacteria and this may be due to the ability of the former to develop as a surface film. On the other hand, pili may serve as a means of transporting certain materials directly into the cell, thus bypassing the wall and membrane barriers, but, in any case, they increase greatly the potential absorption area of the organism.

2.5 | Flagella and Bacterial Behavior

The organelle for propelling bacteria through a fluid medium is the flagellum (Fig. 2.7A). Flagella may occur singly or in arrangements of up to 100 or more per organism. Few spherical bacteria have flagella, whereas nearly all rigid curved and spiral forms do. Rod forms isolated from soil or water are more likely to possess flagella than those from skin surfaces or mucous membranes.

The structure of bacterial flagella is quite different from that of the analogous organelle in eukaryotic forms and provides another example of the dichotomy between structures used for similar functions in the two groups which have been given the same name. Bacterial flagella are hollow, proteinaceous, usually helical tubes, from 12 to 20 nm in diameter and from 2 to 5 μm in length, however, some up to 50 μm in length have been reported. Bacteria may lose their ability to produce flagella by mutation and here again, as with capsules and pili, presence of the structural element is not necessary for continued cell life. Flagella are arranged on bacteria in patterns, each of which is characteristic and constant for a given motile group. In rod-shaped organisms one or more are inserted polarly into the cell, or else they are uniformly distributed around it, though sparse or absent at the poles. The former mode is termed *monotrichous* (Fig. 2.7B) if only one flagellum is present, or *lophotrichous* (Fig. 2.7C) if there is a group or *fascicle* (tuft) of several flagella (Fig. 2.7D). The lateral arrangement of numerous flagella around an organism is called *peritrichous* flagellation (Fig. 2.7A).

Bacterial flagellar filaments consist almost entirely of a single protein given the generic name *flagellin*. The molecular weight of flagellins is around 4×10^4 and they occur as globular subunits of 4.5 nm

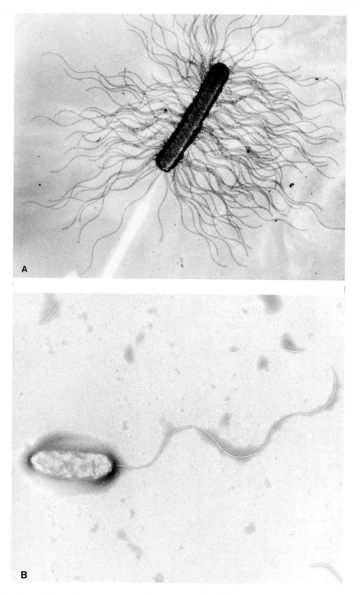

Fig. 2.7. *A,* Peritrichously flagellated *Proteus mirabilis.* (From J. M. F. Hoeniger, Development of flagella by *Proteus mirabilis, J. Gen. Microbiol.* **40,** 29 (1965). Courtesy J. M. F. Hoeniger and permission of the Society for General Microbiology.)

B, Monotrichously flagellated *Thiobacillus thiooxidans.* (Courtesy Z. Vaituzis.)

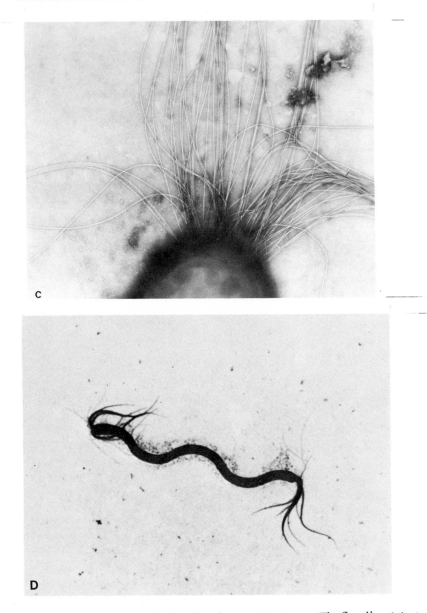

Fig. 2.7 *(Cont'd). C,* Lophotrichously flagellated *Thiospirillum jenense.* The flagella originate at one end of the organism and occur as a fascicle of over 80 individual strands. (Courtesy Z. Vaituzis.)

D, Spirillum sp. showing a fascicle of flagella at each pole. (Courtesy G. J. Hageage.)

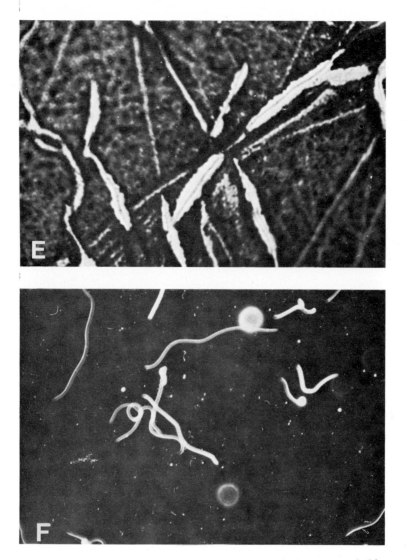

Fig. 2.7 *(Cont'd)*. *E, Proteus mirabilis* suspended in India ink. The bacteria are held station-ary while flagellar motion sweeps out clear fields around them. (From T. Y. Kingma-Boltjes, Function and arrangement of bacterial flagella, *J. Pathol. Bacteriol.* **60,** 275 (1948). Courtesy T. Y. Kingma-Boltjes and permission of Oliver and Boyd, Publishers.)

F, "Motility tracks" or trails of swimming bacteria as revealed by continuous exposure dark-field photography. (From Z. Vaituzis and R. N. Doetsch, Motility tracks: technique for quantitative study of bacterial movement, *Appl. Microbiol.* **17,** 584 (1969). By permis-sion of the American Society for Microbiology.)

in diameter, of which 40% are in the α-helix form. Flagellin molecules are believed to be arranged in either helical or longitudinal strands. Nonproteinaceous elements are found associated with purified flagella preparations, and it is difficult to decide whether these are tightly bound contaminants, or whether they constitute an integral part of the organelle. Different parts of the flagellum vary in chemical composition as well as structure, and on the basis of electron microscopic studies bacterial flagella appear to possess a long *filament,* a short *hook,* or more accurately a bend, and possibly a portion containing paired discs attached to the cell wall and cytoplasmic membrane. Flagella terminate at a differentiated portion of the cytoplasmic membrane and although they traverse the cell wall, they do not enter the cytoplasm proper. Flagella are fragile structures and easily become detached from the cell body leaving short "stubs" behind, from which new whole flagella may be regenerated.

Some "flagellatropic" bacteriophages are known, for example, those designated as $\phi\chi$ and PBS-1, attacking *Escherichia coli* and *Bacillus subtilis,* respectively. Their primary attachment site appears to be the functioning flagellum. It was once believed that the phage nucleic acid was conducted down through the flagellum but now it has been proposed that the attached phage particle is led down the flagellum to the cell surface, where its nucleic acid is injected at receptor sites situated there. The details of flagellatropism remain to be clarified.

On the basis of x-ray diffraction studies, it was concluded that flagellin is a fibrous protein of the keratin-myosin-epidermin-fibrinogen group. This contention, together with subtle extrapolations from studies on flagellar mechanics of large eukaryotic forms, has reinforced the view, which never has been experimentally demonstrated, that bacterial flagella are, in fact, contractile elements. There are several objections to this: (1) flagella appear too simple in cross-sectional structure to contain the necessary machinery for a complete contractile system; (2) an adenosine triphosphatase (ATPase) system is not present in them, and isolated flagella do not contract in the presence of ATP; (3) paralyzed (nonmotile) flagella have the *same* x-ray diffraction patterns as nonparalyzed, functional flagella, and even the gross morphology may appear to be the same; (4) there is no evidence of actin or actin-like materials in flagella, and studies on isolated flagella reveal that they act as if they were semirigid structures; and (5) a contractile flagellum would function only if flagellin molecules could be displaced sequentially. A "wave" of contraction generated only at the flagellar base would not be transmitted distally because of immediate viscous damping by the medium. The helical form of some, but not all, bacterial flagella as seen in fixed or detached preparations, suggests the passage of contractile waves down them, but this may merely reflect tertiary protein structure.

An alternative view concerning bacterial flagellar function, especially in spiral lophotrichous forms, is to consider them as helices or whips activated by a specialized area of the cytoplasmic membrane, and anchored there and at sites in the cell wall (Fig. 2.8). Sequential contractions around this specialized area would "wobble" the flagellum, or a group of flagella, which then would swivel around at the cell wall anchors. The bend of the hook would cause the filament to describe a cone of revolution, and this rotation, occurring at rates of up to 3,000 revolutions per minute, would serve as an efficient "propeller" (Fig. 2.7E). A fascicle of flagella would come into phase hydrodynamically, thus acting as a single giant multistranded flagellum. The most efficient operation of flagella occurs when they are inserted polarly into cylindrical or spiral cells. It may be that peritrichously distributed flagella are swept back to act in this fashion, or

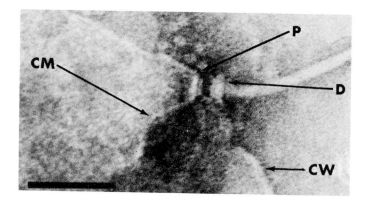

Fig. 2.8. Attachment of a flagellum to the cytoplasmic membrane and cell wall in *Vibrio metschnikovii.* A set of proximal (P) discs connects the flagellum to the cytoplasmic membrane (CM) and a second distal (D) set connects it to the cell wall (CW). These discs are firmly attached to the basal portion of the flagellum below the hook. They are about 25 nm in diameter and the two sets are about 15 nm apart. The double discs themselves are separated by a 4 nm space. (From Z. Vaituzis and R. N. Doetsch, Relationship between cell wall, cytoplasmic membrane, and bacterial motility, *J. Bacteriol.* **100,** 512 (1969). By permission of the American Society for Microbiology.)

that most of the lateral flagella are inoperative, moving for only a certain fixed period in the life of the bacterium. Straight or nonhelical flagella, although not paralyzed, may cause cell rotation, but not translation, if attached to spherical or cylindrical shaped organisms. Spherical, flagellated bacteria move in an awkward tumbling fashion, and cylinders and spirals are much more efficient swimming forms, tending to describe long curving arcs or straight lines as they move (Fig. 2.7F).

Bacteria 2 μm long have been observed to swim at velocities up to 50 μm/sec, or 25 times their cell length per second, and a six-foot man would have to swim in excess of 100 miles per hour to equal this rate, but in absolute terms bacteria do not cover much distance per unit time. Bacteria do not move with a constant velocity and they may be observed to stop, start, and move slowly or rapidly in their medium. Hydrodynamic analysis of swimming in bacteria shows that viscous forces are predominant. The *Reynolds number* (R), that is, the ratio of inertial to viscous forces acting on the steady translational movement of the organism, is calculated from the equation $R = [LV \rho/\mu]$, where L equals the length of the moving system, V is the velocity in μm/sec, ρ is the medium density, and μ is the medium viscosity. In the case of a bacterium 2 μm in length swimming at a velocity of 20 μm/sec in water, R is 4×10^{-6}, that is, inertial forces are a negligible factor. The energy expenditure required to propel a bacterium, such as *B. subtilis,* at a velocity of 10 μm/sec is roughly equivalent to that resulting from the hydrolysis of 10^2 molecules of ATP/sec, a small expenditure of the total generated per unit time. The conversion of chemical to mechanical energy is a fascinating biological problem on which the study of bacterial flagellar mechanics might shed some light.

The structural and physicochemical mechanisms by which motile bacteria "monitor" their environment are, at present, unknown. Bacteria are believed to respond to environmental discontinuities (gradients rather than sharp boundaries) either positively (moving toward) or negatively (moving away), depending upon whether these are physiologically favorable. In the absence of stimuli bacteria swim in a random manner, tracing patterns of relatively straight lines or graceful arcs, unless disturbed by effects of currents, particulate debris, other organisms, and brownian agitation. Upon entering an "unfavorable" zone, a given bacterium will display a strikingly different pattern characterized by a series of short runs, reversals, stops and starts, as though "hunting," and this behavior continues until the organism again moves into a more favorable area. Responses of this sort are said to be *tactic reactions* and, depending upon the nature of the stimulus, are termed *chemotaxis* (response to chemical agents), *phototaxis* (to light), or *aerotaxis* (to gases, Fig. 2.9). Positive tactic responses are those in which organisms gather in a favorable area, reversing direction if they move too far away from it. However, it seems reasonable to look upon all tactic responses as cases of *negative chemotaxis* wherein the organism moves *away* from an unfavorable location, hence remains in a favorable one. In this case a stimulus response reaction $(S \rightarrow R)$ may be invoked to explain movement of the organism away from the unfavorable environment. Explanations of tactic responses of motile bacteria are tinctured with anthropocentric overtones, and their study, although initiated in the early days of bacteriology, is still in a primitive state. The link between flagellar operation and tactic response is likely to be the cytoplasmic membrane, and it has been observed that flagella seem intimately associated with it. Possibly alterations of membrane potential and conformation are "signals" governing flagellar operation, but the mechanisms by which this could be accomplished are still speculative.

Several kinds of mutations affecting flagella occur; there may be loss of ability to produce flagella ($fla^+ \longrightarrow fla$); or flagella may be formed that are inoperative or "paralyzed" ($mot^+ \longrightarrow mot^-$); or flagella are formed that have abnormal shapes such as "curly," "coiled," or "straight" rather than helical, resulting in defective cell movement.

Although a variety of nonmotile bacteria abound, it appears that ability to move in the environment confers "survival advantage" upon an organism. A motile cell may not only avoid predators, but also "explore" its surroundings to a much greater extent than organisms carried along passively by external forces. The latter, of course, if residing on or in a host of more or less constant properties would not ordinarily need to move about in order to survive, and many bacterial parasites dwelling on the skin or mucous membranes of man and animals are, in fact, nonmotile. The relationship between virulence and motility in pathogenic bacteria has not been explored, but it seems that an organism able to escape ingestion by macrophages by moving away from them, or able quickly to contact the "target organ," would be more likely to breach host defenses.

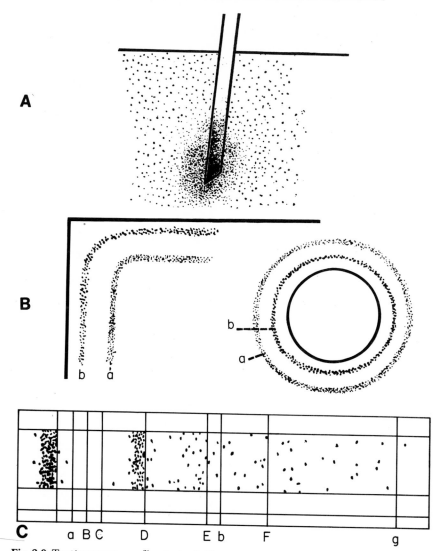

Fig. 2.9. Tactic responses of bacteria. *A,* Chemotaxis—*Chromatium weissii* around a capillary filled with 0.3% ammonium nitrate; *B,* Aerotaxis—collections of spirilla (*a*) and a ciliate, *Anophrys (b),* at the corner of a cover-glass and around a bubble of air under a cover-glass; *C,* Phototaxis—distribution of *Chromatium photometricum* in a microscopic spectrum. The largest number of bacteria are in the infrared region, and others are in the yellow-orange region (line *D*). (From H. S. Jennings, *Behavior of the Lower Organisms,* Indiana University Press, Bloomington, Indiana, 1962, reprint of 1906 edition. By permission of Indiana University Press.)

2.6 | Cell Walls

The shapes of many kinds of bacteria are governed by the presence of a rigid, boundary wall. The wall has a number of functions, one of which is to prevent the bursting of the cell by osmotic forces. The internal osmotic pressure of gram-positive bacteria is around 20–25 atm (atmospheres), and in gram-negative bacteria it is between 5 and 6 atm. Consequently, a covering structure of some mechanical strength is required to maintain cell integrity and shape. Wall components of bacteria constitute a large portion of the cell dry weight, and they have been found to range from 10 to 40% of the total. In addition to this "corseting" function, the cell wall has other roles which will be taken up later in this section.

The rigid part of bacterial cell walls is unique in that it consists of a macromolecular network sometimes referred to as a "murein sacculus," composed mainly of *N-acetylmuramic acid* (AM) and *N-acetylglucosamine* (AG) (Fig. 2.10A). The saccharide component consists of alternating AG and AM units joined by $1 \rightarrow 4\beta$ bonds, reminiscent of the cellulose of plant cell walls and insect chitin (Fig. 2.11), except for the fact that here each AM contains a 3-O-D-lactyl group, and the average chain length is small (20–140 hexosamine residues). Short peptides are attached to the AM units and contain D-glutamic acid and D-alanine, and in many cases L-alanine, and either L-lysine or α,ϵ-diaminopimelic acid (DAP) (Fig. 2.10B). In *Staphylococcus aureus* the tetrapeptide chains (L-alanine-D-glutamic acid-L-lysine-D-alanine) are cross-linked by a pentaglycine bridge, thereby forming the peptidoglycan* wall network (Fig. 2.12). In gram-negative cell walls about 10% is peptidoglycan, but there is from 40 to 90% in the walls of various gram-positive bacteria.

That cell walls are complex both in structure and chemical composition may be inferred from the presence of various layers seen in wall profiles of thin-sectioned bacteria viewed with the electron microscope. Measurements indicate that cell wall profiles range in thickness from 7.5 to 80 nm, and although numerous variations have been observed, there are four main types: (1) In extreme halophiles, such as *Halobacterium halobium*, there are an inner membrane and an outer cell wall. The latter is about 13–15 nm thick, and along its outer

---

*Peptidoglycan cell walls have never been reported to be present in eukaryotic forms.

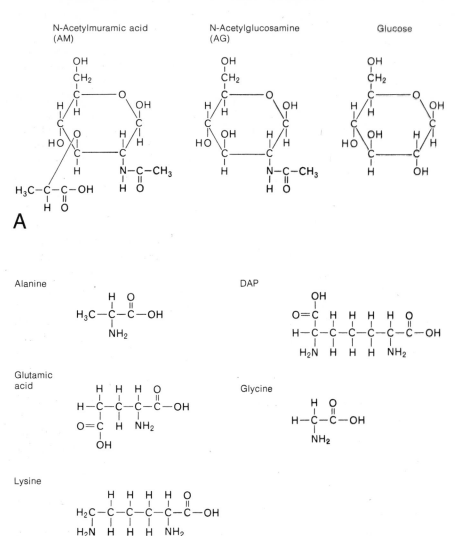

Fig. 2.10. *A,* Comparison of the structures of N-acetylmuramic acid (AM), N-acetylglucosamine (AG), and glucose.

B, Principal amino acids comprising the peptide chain of a peptidoglycan. Half of the alanine and all of the glutamic acid residues are D-isomers. Naturally occurring proteins consist of L-isomers of amino acids. Diaminopimelic acid does not occur in proteins and is found exclusively in bacterial cell walls.

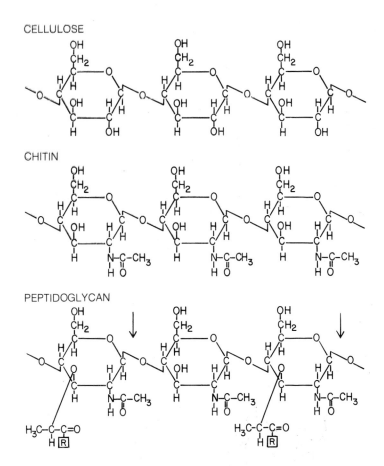

Fig. 2.11. Comparison of the structures of cellulose, chitin, and the peptidoglycan of bacterial cell walls. Arrows indicate locations of bonds hydrolyzed by muramidase.

edge there sometimes appear quite dense places, having a periodicity of 15 nm along its length. Extreme halophiles (Section 3.20) are fragile and lyse readily if the sodium chloride content is lowered below 8%. Indeed, membrane stability is dependent on a high osmotic pressure and the specific presence of sodium ions. (2) Gram-positive bacteria have walls from 15 to 80 nm thick. The character of the profile may depend upon the staining agent employed (Fig. 2.13). Associated with cell walls of some gram-positive and gram-negative bacteria are unique

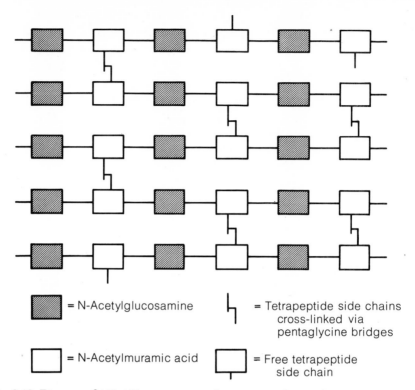

▨ = N-Acetylglucosamine	⌐ = Tetrapeptide side chains cross-linked via pentaglycine bridges
□ = N-Acetylmuramic acid	⌐ = Free tetrapeptide side chain

Fig. 2.12. Diagram of AG, AM arrangement showing cross-linking by tetrapeptide side chains. In *Staphylococcus aureus* the AM peptide residues are almost all cross-linked to one another by means of pentaglycine bridges.

A

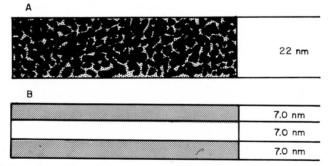

22 nm

B

7.0 nm

7.0 nm

7.0 nm

Fig. 2.13. Cell wall profiles of thin sections of *Bacillus megaterium* as observed in the electron microscope after post-staining with: *(A)* 1% uranyl acetate (pH 4.5), 1% phosphomolybdate (pH 7.2), 2% phosphotungstate (pH 7.2), or 1% ruthenium red; or *(B)* 1% osmium tetroxide, 0.5% permanganate, or Reynold's lead citrate. The triple-layered cell wall is seen only when the stains listed in *(B)* are used. In *(A)* there appear dense granules from 2.5 to 3.5 nm in width, but it is not known whether these represent polymeric cell wall material or precipitated stain where no layered arrangement is observed. (From M. V. Nermut, The ultrastructure of the cell wall of *Bacillus megaterium, J. Gen. Microbiol.* **49,** 503 (1967). Courtesy M. V. Nermut and permission of the Society for General Microbiology.)

polymers, namely, *teichoic* and *teichuronic acids* (Fig. 2.14). The former are ribitol phosphate or glycerol phosphate polymers together with alanine, N-acetylglucosamine, glucosamine, and glucose, and are of the following general structures:

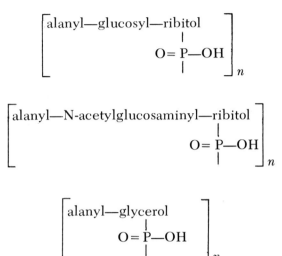

Teichuronic acids (uronic acid polysaccharides) also are present in walls of some bacteria. In *Bacillus subtilis* there is one composed of N-acetylglucosamine and glucuronic acid units, and in *Micrococcus lysodeikticus* walls there is a polymer composed of equimolar amounts of glucose and aminomannuronic acid. (3) The basic gram-negative wall structure (Fig. 2.15 A and B) comprising an intermediate dense layer (peptidoglycan) 3–8 nm thick is sandwiched between the cytoplasmic membrane and a wavy outer membrane 6–10 nm thick which may appear as a dense-light-dense profile. In general, cell walls of gram-negative bacteria are thinner in profile, more complex in structure and composition, and mechanically weaker than those of gram-positive bacteria. The intermediate layer may sometimes be so closely

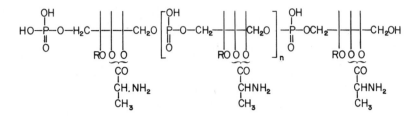

Fig. 2.14. Structures of ribitol teichoic acids from the cell walls of various bacteria. In *Bacillus subtilis*, R = β-glucosyl; *n* = 7; in *Staphylococcus aureus* H, R = α and β-N-acetyl glucosaminyl, *n* = 6; in *Lactobacillus arabinosus* 17-5, R = α-glucosyl, and alternate ribitol residues may have α-glucosyl at the 3 position; *n* = 4–5.

applied as to be indistinguishable from the outer membrane, and this association has been observed in *Escherichia coli, Nitrosocystis oceanus, Shigella flexneri,* and *Proteus mirabilis,* among others. More complex layerings over the basic gram-negative wall structure are seen in such bacteria as *Ectothiorhodospira mobilis, Beggiatoa,* and *Spirillum,* but the function of these additional layers is not clear. (4) Extremely complex walls and outer layers occur in *Lampropedia hyalina,* which divides in such a way that square "tablets" of several hundred bacteria result. Although the inner wall is that of a typical gram-negative organism, there is superimposed on this a "perforate" layer, attached to it by means of fine fibers, and a "punctate" or "echinulate" layer which in turn overlays it. The latter appears in cross-section as an array of spines projecting from hollow bulbs connected together at their their bases.

The organism known as *Micrococcus radiodurans* was originally isolated from canned meat that had been exposed to approximately 2–3 Mrads of gamma radiation and in fact, 5–6 Mrads are needed to kill it. The mechanism of such radiation resistance is unknown but the wall layers of this bacterium show very complex profiles (Fig. 2.17 A-E). There is a dense layer of cell wall 8–20 nm thick which shows transverse striations having a periodicity of 17.5–20 nm. The outer layer is a wavy membrane 7.5–10 nm thick and this overlays two other intermediate layers, namely, the "compartment" and the "hexagonal." The former appears as pairs of light lines around irregular dark spaces (compartments), and the latter shows a regular hexagonal pattern with center-to-center spacings of 17.5 nm between ring-shaped units joined by fine spokes.

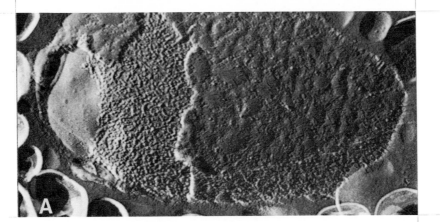

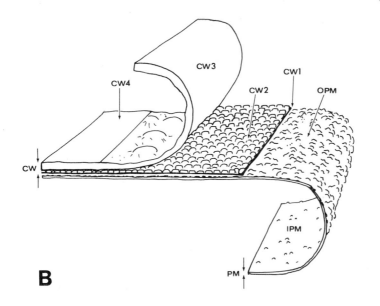

Fig. 2.15. *A,* Freeze-etching preparation of *Escherichia coli* showing the cell wall with a wavy appearance, and some fibrils apparent; the outer surface of the cytoplasmic membrane bears blunt protrusions.

B, Schematic diagram of the outer layers of *Escherichia coli*. The cell wall shows the peptidoglycan layer (CW 1), the globular protein layer (CW 2), the inner surface (CW 3), and two aspects of the outer surface (CW 4) presumably of the lipoprotein-lipopolysaccharide complex. The cytoplasmic membrane outer (OPM) and inner (IPM) surfaces also are depicted. (From N. Nanninga, Ultrastructure of the cell envelope of *Escherichia coli* B after freeze-etching, *J. Bacteriol.* **101,** 297 (1970). Courtesy N. Nanninga and permission of the American Society for Microbiology.)

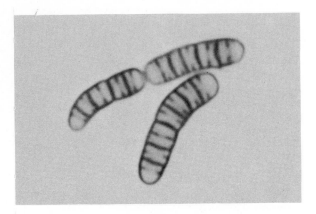

Fig. 2.16. *Caryophanon latum* stained with phosphomolybdic acid and methyl green after Bouin fixation. The numerous cell wall septa laid down by this organism are clearly shown. (Courtesy G. J. Hageage.)

Surface features of intact bacteria and various wall component preparations have been revealed using shadowing, carbon-replica, and freeze-etching techniques. Regular patterns occur in the peptidoglycan and outer layers of some, but not all, bacteria (Fig. 2.18A). Cell wall fragments of *Spirillum serpens* show hexagonal arrangements of spherical units having a center-to-center distance of 12–14 nm. Similar patterns also have been observed in cell wall preparations of *Rhodospirillum rubrum, Azotobacter agilis, Ectothiorhodospira mobilis, Lampropedia hyalina,* and *Halobacterium halobium.* In the cell walls of gram-positive bacteria, such as *Bacillus subtilis, B. polymyxa, B. megaterium,* and *Clostridium sporogenes,* the pattern appears tetragonal rather than hexagonal but the latter pattern has been observed in *B. alvei* and *B. anthracis.* Any special significance of these structures has yet to be ascertained (Fig. 2.18 A and B).

Bacterial growth implies synthesis of all cell components including those of the wall, and one problem that arises is that peptidoglycan must be inserted into a rigid macromolecular network. Cell wall growth has been studied using immunofluorescence techniques. Fluorescent-labeled antibodies directed against specific cell wall antigens show that in *Streptococcus pyogenes* and *S. faecalis* new material is either brought from the cytoplasmic membrane in the region of a septal mesosome (Section 2.7) at the edge of the closing septation, or that both septum and peripheral wall are supplied by

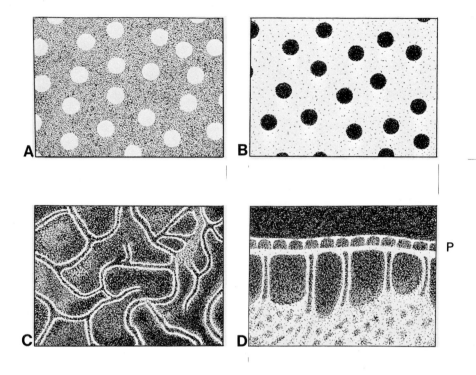

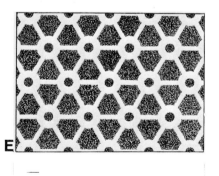

Fig. 2.17. Diagram of the cell wall structure of *Micrococcus radiodurans. A,* Appearance of the dense layer in tangential section; *B,* the "holey" layer as seen in negatively-stained preparations; *C,* surface view of the "compartment" layer; *D,* folded edge of cell wall fragment showing a side view of the compartments and the projecting subunits (P) of the hexagonal layer on the outer surface (*E*). (From A. M. Glauert and M. J. Thornley, The topography of the bacterial cell wall, *Ann. Rev. Microbiol.* **23,** 159 (1969). Courtesy of M. J. Thornley and permission of Annual Reviews, Inc.)

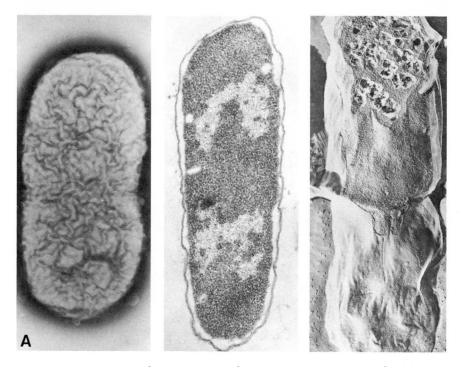

Fig. 2.18. *A,* Comparison of *Escherichia coli* B after *(left)* negative staining in silicotungstate; *(center)* fixation in osmium tetroxide, dehydration, embedding, and ultra-thin sectioning (wall and membrane separated); and *(right)* freeze-etching. The intact surface is seen at left, the cytoplasmic membrane is in the center, and the protoplasm is seen at the right. These photographs represent three principal methods of preparing and viewing bacteria for electron microscopic studies. Note that the surface appearance of the organisms on the left and right are not quite the same. (From M. E. Bayer and C. C. Remsen, Structure of *Escherichia coli* after freeze-etching, *J. Bacteriol.* **101,** 304 (1970). Courtesy M. E. Bayer and permission of the American Society for Microbiology.)

a ring of membrane at the edge of the septum. The mechanisms of cell wall growth may differ in gram-positive and gram-negative bacteria but little is really known about the process, and results obtained by fluorescent antibody studies permit a degree of latitude in interpretation. As a rule, gram-positive rod-shaped bacteria divide by laying down a wall septum or cross wall without constriction (Fig. 2.16), whereas gram-negative forms divide by constriction (Fig. 2.21). This suggests that there are fundamental differences. Bacteria that constrict during division generally separate shortly thereafter, whereas those

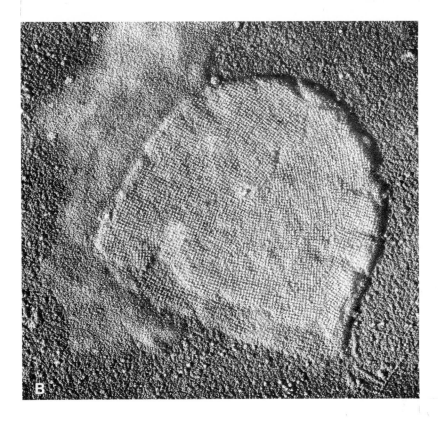

Fig. 2.18 *(Cont'd).* B, Fragment of cell wall from *Bacillus polymyxa* showing a regular globular structure of 7 nm diameter with an average interspace of 3 nm. (From M. V. Nermut and R. G. E. Murray, Ultrastructure of the cell wall of *Bacillus polymyxa*, *J. Bacteriol.* **93**, 1949 (1967). Courtesy R. G. E. Murray and permission of the American Society for Microbiology.)

that form septa during division without constriction may form long chains of individual organisms (Fig. 2.19A and B).

In *S. faecalis* a mucopeptidase located in the septal region and in the most recently deposited wall, when activated by proteolytic enzymes, hydrolyzes the $1 \rightarrow 4\beta$ bonds between AG and AM, and in this way allows for insertion of new units. Hydrolases in the walls of *Escherichia coli* and certain bacilli and staphylococci release reducing groups from AG rather than AM. Wall lysins provide a precisely desig-nated locus for the insertion of new wall material, rather than permit-

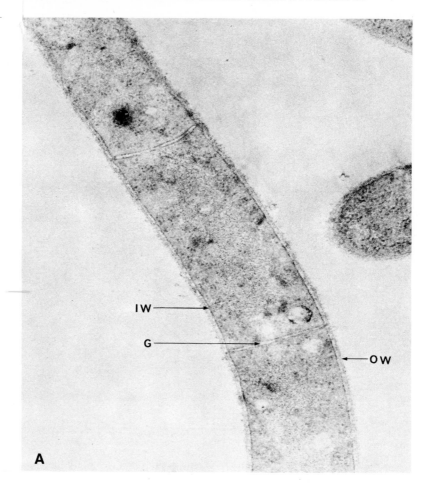

Fig. 2.19. *A,* Thin section of *Saprospira grandis* showing division by septation. Outer wall (OW), inner wall (IW), and gap (G) are clearly depicted. (Courtesy G. J. Hageage.)

ting nonspecific distribution of it. This is accomplished by providing acceptor ends of peptidoglycan, or by opening bonds to permit rearrangement of existing peptidoglycan and positioning of new material.*

---

*Peptidoglycan and polysaccharide components of lipopolysaccharides are formed on specific polyisoprenoid carriers located in the cytoplasmic membrane and these units are transferred to the growing portion of the cell wall.

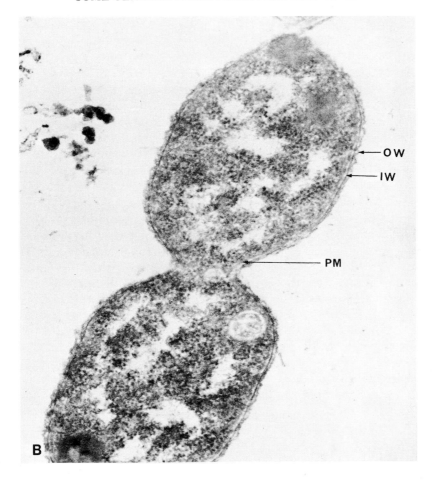

Fig. 2.19 *(Cont'd)*. *B,* Thin section of *Vitreoscilla* RCM I, showing folded outer wall (OW), inner wall (IW), and cytoplasmic membrane (PM). This organism divides by constriction, and all cell wall elements participate in the process. (Courtesy G. J. Hageage.)

Lytic enzymes are required for cell separation after division is completed, and it thus appears that wall lysins serve several functions in growing and dividing cells.

If gram-positive bacteria are exposed to the enzyme muramidase (lysozyme) while being maintained in an osmotically stabilizing medium such as 0.2 M sucrose, their peptidoglycan layer may be entirely hydrolyzed into low molecular weight fragments. It follows that some organisms such as *Bacillus megaterium* KM, *Sarcina ureae,* and *Micrococcus lysodeikticus,* whose walls are composed mainly of peptido-glycan, may be converted into bodies essentially without cell walls. These have been named *protoplasts,* although the word *gymnoplast* is more precise. (Protoplast is derived from the Greek *protos*—first, and *plastos*—form or thing, hence, "first formed"; on the other hand, gymnoplast is from the Greek *gymno*—naked, and *plastos*—form, hence, naked form, that is, a form lacking a cell wall.) Protoplasts are defined as spherical (usually), osmotically unstable forms that are unable to divide, although they may grow and increase in volume for some time before bursting. Protoplasts possess most biochemical capabilities of their cell-walled counterparts, but they appear gener-ally unable to synthesize new cell walls if the peptidoglycan com-ponent is completely removed. Flagellated protoplasts are nonmotile, clearly indicating a role of the cell wall for both function of flagella, and, indeed, their polymerization from flagellin molecules. *Sphero-plasts* are defined as those spherical bodies that develop upon ex-posure to treatments that modify the cell wall, generally by inter-fering with the peptidoglycan layer, but which retain the ability to revert to their original form when conditions are normalized (Fig. 2.20). If treatment of a bacterium with penicillin in an osmotically stabilized medium, for example, results in the development of spherical forms lacking a rigid layer, inactivation of this antibiotic with penicillinase allows the spheroplasts to revert to their original bacterial form. D-Cycloserine, bacitracin, vancomycin, ristocetin, the penicillins, and the cephalosporins are antibiotics that affect the peptidoglycan formation, and may, under appropriate condi-tions, lead to the production of osmotically-sensitive forms.

The investigation of protoplasts and spheroplasts has been useful in determining the functions of cell wall components. The general conclusion is that their main function is mechanical, involving shield-ing against osmotic and environmental damage. The bacterial cell wall, however, possesses a variety of chemical structures interacting with elements of the environment. These include: virulence and toxicity fac-tors of wall components of gram-negative forms such as the lipo-polysaccharides, some of which act as endotoxins (Section 4.18); the sites for bacteriophage adsorption such as the O-antigen (a repeating trisaccharide chain composed of mannose, rhamnose, and galactose) to which the ϵ_{15}-phage attaches in *Salmonella anatum;* cell "recog-nition" sites required for conjugation; and antigenic sites reacting with specific antibodies which may cause clumping together of organ-

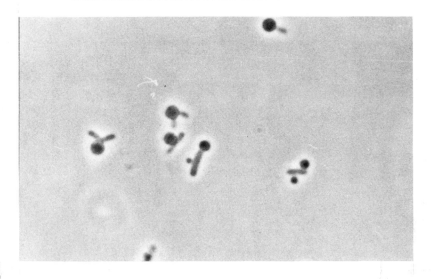

Fig. 2.20. Spheroplast formation in *Salmonella typhimurium*. In this phase-contrast photograph the rounded spheroplasts are seen developing both at the ends and centers of the organisms, and in a few instances more than one per cell are formed. (From Z. Vaituzis and R. N. Doetsch, Flagella of *Salmonella typhimurium* spheroplasts, *J. Bacteriol.* **89,** 1586 (1965). By permission of the American Society for Microbiology.)

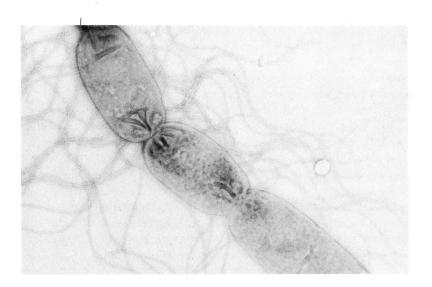

Fig. 2.21. *Escherichia coli* shown dividing by constriction. (Courtesy Z. Vaituzis.)

isms. The wall also serves as a coarse molecular sieve, and it is likely that substances of molecular weight greater than 10^4 are excluded by it except in special circumstances.

Although the "gram procedure" is an ancient laboratory staining technique, only now have the results of this operation been explained in terms of cell wall composition and structure, following countless other explanations invoking almost every possible cell constituent. The gram stain differentiates bacteria into two main groups, the gram-positive and the gram-negative. In essence, a dried film of bacteria contained on a glass slide is stained first with a solution of crystal violet and then with an iodine solution. The stained film is exposed to a decolorizing agent such as acetone or alcohol, and then it is counter-stained with another dye solution, usually safranine. Gram-negative bacteria do not retain the primary crystal violet stain when treated with a decolorizing agent; hence they take up the counterstain, saf-ranine, and appear reddish in color when examined with a microscope. Gram-positive bacteria retain the primary stain and thus are deep purple in color. The failure of gram-negative bacteria to retain the crystal violet-iodine complex is ascribed to their high cell wall lipid content and its solubilization upon treatment with organic solvents such as acetone or alcohol. In gram-positive bacteria the dye complex remains trapped in the cell, because of the barrier formed by the thicker, dehydrated peptidoglycan cell wall. Support for this explana-tion is supplied by results of various experiments on the removal of cell wall of a gram-positive bacterium with muramidase *after* the gram-staining procedure has been applied to it. Treatment with a de-colorizer completely removes the crystal violet-iodine complex and the organisms are converted into gram-negative forms.* A number of biochemical and physiological characteristics appear correlated with the gram-positive and gram-negative reactions of bacteria, and their cell wall structures may be a reason for, or a consequence of, their marked differences (Table 2.2). As with so many other laboratory procedures, some organisms appear to be neither altogether gram-positive nor gram-negative, and these are referred to as "gram-variable" forms. Presumably the peptidoglycan layers in these in-stances would be at a critical thickness in terms of the explanation given above. A definitive theory of the gram stain, that explains all of the observations reported, has yet to be propounded.

---

*It has been shown that the gram-negative *Escherichia coli* can be induced to become gram-positive by treatment with chloroform and methanol. The stability of the cell component-dye-iodine complex may also be a determinant of the gram characteristic.

Table 2.2.   Differences between Gram-Positive and Gram-Negative Bacteria*

Property	Gram-positive	Gram-negative
Cell wall thickness	Greater	Less
Cell wall ruptured by heat	No	Yes
Amino acids in cell wall	Few kinds	Numerous kinds
Separation of plasma membrane from cell wall	Difficult	Relatively easy
Hexosamine content of cell wall	More	Less
Magnesium content	More	Less
Excretion of putrescine	Slight or none	Present
Digestion by gastric juice	Resistant	Less resistant
Digestion of killed organisms by trypsin or pepsin	Resistant	Not resistant
Resistance to strong alkali (1% KOH)	Not dissolved	Dissolved
Solubility of lipids in fat solvents	Resistant	Less resistant
pH of optimal growth	Relatively high	At acidic values
Apparent isoelectric point by staining	pH 2–3	About 4–5
Amounts of acid and basic dyes retained at apparent isoelectric point	More	Less
Susceptibility to acriflavine dyes	Marked	Less marked
Bacteriostatic effect of triphenylmethane dyes	Very susceptible	More resistant
Inhibition by sodium azide	Resistant	Less resistant
Permeability to dye in living state	More permeable	Less permeable
Bacteriostatic action of iodine	More susceptible	Less susceptible
Autolysis	Less common	More commonly noted
Lysis by specific complement fixation	Not readily apparent	Often observed
Nature of toxins produced	Exotoxins	Endotoxins and antigens of Boivin type
Nutrient requirements	Generally complex, none autotrophic	Relatively simpler, many species autotrophic
Susceptibility to penicillin	More	Less
Susceptibility to low surface tension	More	Less
Susceptibility to anionic detergents	Very susceptible	Much less; susceptible only in acid media
Bacteriostatic action of tellurites	More resistant	More sensitive
Acid-fastness	Some species	None
Rupture by sudden changes in pressure and supersonic energy	More resistant	Less resistant
Optical density changes of suspensions with increasing solute concentration	None or little	Extinction increases

*From C. Lamanna and M. F. Mallette, *Basic Bacteriology*, 3rd ed. Williams and Wilkins, Baltimore, 1965. Courtesy C. Lamanna and permission of Williams and Wilkins Co.

2.7 | The Cytoplasmic Membrane and Its Ramifications

In thin-sectioned bacteria the cytoplasmic membrane, when observed with the electron microscope after staining with osmium tetroxide, appears as a structure consisting of two dense profiles each approximately 2 nm thick separated by a light band 3 nm thick. Membranes therefore appear as approximately 7 nm "dense-light-dense" layers. The gross chemical composition of these membranes is similar to that of other biological membranes, that is, about 50 to 75% protein and 20 to 35% lipid. The composition of purified membranes of *Micrococcus lysodeikticus,* for example, is reported to be from 65 to 68% protein, and from 23 to 26% total lipid. Of this the main phospholipids (from 75 to 80% of total lipids) are diphosphatidyl glycerol, phosphatidyl glycerol, and phosphatidyl inositol. Menaquinone-45 comprises from 4 to 5%, and carotenoids 0.5% of the remainder. Trace amounts of RNA, DNA, and carbohydrate occasionally are found, but these may be cytoplasmic contaminants. Investigations of properties of bacterial membranes usually have been made using preparations obtained from gram-positive bacteria. This is because the cell wall may be easily removed enzymatically, and the resulting protoplast membranes isolated after osmotic lysis.

In addition to the limiting cytoplasmic membrane, other intracellular membrane systems are present in bacteria. These include vesicular, lamellar, and tubular *mesosomes* (Fig. 2.22) resulting from ingrowth of the cytoplasmic membrane. Various gram-negative bacteria—including nitrogen-fixing, photosynthetic, nitrifying, and marine forms—possess extensive membranous systems. Lamellated membranes in *Nitrosocystis* (Fig. 2.23A) are probably sites of linked biochemical reactions leading to oxidations of ammonia to nitrite (Section 5.9).

The cytoplasmic membrane plays a principal role in cell division, and cell permeability, and it is a structure upon which functional macromolecules may be situated in orderly arrays. The cytoplasmic membrane is also the agent for differentiating subcellular structures in prokaryotic forms (Figs. 2.22 and 2.26 B and C). Cytochromes, dehydrogenases, ATPase, and NADH$_2$ oxidase (Section 5.1) are among the enzymes associated with membranes of gram-positive bacteria.

In cell division, ingrowth of the cytoplasmic membrane occurs at the approximate mid-point along the longest axis of rod-shaped cells, apparently because this is the point of least mechanical stability. Here

Fig. 2.22. Mesosome in *Escherichia coli* showing the close relationship between it and the nucleus. (From A. Ryter, Association of the nucleus and the membrane of bacteria: A morphological study, *Bacteriol. Rev.* **32,** 39 (1968). Courtesy A. Ryter and permission of the American Society for Microbiology.)

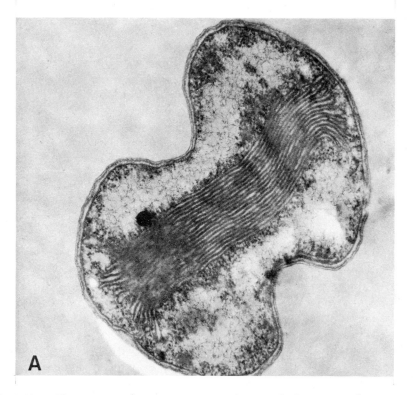

Fig. 2.23. *A,* Thin section of *Nitrosocystis oceanus* showing the beginning of constrictive division and a membranous organelle consisting of 20 vesicles so flattened that the lumen is only 10 nm thick; these lamellae almost completely traverse the organism cutting across both cytoplasm and nucleoplasm. Such membrane systems may be specialized mechanisms for acquiring energy from conversion of NH_4^+ to NO_2^-. (From R. G. E. Murray and S. W. Watson, Structure of *Nitrosocystis oceanus* and comparison with *Nitrosomonas* and *Nitrobacter, J. Bacteriol.* **89,** 1594 (1965). Courtesy R. G. E. Murray and permission of the American Society for Microbiology.)

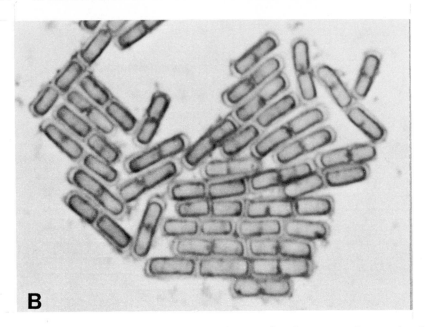

Fig. 2.23 *(Cont'd)*. *B,* Plasmolyzed *Bacillus cereus* stained with tannic acid—crystal violet after Bouin fixation to reveal the cytoplasmic membrane withdrawn from the cell wall. (Courtesy G. J. Hageage.)

the membrane is at its maximum distance from the curved cell surfaces at the poles, and continued membrane growth thus leads to an infolding (Section 2.6). The membrane, and *not* the cell wall, initiates the processes leading to cell division, but the specific signal for doing so, probably connected in some way with DNA replication, is not known. During division in gram-negative bacteria, membrane constriction rather than septum formation is observed, except under unusual circumstances. Consequently, the cell is "pinched off" from the daughter cell rather than "walled off" as in gram-positive forms.

The membranous invaginations of gram-positive and some gram-negative bacteria that are referred to above as mesosomes or "chondroids" have had a number of functions ascribed to them: (1) they are functionally equivalent to mitochondria, based upon evidence which suggests that they are active reduction sites of tellurium and tetrazolium salts; (2) they are electron transport organelles (for example, the mesosome fraction of *Bacillus subtilis* contains *all* of its cell cytochrome); (3) they serve to anchor the nucleus and mediate its replication; (4) they function in the synthesis of the cell wall septum

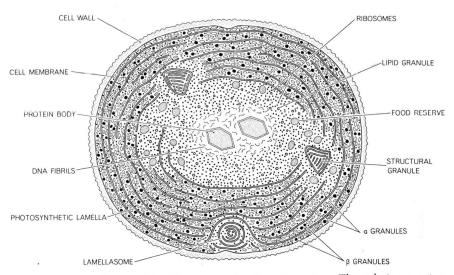

CELL WALL

RIBOSOMES

CELL MEMBRANE

LIPID GRANULE

PROTEIN BODY

FOOD RESERVE

DNA FIBRILS

STRUCTURAL GRANULE

PHOTOSYNTHETIC LAMELLA

α GRANULES

LAMELLASOME

β GRANULES

Fig. 2.24. Principal parts of the blue-green alga *Anacystis montana*. The gelatinous outer sheath is not shown here, and as in other blue-green algae, the highly specialized mitochondria and chloroplasts are lacking. (From P. Echlin, The blue green algae, *Scient. Amer.* **214,** 74 (1966). Courtesy P. Echlin and permission of the *Scientific American.*)

in gram-positive bacteria, and hence, in new cell wall formation; (5) they are centers for formation and secretion or transfer of exocellular proteins; (6) they are the site of fatty acid biosynthesis; and (7) they "guide" part of the DNA into the forespore of a sporulating organism (Section 2.10). Thus, mesosomes may serve a variety of structural and physiological functions. In some cases the evidence is inferential and more direct experimental results are needed to prove these suggested functions.

In penicillin-treated bacteria, mesosome vesicles, following effects of the drug on the growing transverse septum, are eventually extruded into the cell wall-cytoplasmic membrane interspace. The same phenomenon occurs in bacteria following osmotic shock or upon treatment with muramidase.

Fundamental differences in structures employed for photosynthetic processes sharply demarcate prokaryotes from eukaryotes. The detailed structure of the prokaryotic photosynthetic apparatus has yet to be worked out, but it appears to be a pigment-containing system of membranous tubes and vesicles. Three groups of photosynthetic prokaryotes are known; these are the blue-green algae (Fig. 2.24), the green bacteria, and the purple bacteria (Section 3.19).

A differentiated portion of the cytoplasmic membrane may be involved in the operation of bacterial flagella. It is known that the base (proximal) portions of flagella are intimately associated with the membrane (Fig. 2.8), and that chemical agents active against the membrane interfere with motile function, but this function for membranes has not been unequivocally proven.

In the presence of a high concentration of impermeable solutes (a *hypertonic* solution) the cytoplasmic membrane of gram-negative organisms may be observed to withdraw inward from portions of the cell wall, leading to the formation of vacuoles or gaps due to the outward flow of water. This phenomenon is known as *plasmolysis* and is ordinarily a nonlethal event (Fig. 2.23B). Eventually the osmotic pressure is equilibrated and the organisms again appear normal. Interestingly, strongly plasmolyzed bacteria always become nonmotile if originally motile. Plasmolysis in gram-positive bacteria is difficult to demonstrate except in the presence of very high concentrations of external solutes such as 1 M KNO_3. It may be that in these organisms the whole cell shrinks, and in any event, the cytoplasmic membrane is firmly attached to the cell wall. *Plasmoptysis* is a term used to describe the bursting of cells as a result of increased internal pressure caused by water intake in solutions of osmotic pressure less than that of the cell contents (*hypotonic* solutions). This would only occur in bacteria lacking strong cell walls.

The development of a limiting membrane was probably an early event in the history of life on earth. Organisms capable of growth and multiplication must have been organized within a definite boundary area. Dilute essential solutes of the environment must be concentrated within the cell, and this is accomplished by reactions mediated by membrane-bound enzymes. These are responsible not only for the specificity of solute selection, but also for maintenance of high internal levels against a concentration gradient. Such solutes then are vectorially processed within the cell by other membrane systems. The membranous organelles in eukaryotic forms, such as the nuclear membrane, mitochondrial membrane, rough and smooth endoplasmic reticulum, chloroplasts, and the like, seem to attest to this. Perhaps improvements in delicacy and specificity of fixation and staining techniques, in conjunction with electron microscope observations, will reveal further membranous organization in the prokaryotic cytoplasm in the form of a fine network or "spongy" matrix upon which the many complex multienzyme systems are arrayed.

2.8 | The Nucleus

Bacterial genetic material resides mainly in a *nucleus,* but this does not imply identity with nuclei of eukaryotic forms. Bacterial nuclei* (Fig. 2.1A and B) have great plasticity of form, though not of texture, characterized by absence of many of the hallmarks of eukaryotic nuclei, namely, a nuclear membrane, associated proteins (histones), staining reactions, a resting stage, and the ordinary features of mitotic division.

The nucleus of a bacterium, such as *Escherichia coli,* is comprised of a single closed double-stranded DNA loop with a total length of 1×10^3 to 1.4×10^3 μm. During exponential growth a single cell may possess from two to four nuclei, and at lower growth rates, either one or two. A single organism contains about 10^{-14} g of DNA during the exponential growth phase. The nucleus must be tightly coiled to fit into the small volume of the cell, and it is difficult to understand how the DNA is, in fact, fitted in and replicated. The closed bacterial nucleus approximately 1 mm long must be duplicated in fast growing bacteria within a division time of 15 minutes. Topological considerations aside, the double helix, at this rate of replication, must unwind at the rate of 5,000 revolutions per minute! The mechanics of this process are still unknown although various theories have been put forward. It has been calculated that only 20% or less of a generation time would be available for *separation* of DNA strands if this occurred after and not during replication. From available cytological investigations it seems reasonable to assume that replication and separation of DNA occur simultaneously.

Prokaryotic organisms are as genetically stable as eukaryotic ones, and one must conclude that, although their mode of nuclear replication is indeed distinctive, there is no disadvantage in its operation. However, mutations to a recessive characteristic such as streptomycin resistance are immediately expressed; consequently, bacteria appear to be haploid,† not diploid; otherwise remaining unmutated dominant genes would still be expressed.

---

*The bacterial *genophore* is a discrete unit of genetic material that corresponds to a linkage group and this designation is preferable to nucleus.

†A chromosome may be defined as a linkage structure consisting of a linear arrangement of genes. In bacteria, in addition to a "standard" chromosome, a number of independent genetic elements (*episomes* and *plasmids*) may exist in the cytoplasm; consequently, the genome of bacteria may consist of any number of chromosomes.

In electron micrographs of thin sections, the area of the bacterial nucleus appears as a ribosome-poor region characterized by a parallel arrangement of DNA fibers or aggregates. These 2-nm thick fibers appear tightly folded back and forth over one another. Mesosomes also appear to have a role in nuclear replication by mediating a process involving the cytoplasmic membrane. Mesosomes appear to contact the DNA over a fairly wide area, and are probably attached to it.

2.9 | Ribosomes

The cytoplasm of most nonphotosynthetic, nonsporulating bacteria is comparatively featureless, that is, vacuoles of the kinds seen in eukaryotic forms, cytoplasmic streaming, and membrane-bounded nuclear and respiratory elements are lacking. One characteristic feature seen in stained thin sections of bacteria is the abundance of *ribosomes,* scattered throughout the cytoplasm (Fig. 2.1A and B). Isolated bacterial ribosomes appear as particles approximately 20 nm in size, but contrary to the picture observed in eukaryotic cells there is no ribosome-studded endoplasmic reticulum.

Ribosomes of 70 S (S stands for *Svedberg unit;* it equals a sedimentation rate in the ultracentrifuge of 10^{-13} cm/sec/unit field) size, which are the functional forms in bacteria, characteristically are composed of two chemically nonidentical subunits, 30 S and 50 S, one of which is twice the size of the other. These subunits, which may be obtained by dissociation *in vitro* of 70 S ribosomes in a Mg^{2+} poor medium, recombine to make the 70 S form when the Mg^{2+} concentration is brought to 10^{-2} M. In *Escherichia coli,* ribosomes have been found to consist of 63% ribonucleic acid (RNA) and 37% protein. The 30 S subunit has a molecular weight of ~0.9×10^6, and the 50 S subunit of ~1.8×10^6. The proteins of ribosomes are complex, and more than 30 different kinds, mostly basic, have been obtained from 70 S ribosomes of this bacterium. The average molecular weight of these proteins is around 25×10^3; it is estimated that in *E. coli* there is sufficient protein to allow for one molecule of each kind in each ribosome.

Ribosomal aggregates called *polysomes* are ribosomes held together by a filamentous thread, probably messenger RNA (mRNA). Each 70 S ribosome moves along the mRNA and forms a polypeptide chain as it travels down its length. When one 70 S ribosome has gone sufficiently far from the initiation point, another can be added onto the mRNA. When the ribosome reaches the end of the mRNA it is detached and the polypeptide is released.

2.10 | Endospores, Cysts, Conidia, and Myxospores

The principal differentiated structure of eubacteria is the *endospore;* however, *cysts, conidia,* and *myxospores (microcysts)* are produced in other prokaryotes. Endospores develop in *Bacillus* and *Clostridium,* cysts in *Azotobacter,* conidia in various actinomycetes, and myxospores in myo-bacteria. Endospore formation is not a widely distributed trait among prokaryotic organisms, and although endospores may confer certain advantages, numerous kinds of bacteria have maintained themselves successfully without them.

The endospore is dense, highly refractile, difficult to stain, and very resistant to inimical external environmental conditions (Fig. 2.25). Although the functional reason for endospore heat resistance is not known, this property has been useful experimentally in sporulation and germination studies. Cysts, conidia, and myxospores are at best only weakly heat resistant.

In *Bacillus* and *Clostridium, Sporosarcina,* and *Spirillum,* one endospore is formed which is generally liberated upon lysis of the *sporangium,* or mother cell; hence, the process is not reproductive. Endospore formation may guarantee survival of a sporulated organism, but either the environment has to become favorable for germination or the endo-spore must be carried to one that is. This point should be noted since aerial distribution of fungal spores is a method for making this possible.

The *biological* role of endospores has been speculated upon exten-sively, but remains unknown. Bacteria do not form endospores in adverse conditions. In fact, the environment in which sporogenesis occurs must be physiologically favorable for growth and biosynthesis. Once formed, however, an endospore is resistant not only to a high degree of heating, even boiling in some cases, but also to desiccation, freezing, radiation, and numerous toxic and lethal chemical agents. Bacterial endospores may continue to remain viable, or retain ability to germinate under appropriate conditions for an indefinite period, and therefore, no time limit may be placed on the sporogenous state.

The development of a bacterial endospore is a complex cytological event, and current ideas on the sporulation process are derived from electron microscopic studies of thin-section preparations. Bacterial sporulation is a useful process for investigating differentiation at the unicellular level (Fig. 2.26B and C).

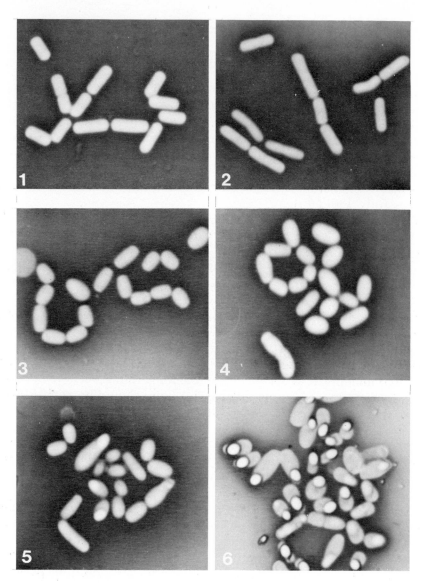

Fig. 2.25. Development of endospores in *Clostridium perfringens* as seen using light micros-copy of nigrosin films. (1) 0 hr; (2) 1.5 hr; (3) 3 hr; (4) 4 hr; (5) 5 hr; and (6) 6 hr. (From J. F. M. Hoeniger, P. F. Stuart, and S. C. Holt, Cytology of spore formation in *Clostridium perfringens,. J. Bacteriol.* **96**, 1818 (1968). Courtesy J. F. M. Hoeniger and permission of the American Society for Microbiology.)

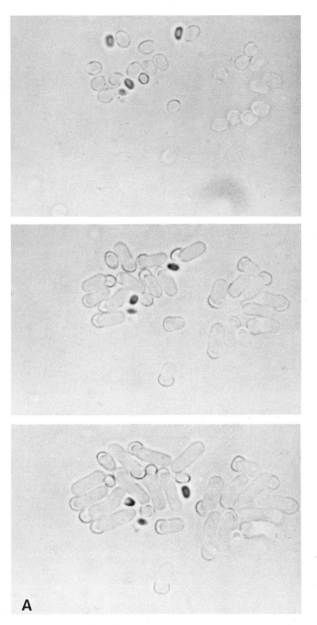

Fig. 2.26. *A*, Morphological changes during the germination of *Bacillus* endospores. (Courtesy C. F. Robinow.)

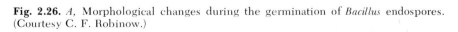

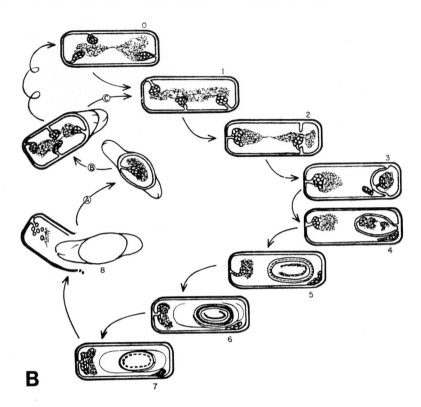

Fig. 2.26 *(Cont'd) B,* Diagrammatic summary of sporulation in a *Bacillus* species. Membrane and associated mesosomes are shown. 0–1, transition from replicating cell to axial stage; 2, 3 and 4, stages in forespore development. At stage 4 the cell becomes "committed" to proceed to 8. Thus a cell at stage 3 can be returned to the vegetative form by fresh medium. Stage 5, cortex development commences *(dotted line)* and continues through 6 when the coat protein is deposited. Stage 7 is characterized by a dehydration of the spore protoplast and an accumulation of DPA (dipicolinic acid) and calcium in the spore. In stage 8 (complete refractility), a lytic enzyme acts to release the spore. Also shown are germination *(A);* outgrowth to a primary cell *(B);* from which the cell may, under special conditions, enter sporulation by a shortcut *(C),* "the microcycle", but normally undergoes logarithmic growth *(spiral arrow).* (From P. Fitz-James and E. Young, Morphology of sporulation, in G. W. Gould and A. Hurst (eds.), *The Bacterial Spore* (1969). Courtesy of P. Fitz-James and permission of Academic Press.)

The cytoplasmic membrane participates in the unfolding, via mesosomes, of a spore septum which eventually carves out a *forespore* containing the nuclear material of the future endospore. Mesosomes are not absolutely essential to the process, since protoplasts which are devoid of them will sporulate, if the organism has developed a fore-

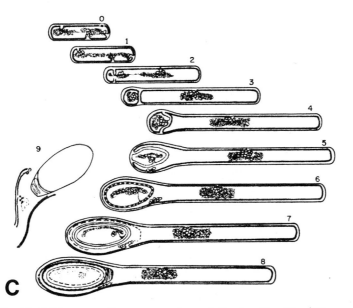

Fig. 2.26(*Cont'd*). *C*, Diagrammatic summary of sporulation in a *Clostridium*. The processes, although basically those shown in Fig. 2.26*B* for a *Bacillus*, are accompanied by considerable growth in cell length and a clubbing to accommodate first the forespore (steps 4 and 5), then the cortex (steps 6 and 7), and finally coats. Refractility (step 8) glazes the electron microscopic detail, and lysis (step 9) liberates the completed spore. Storage substance fills the stem of the cell but is not drawn in. (From P. Fitz-James and E. Young, Morphology of sporulation, in G. W. Gould and A. Hurst (eds.), *The Bacterial Spore* (1969). Courtesy of P. Fitz-James and permission of Academic Press.)

spore prior to removal of the cell wall. Subsequent differentiation of the membrane surrounding the endospore cytoplasmic material leads to the formation of a multi-layered body of complex structure (Fig. 2.27A and B). The finished endospore is liberated from the sporangium as an elliptical, cylindrical, or spherical body. While still in the sporangium the spore may cause it to bulge at the center or the end, resulting in "racket" or "snowshoe" forms (Fig. 2.26C).

Surrounding the endospore is a membranous envelopment called the *exosporium* (Fig. 2.27A). In some bacteria it lies loosely draped over the endospore, as in *Bacillus cereus,* whereas in others it adheres tightly to the endospore surface, as in *B. megaterium.* In *B. anthracis* the exosporium has a "nap" structure of unknown function projecting from its surface (Fig. 2.27B). The exosporium is formed *within* the vegetative cell or sporangium at the time of sporulation and is not a residual product.

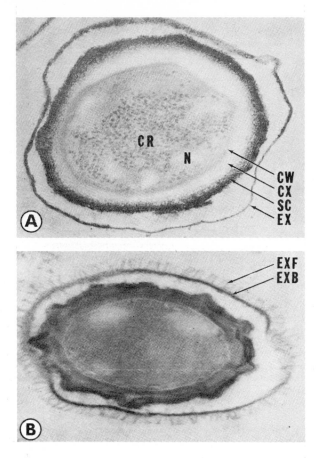

Fig. 2.27. *A,* Dormant endospore in cross section stained with permanganate, uranyl, and lead ions. The main structural components of the spore are evident: exosporium (EX), spore coat (SC), cortex (CX), core wall (CW), and nucleoplasm (N), and cytoplasm with ribosome granules (CR).

B, Endospore showing a fringe (EXF) of hairlike projections on the outer surface of the exosporium basal layer (EXB). (*A* and *B* from B. J. Moberly, F. Shafa, and P. Gerhardt, Structural details of anthrax spores during stages of transformation into vegetative cells, *J. Bacteriol.* **92,** 220 (1966). Courtesy P. Gerhardt and permission of the American Society for Microbiology.)

The endospore consists of a series of layers, the outermost of which has a characteristic surface distinctive for each species. Grooved, spiked, dimpled, and corrugated endospore surfaces of great beauty are revealed in carbon replicas examined with the electron microscope (Fig. 2.28). Beneath one or two of the *spore coats* there occurs

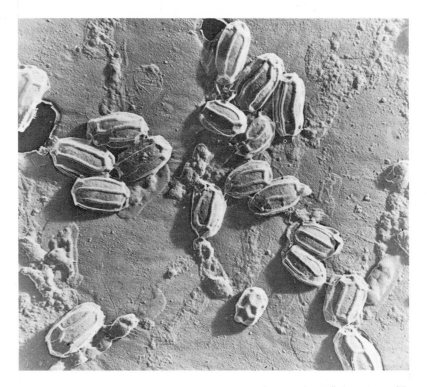

Fig. 2.28. *Bacillus polymyxa* endospores as seen in carbon-replica preparations. (From D. E. Bradley and J. G. Franklin, Electron microscope survey of the surface conformation of spores of the genus *Bacillus*, *J. Bacteriol.* **76**, 618 (1958). Courtesy D. E. Bradley and permission of the American Society for Microbiology.)

a layer called the *cortex,* a region comprised of concentric layers of fibrils. It is believed that here the major portion of *dipicolinic acid* (DPA), a compound found in all endospores, is located. DPA is not detectable in the vegetative bacterial cell, but as a complex with calcium, it constitutes up to 10% of the endospore dry weight. If sporulation proceeds in a calcium-deficient medium, aberrant endospores result. These have lessened heat resistance and much less refractility than normal. Next, there is the *spore wall,* a relatively thin structure separating the cortex from the spore cytoplasm or *core.* This spore wall eventually forms the cell wall of the germinating organism (Fig. 2.26A).

Sporulation may occur in the complete absence of nutrients; in distilled water or buffer. This process is called *endotrophic sporulation.*

In this case the materials for endospore formation are derived exclusively from internal or endogenous sources.

After commitment to sporulation, some members of the genus *Bacillus,* for example *B. thuringiensis, B. medusa, B. popilliae,* and *B. laterosporus,* produce crystalline inclusions, usually one per organism, termed *parasporal bodies* (Fig. 2.29A and B). These crystalliferous bacteria form diamond-shaped, cuboidal, rhombiodal, canoe-shaped, and irregularly shaped inclusions, which in nearly all cases lie *outside* the exosporium. When endospores are released upon sporangium lysis, crystals are seen lying among them. The crystal formed by *B. thuringiensis* is lethal to insects and, hence, is a microbial insecticide. It is inert to proteolytic enzymes such as trypsin, chymotrypsin, and pronase, and is insoluble in water and dilute nitric and hydrocholoric acids, but soluble in alkaline solvents, like sodium or ammonium carbonate, and sodium hydroxide. It is also a protein, formed from amino acids resulting from the breakdown of intracellular protein, and it consists of 18 amino acids of which aspartic and glutamic acids constitute 25% of the total. This parasporal body has a silicon-containing framework, and is a face-centered bipyramidal crystal possessing a finely striated surface structure.

In the alkaline gut of the silkworm, the toxic component of the crystal is released by the action of nonenzymatic alkaline factors, resulting in gut paralysis within minutes. Naturally occurring crystalline toxins are rare and their pathogenicity for lepidopterous insects is difficult to explain in functional terms. It may be that crystal formation is an organism's way of dealing with certain materials which are, in fact, toxic to itself. However, not all strains of organisms of the same species form them. The involvement of silicon is noteworthy, but its role in the process of toxic crystal formation is obscure.

Among some of the sporogenous soil clostridia, attached endospore appendages are visible while the endospore is in the sporangium and when it is liberated. These appendages are of great structural and morphological diversity, and various hypotheses have been forwarded explaining their probable function. The appendages are generally tubular, proteinaceous, and appear to be derived from the outer spore coat rather than the exosporium (Fig. 2.30). Laboratory cultures may lose the ability to form endospore appendages, whereas freshly isolated strains generally possess it. It may be that these structures function as a means of attaching the free endospore firmly to a surface, although too firm an attachment might ultimately prove disadvanta-

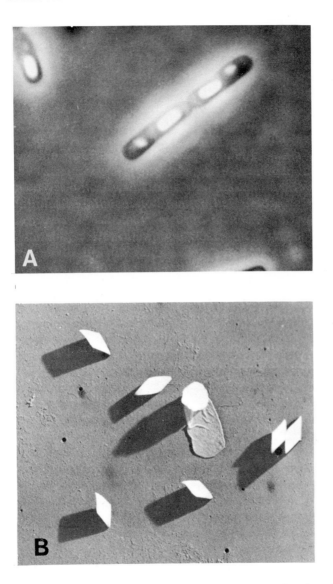

Fig. 2.29. *A,* Phase-contrast view of *Bacillus thuringiensis* showing the refractile endospore and accompanying parasporal body. (Courtesy C. Reichelderfer.)

B, Shadowed preparations of the parasporal bodies of *Bacillus thuringiensis.* An endospore and its accompanying exosporium are also shown. (From C. L. Hannay and P. Fitz-James, The protein crystals of *Bacillus thuringiensis* Berliner, *Canad. J. Microbiol.* **1,** 694 (1955). Courtesy C. L. Hannay and permission of The National Research Council of Canada.)

Fig. 2.30. Free endospore of *Clostridium taeniosporum* showing ribbon-like appendages flared out from the spore body. (From L. J. Rode, M. A. Crawford, and M. G. Williams, *Clostridium* spores with ribbon-like appendages, *J. Bacteriol.* **93,** 1160 (1967). Courtesy L. J. Rode and permission of the American Society for Microbiology.)

geous. They may serve as chemosensory organelles from which signals could be transmitted concerning suitability of the external environment for germination. Perhaps they are involved in transport (*see Pili,* Section 2.4) of materials into the endospore itself. The stability of clostridial endospore appendages varies with cultural conditions, and thus far, strictly anaerobic environments have been found necessary for their formation. The appendages are not absolutely necessary for germination, since not all endospores possess them.

Various spore appendages have been described as "ribbon-like," "pin-like," "spine-like," "hirsute," and so forth. They all have marked structural detail when examined with the electron microscope. Some *B. anthracis* endospores have been described as surrounded by a nap of filaments, but this has been confined mainly to clostridial endospores. In any case, endospore appendages are representative of "structures in search of a function."

Cysts formed by *Azotobacter* differ markedly from endospores in structure (Figs. 2.31 and 2.32). Cysts lack heat resistance almost without exception, but as in endospores, they are resistant to desiccation. Cysts have a lamellar outer part, or *exine*, which has a bark-like surface. Within the exine is a fibrillar layer called the *intine*, which surrounds a contracted cytoplasm in which prominent globules of poly-β-hydroxybutyrate are embedded (Section 2.11). The nuclear material may be seen directly in thin sections and appears similar in texture to that in the vegetative cell.

Conidia are weakly heat-resistant reproductive structures produced by actinomycetes (Section 3.8). They develop by constriction and fragmentation of a specialized part of the aerial mycelium. These oval or spherical bodies possess a thick wall and vacuolated cytoplasm. The conidial surfaces are either smooth or covered with protuberances, spines, and other distinctive details. Since many conidia are formed from a single mycelium, they are reproductive in function (Fig. 3.16A and B).

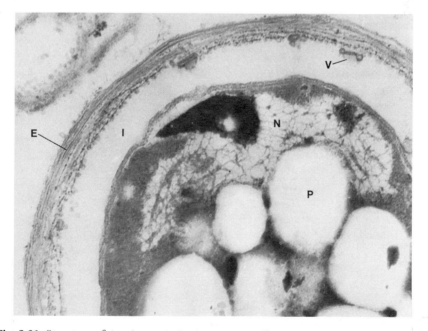

Fig. 2.31. Structure of *Azotobacter vinelandii* cyst. Lamellar exine (E), empty intine space (I), intine vesicles (V), nuclear area (N) and poly-β-hydroxybutyrate (P) inclusions are shown. (From G. R. Vela, G. D. Cagle, and P. R. Holmgren, Ultrastructure of *Azotobacter vinelandii, J. Bacteriol.* **104**, 933 (1970). Courtesy G. R. Vela and permission of the American Society for Microbiology.)

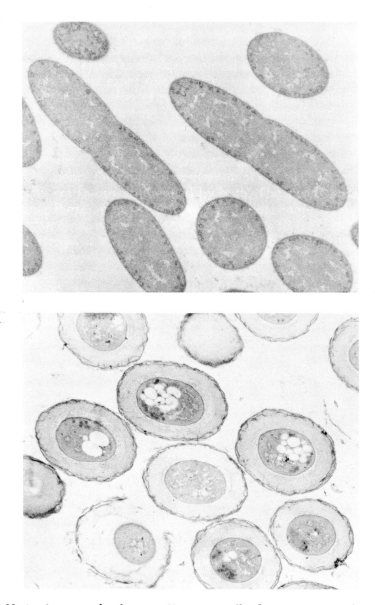

Fig. 2.32. *Azotobacter* cyst development. Vegetative cells of *Azotobacter (top)* and sectioned cysts *(bottom)* showing the complex structure of these bodies. Note the bark-like exine and the numerous inclusions of poly-β-hydroxybutyrate. (From O. Wyss, M. G. Neumann, and M. D. Socolofsky, The development and germination of the *Azotobacter* cyst, *J. Biophys. Biochem. Cytol.* **10**, 555 (1961). Courtesy O. Wyss and permission of the Rockefeller University Press.)

Members of the genus *Myxococcus* (Section 3.4) form spherical myxo-spores. These are thick-walled resting cells that are resistant to drying, and they may remain dormant for long periods. As in the case of endospore and cyst formation, one vegetative cell gives rise to one myxospore (Fig. 2.33).

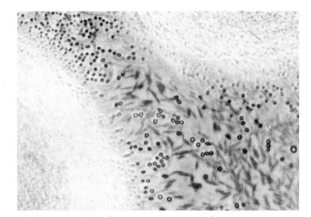

Fig. 2.33. Two fruiting bodies of *Myxococcus xanthus* showing vegetative organisms and myxospores (dark spheres) lying between them. (Courtesy R. Burchard.)

2.11 | Cytoplasmic Inclusions

In nonsporeforming, nonphotosynthetic bacteria, the cytoplasm is differentiated mainly into a nuclear and a nonnuclear region. However, in many organisms characteristic, generally spherical, accumulations of nonprotoplasmic material occur, to which the name *cytoplasmic inclusion* or merely "inclusion" has been given. Some inclusions are energy reserves which may be utilized when exogenous sources are depleted. Others probably represent accumulated waste materials de-toxified to a polymeric form. Bacterial cytoplasm is not vacuolated.* It seems likely that all inclusions are membrane-bounded, although this has yet to be generally demonstrated. It is difficult to visualize various inclusions freely floating in the cytoplasm like so many raisins in a bowl of porridge. Inclusions may appear in organisms after maxi-mum growth (Section 4.4) of the culture, or earlier, if all enzyme systems are saturated by an excess of nutrients. The most commonly

---

*It has been pointed out by Stanier (1970) that although the cytoplasmic membrane of bacteria may undergo deep invaginations, it apparently lacks the plasticity "to inter-nalize a relatively large external object."

occurring inclusions in bacteria are *metachromatic granules,* poly-β-hydroxybutyric acid droplets, sulfur droplets, and polysaccharide accumulations.

Metachromatic granules are deposits containing, among other things, unbranched, condensed phosphates (polyphosphates), that occur singly or in various numbers in a given cell (Fig. 2.34). Polyphosphates have an elementary composition of $M_{N+2}P_NO_{3N+1}$. They are roughly spherical, they vary little in diameter, and, in a stained rod-shaped bacterium, they appear similar to "peas in a pod." In the living organism, polyphosphate polymers appear as highly refractile granules or spherules. The amount formed depends upon culture conditions and the nature of the organism. The term "metachromatic granule" arises from the observation that they may be reddish-violet when stained with blue dyes such as aged alkaline methylene blue or toluidine blue, a reaction termed *metachromatism.* The color shift is due either to changes in dye concentration or to the state of dye dispersion. Metachromatic granules may not consist solely of polyphosphate and some work suggests that they include RNA, lipid, protein, and magnesium. Since these granules have never been isolated or chemically purified, conclusions about their composition come from observing staining reactions following treatment with various reagents or exposure to enzymes, and some so-called metachromatic granules may not, in fact, be composed of

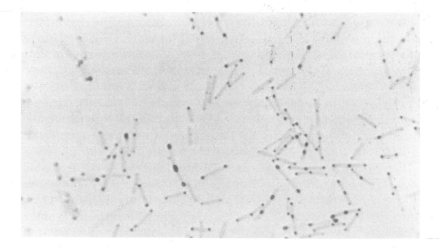

Fig. 2.34. Metachromatic granules in *Lactobacillus* stained with methylene blue. Some of the granules appear to be bulging the cells slightly, and many of them appear at the poles or cell septa. (Courtesy G. J. Hageage.)

polyphosphate exclusively. The former approach exemplifies the aim of "analytical morphology," that is, isolation and chemical analysis of the purified structure; the latter is that of "direct cytochemistry," in which cellular reactions to specific dyes are used to localize and identify structures *in situ*.

Poly-β-hydroxybutyric acid (PHB) droplets are membrane-bounded spherules that are rather smooth-appearing and differ greatly in diameter, even in the same organism. Production of PHB is dependent upon specific culture conditions, and in some instances it may comprise a large proportion of cell dry weight, for example, in cultures of *Caryophanon latum* almost 50% of the total dry weight may be composed of this polymer. If organisms accumulating PHB are placed in a medium lacking an energy source, but satisfactory in other respects, it may be utilized as an internal reserve (Fig. 2.35). PHB is a cell material produced solely by prokaryotic organisms.

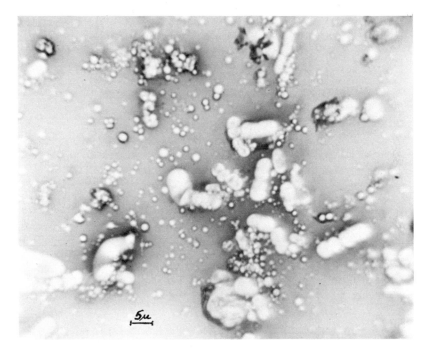

Fig. 2.35. Poly-β-hydroxybutyrate droplets released from lysing cells in a 6-day culture of *Caryophanon latum*. The globules appear to be of various diameters in this nigrosin film preparation. (From P. J. Provost and R. N. Doetsch, A reappraisal of *Caryophanon latum, J. Gen. Microbiol.* **28,** 547 (1962). By permission of the Society for General Microbiology.)

Sulfur droplets occur as spherical, refractile accumulations in the larger sulfur bacteria (Fig. 2.36 A and B). They, too, may be used as an endogenous reserve, and in cases where exogenous sulfide becomes exhausted, internal sulfur deposits may be utilized. The precise form of cytoplasmic sulfur is not known. Generally, sulfur chemically extracted from bacteria and then crystallized is in rhombic form. But inside the organism sulfur may have a quite different physical form (perhaps liquid), if it is to be mobilized for metabolic purposes.

Polysaccharide inclusions are reserve substances stored by some bacteria, and these substances, if they stain blue with iodine, are generally starch-like or amylopectin polymers. In clostridia, iodine stains the deposits reddish-purple, and they have been given the name *granulose*. Polysaccharide accumulations tend to be amorphous and seem more or less generally distributed throughout the cytoplasm. Such cytoplasmic inclusions are seldom utilized by the bacteria that make them if supplied externally in purified form as a sole carbon source.

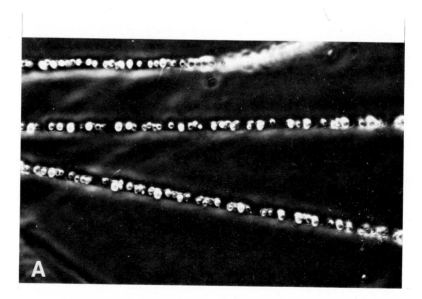

Fig. 2.36. *A, Beggiatoa* trichomes showing stored globules of sulfur. The size of the globules varies considerably and there appears to be no favored location for their deposition. (From S. D. Burton and R. Y. Morita, Effect of catalase and cultural conditions on growth of *Beggiatoa, J. Bacteriol.* **88,** 1755 (1964). Courtesy R. Y. Morita and permission of the American Society for Microbiology.)

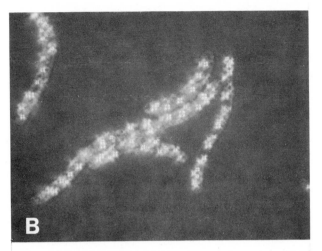

Fig. 2.36 *(Cont'd)*. B, Appearance of *Thiospirillum jenense* when viewed with a polarizing microscope. The sulfur globules have a characteristic "Maltese Cross" appearance suggesting that they are spherically symmetrical aggregates of radially arranged arrays of S^8 molecules. (Courtesy G. J. Hageage.)

2.12 | Rhapidosomes

Bacteria of the genera *Saprospira, Pseudomonas, Proteus,* and *Photobacterium* possess intracellular rod-shaped structures given the name *rhapidosomes,* their number varying from one to twelve per organism. Rhapidosomes are structurally diverse, appearing as hollow or solid cylinders around 25 nm in diameter and 250 nm in length (Fig. 2.37). Their function is unknown, but it has been suggested that they could serve as "anchors" for the nucleus, or play a part in nuclear replication. Gram-negative bacteria such as *Escherichia, Klebsiella,* and *Salmonella,* and gram-positive bacteria such as *Bacillus, Mycobacterium,* and *Micrococcus* have not been shown to contain rhapidosomes. The evidence thus far adduced suggests that they are most likely to be defective bacteriophages of the tailed type.

A number of intracellular tubular structures have been observed in bacteria. Among these, the *bacteriocins* are best known. These particles are recognized mainly by the fact that they specifically lyse susceptible organisms, but not the organism producing them. Bacteriocins do not multiply within the target organisms, but act as highly specific antibiotic substances. A bacteriocin termed "pyocin" (from *Pseudomonas aeruginosa*) consists of particles resembling the tails of certain bacteriophages (Section 4.18).

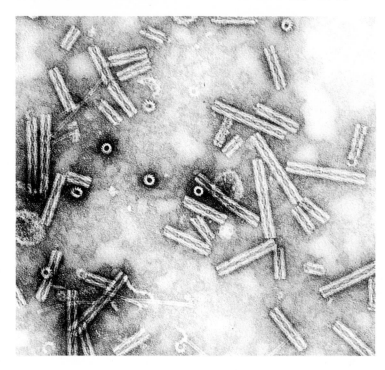

Fig. 2.37. Purified rhapidosomes from *Saprospira grandis;* both cross-sections and longitudinal sections are seen. (From R. E. Reichle and R. A. Lewin, Purification and structure of rhapidosomes, *Canad. J. Microbiol.* **14,** 211 (1968). Courtesy R. E. Reichle and permission of the National Research Council of Canada.)

2.13 | Colony Formation

Repeated division of bacteria at a fixed site in or on a solid medium eventually results in the surface or subsurface development of a macroscopic aggregation of individuals called a *colony* (Fig. 2.38A and B). Differences in cell surface characteristics, mode of cell division (by constriction or septation), and physicochemical conditions of the medium are reflected in the gross topological, textural, and physical appearance of such macroformations. The conventional designations for gross colonial appearance are *mucoid* (M), *smooth* (S), *and rough* (R). A given organism may undergo mutations, particularly those affecting cell surface components, and thereby change from colony appearance S to R, or M to R, or M to S.

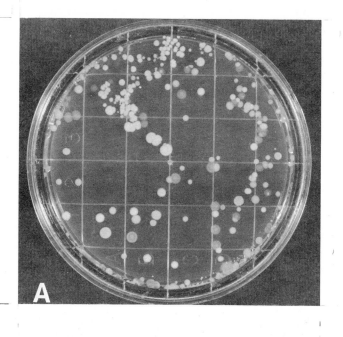

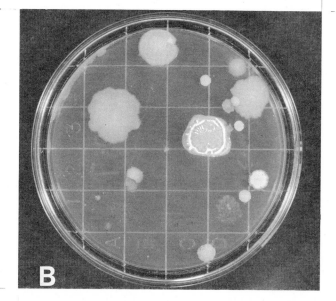

Fig. 2.38. Surface colonies on nutrient agar showing smooth and regular *(A)*, and rough and irregular *(B)* features as described in the text.

The M colony is viscous, almost watery and glistening, and relates to the consistency of the encapsulated forms responsible for its production (Section 2.3). In addition, the organisms comprising the M colony are usually observed to separate completely, immediately following cell division. The S colony is butyrous, smooth, and shiny, and the bacteria therein are not conspicuously capsulated. Cells may adhere in short chains, but never long ones. R colonies appear dull, rugose, and are sometimes infrangible, with usually irregular edges, and the organisms comprising it are arranged in festoons of long chains.

Occasionally colonies are observed that move as a *unit* over the surface of a solid medium. *Bacillus rotans* and *B. alvei* colonies are composed of highly motile organisms arranged in such a manner that the *entire* colony partakes of their movement, which is generally a clockwise or anticlockwise rotation. As the colony progresses and rotates along the surface it leaves a trail of slower organisms behind it. This behavior sometimes has been interpreted as evidence for a cooperative function in bacteria, but the motile colony may be a phenomenon more fortuitous than purposeful. Distribution of independent organisms in a static colonial mass is not advantageous to all, since various individuals are subjected to differences in aeration, nutrient supply, and waste products, and perhaps colonial movement allows for a more uniform exposure of organisms to renewed favorable conditions.

Postfission movements of bacteria may be related to colony characteristics. These movements, following cell division, have been described as slipping, snapping, and whipping. Slipping movement, characterized by bacteria in S colonies, consists of a motion past the daughter cell immediately following division, so that there is a random arrangement of individuals within the colony. Snapping movement, characterized by bacteria of the aerobic, sporogenous group, is a movement whereby the divided cells push against one another, thus "snapping" at the point of division. Generally these organisms form long chains of cells which remain attached after division, and the snapping occurs at points of stress where elements growing in opposite directions meet, thus imparting a force on the cells in the elongating chain. Colonies composed of such organisms are usually R. In the corynebacteria, a whipping postfission movement has been observed; here the divided organisms remain attached and one rapidly whips through about a 180° angle. Thus, colonies of M, S, and R types reflect cell surface features and postfission movements of the bacteria constituting them.

It is uncertain whether bacterial colonies are the result of laboratory manipulations, or whether they actually occur in nature. Colonies of a sort are probably formed in natural instances, such as "micro-colonies" too small to be noticed, or in abscesses, tooth plaques, or masses of organisms embedded in a slimy matrix (*zoogloea*), and in acute localized infections. Conditions in nature only occasionally offer the surfeit of nutrients and absence of competition suitable for colony formation of the kind seen in laboratory cultures. The exquisite and sculptured beauty of these formations may, after all, be only chimeras of the laboratory!

Bacteria cultivated in fluid mediums also reflect the characteristics of the organisms involved. The formation of stable, evenly turbid, or unstable, agglutinable suspensions and the development of surface pellicles or films, or viscous slimes, depends upon numerous inter-related factors. Some of these factors are the presence of capsules and surface polysaccharides, the relative hydrophilic and hydrophobic nature of the surface, the pH of the medium, and the presence and concentration of various inorganic ions. Bacteria are generally negatively charged at physiological pH, and this charge is sufficient to prevent them from autoagglutinating. Surface pellicles and films are formed by obligately aerobic bacteria, such as gram-positive sporogenous rods. The medium beneath the pellicle is usually clear and almost completely devoid of organisms.

2.14 | Gross Chemical Composition of Bacteria

The chemical composition of bacteria varies directly with that of their environment, and it is necessary to specify growth conditions precisely if meaningful statements are to be made. We will merely note here that bacterial cells are about 70–90% water, and they, therefore, resemble other cells in this regard. The dry weight (10–30% of the "wet weight") of bacteria has an approximate elementary analysis (in percentage) of carbon, 50; oxygen, 20; nitrogen, 15; hydrogen, 8; sulfur, 3; and phosphorus, 1. Small amounts of potassium, magnesium, sodium, calcium, and chlorine, and trace amounts of boron, copper, manganese, aluminum, and iron also are found. The organic matter is composed of roughly 40–80% protein, 1–30% carbohydrate, and 1–40% lipid, depending upon cultural conditions.

It is of interest that living matter, including bacteria, contains elements whose order of abundance reflects their distribution in the universe and not their distribution on earth. Thus, hydrogen, oxygen, carbon, and nitrogen comprise more than 90% of the total elements in living forms as well as in the universe, whereas in the earth's crust the major elements, excepting oxygen, are silicon, aluminum, sodium, and magnesium.

References

Brinton, C. C., Jr., Contributions of Pili to the Specificity of the Bacterial Surface, and a Unitary Hypothesis of Conjugal Hereditary, in B. D. Davis and L. Warren (eds.), *The Specificity of Cell Surfaces*, pp. 37-70, Prentice-Hall, Englewood Cliffs, N. J., 1967.

Cosslett, V. E., *Modern Microscopy*, Cornell University Press, Ithaca, N.Y., 1966.

Doetsch, R. N. and G. J. Hageage, Motility in Procaryotic Organisms; Problems, Points of View, and Perspectives, *Biol. Rev.* **43,** 317 (1968).

Doetsch, R. N. Functional Aspects of Bacterial Flagellar Motility, *Crit. Rev. Microbiol.* **1,** 73 (1971).

Fitz-James, P., and E. Young, Morphology of Sporulation, G. W. Gould and A. Hurst (eds.), *The Bacterial Spore*, pp. 39-72, Academic Press, New York, 1969.

Fuhs, W. G., The Nuclear Structures of Protocaryotic Organisms (Bacteria and *Cyanophyceae), in Protoplasmalogia, V. Karyoplasma (nucleus),* Springer-Verlag, Vienna, 1969.

Glauert, A. M. and M. J. Thornley, The Topography of the Bacterial Cell Wall, *Ann. Rev. Microbiol.* **23,** 159 (1969).

Higgins, M. L., and Shockman, G. D., Procaryotic Cell Division with Respect to Wall and Membranes, *Crit. Rev. Microbiol.* **1,** 29 (1971).

Pollack, M. R., and M. H. Richmond (eds.), *Function and Structure in Microorganisms*, 15th Symp. Soc. Gen. Microbiol., Cambridge University Press, Cambridge, 1965.

Rogers, H. J., and H. R. Perkins, *Cell Walls and Membranes*, E. and F. N. Spon Ltd., London, 1968.

Ryter, A., Structure and Function of Mesosomes of Gram Positive Bacteria, in *Current Topics in Microbiology and Immunology,* **49,** 151.

Salton, M. R. J., Structural and Functional Properties of Bacterial Cell Membranes, in F. Snell, J. Wolken, G. Iverson, and J. Lam (eds.), *Physical Principles of Biological Membranes,* pp. 259-284, Gordon and Breach Scientific Publishers, New York, 1970.

Schlessinger, D., Ribsomes: Development of Some Current Ideas, *Bacteriol. Rev.* **33,** 445 (1969).

Sjöstrand, F. S., *Electron Microscopy of Cells and Tissues,* 2 vols., Academic Press, New York, 1966.

Stanier, R. Y., and C. B. van Niel, The Concept of a Bacterium, *Arch. Mikrobiol.,* **42,** 17 (1962).

Stanier, R. Y. Some Aspects of the Biology of Cells and Their Possible Evolutionary Significance, in *Organization and Control in Prokaryotic and Eukaryotic Cells,* 20th Symp. Soc. Gen. Microbiol., Cambridge University Press, Cambridge, 1970.

Thompson, D., *On Growth and Form* (J. T. Bonner, ed.), Cambridge University Press, Cambridge, 1966.

Walsby, A. E., Structure and Function of Gas Vacuoles, *Bacteriol. Rev.* **36,** 1 (1972).

three | Heterogeneity of Form and Function

"No pleasure endures unseasoned by variety."

Maxim 406, Publius Syrus (ca. 42 B.C.)

Devotees of the "coccus-rod-and-spiral" school of microbiology may be unaware of the extent of morphological innovation in prokaryotic organisms. The almost exclusive use of *Escherichia coli* in studies of biochemical transformations and molecular genetics has convinced many practitioners that other bacteria are similar to it. How misleading this attitude is will be seen by noting the groups of bacteria now to be described. In some instances students of "odd" bacteria are confronted with a structure whose function is not apparent. Such cases constitute novel problems whose solutions may have important lessons for general biology. It is obvious that the time-honored laboratory methods of culturing bacteria impose limits on the kinds likely to be uncovered in a given environment. This was demonstrated long ago by Winogradsky who isolated autotrophic nitrifying bacteria (Section 4.17) only after discarding the traditional organic-based concoctions used by medical bacteriologists, and substituting completely inorganic silica gel-based media. This discovery pointed the way for tactics since employed in isolating novel physiological types varying greatly from known forms. Growing bacteria in laboratory-concocted media is a means to an end, and should not be confused with the aim of ecobiology, which is to discover how organisms behave under natural conditions. The ecological optima in nature and the physiological optima *in vitro* are likely to be quite different.

Development of inorganic mineral salts media for bacteria capable of metabolizing sulfur compounds (H_2S, $S_2O_3^{2-}$) anaerobically, autorophically, and photosynthetically, as well as aerobically, heterotrophically, and in the dark, has resulted in the isolation and characterization of morphologically diverse bacteria. Furthermore, use of glass slides, capillary tubes, and other devices, suspended in ponds and lakes, or buried in soil, and thereafter examined directly with a microscope, has revealed an even greater variety of organisms, many of which are not yet amenable to laboratory cultivation (Section 1.3). Investigation of environments such as marine muds, salt brines, sulfur springs, fresh water lakes, and the like, has revealed a world of budding, stalked, ribbon-spinning, fruiting, filamentous, and trichome-forming bacteria that proliferate in their particular habitats and presumably function optimally, aided by structural adornments which distinguish them sharply from the familiar coccus-rod-and-spiral forms.

Prokaryotic organisms appear restricted in the degree to which individuals are able to develop into cooperative multicellular aggregations. It was once thought that although prokaryotes were biochemically versatile, they were limited in morphological diversity. Now this judgment should be revised in light of present knowledge.

The morphological features observed in bacteria in their natural environments may or may not persist when they are isolated and grown under laboratory conditions, so that descriptions based on laboratory observations are not always valid for the natural forms.

Bacteria are classified, in principle, according to rules originally devised by zoologists and botanists. There are "species" of bacteria, so designated by bacteriologists, but it is doubtful that they are equivalent to those described from the study of animals and plants. In any case, there is no valid evolutionary implication in the present-day classification of bacteria, aside from recognizing distinct groups. However, nucleic acid reassociation experiments offer some promise of establishing "natural relationships." In any case, use of generic and species names is a convenient way of referring to bacteria and suggests something about the general characteristics of a particular organism, but reference to higher taxonomic groups, such as families and orders, will not be made here.

3.1 | Budding Bacteria

Budding bacteria are so called because of the manner in which they reproduce (Fig. 3.1A and B). *Hyphomicrobium,* and its photosynthetic, anaerobic counterpart, *Rhodomicrobium,* after attaching to a surface, have a "mother cell," a small rod-shaped organism (0.5–0.75 μm × 0.6–5.0 μm) which produces a thin (0.2 to 0.3 μm wide) stalk-like axial protrusion, or hypha. These may then grow up to several micrometers long. The nuclear material divides and one of the nuclei remains in the mother cell while the other migrates with the growing tip of the stalk. Eventually a "bud" equal in size to the mother cell forms at the terminus of the stalk and may, at the same time, develop a single flagellum, usually at the polar end. Most buds become uniflagellated "swarmers" (peritrichously flagellated in *Rhodomicrobium*), which, after a cross wall is completed, separate from the stalk and swim away. A swarmer may then attach to a surface, lose its flagellum, and repeat the cycle. Frequently, large irregular branched networks of cells and tubes attached by "holdfasts" at the ends opposite the developing stalks are observed. Buds sometimes form directly on the mother cell without developing a stalk, and various other arrangements of hyphal attachments to cells have been described, but in any event, budding results in the formation of two cells, usually by a process of transverse but unequal division—there is no cell division without bud formation. Usually one cell will be stalked and nonmotile while the other is stalkless but motile for a while. The process whereby a sessile organism gives rise to a motile daughter cell which then swims away to another, perhaps more suitable, environment is commonly seen among aquatic bacteria (Fig. 3.2).

Hyphomicrobium and *Rhodomicrobium* are found both in fresh and sea water and in soil. The mechanisms governing developmental events such as nuclear migration, stalk formation, wall formation in the budding daughter cell, appearance and subsequent loss of the flagellum, and the aggregation of rosettes and nets are all phenomena worthy of further investigation.

Cellular extensions also occur in other bacteria (see below) and a varied terminology has arisen to describe them. Historically, structures designated as "hyphae" (as in *Hyphomicrobium*) and "stalks" (as in *Caulobacter*) were described earlier than *prosthecae* (as in *Ancalomicrobium*). The term prosthecae may be applied to all such blunt or filiform extensions of the cell wall and membrane. While these all are cellular extensions, they may be functionally dissimilar.

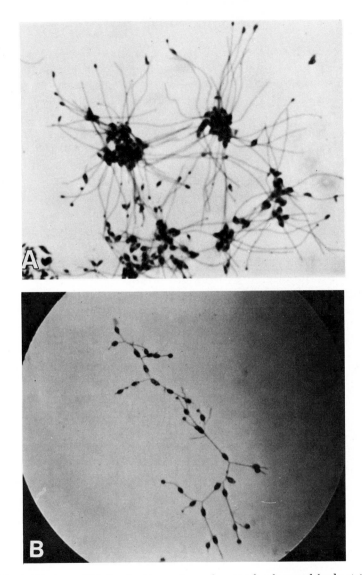

Fig. 3.1. *A,* Clusters of *Hyphomicrobium vulgare* showing hyphae and buds rising from a dense mass of organisms. (From T. Y. Kingma-Boltjes, Über *Hyphomicrobium vulgare Stutzer et Hartleb, Arch. Mikrobiol.* **7,** 188 (1936). Courtesy T. Y. Kingma-Boltjes and permission of Julius Springer, Berlin.)

B, Rhodomicrobium vannielii. In rhodomicrobia the bud generally remains attached to the mother cell, and microcolonies consisting of extensive cell masses attached by means of slender tubes are formed. If "swarmers" develop, they are peritrichously flagellated. (Courtesy H. C. Douglas.)

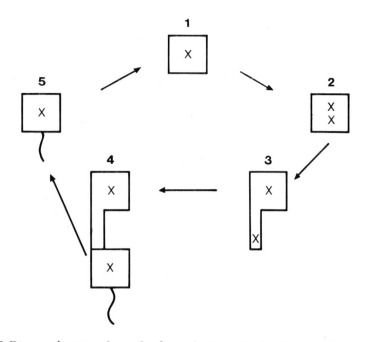

Fig. 3.2. Diagram depicting the mode of reproduction in hyphomicrobia. When an organism reproduces by budding, the mother cell is generally larger than the daughter cell at the time of cell separation, but this is not shown here. The nuclear material (X) divides (1, 2) and one nucleus migrates with the developing hypha (3); a cross-wall is laid down and a uniflagellate swarmer (4) is formed which detaches from the tip of the hypha (5) and swims away to repeat the cycle.

3.2 | Stalked Bacteria

Similar to the budding bacteria in some respects are those organisms that form a stalk, but do not thereafter further develop it as part of a reproductive cycle. In *Caulobacter* (Fig. 3.3), as in the hyphomicrobia, the stalk is an integral part of the cell and in aquatic environments the organism attaches to surfaces by means of sticky holdfasts located at the distal end of the stalk. Holdfasts consist of blebs of extracellular material without a definite form. The cells of *Caulobacter,* apart from the stalk, generally appear "vibrioid" in shape, but cylindrical, lemon-shaped, and subvibrioid forms have been described. Most physiological properties of caulobacters are like those of *Pseudomonas,* and they are aerobes able to utilize a number of organic carbon sources including carbohydrates, amino acids, alcohols, and aromatic compounds.

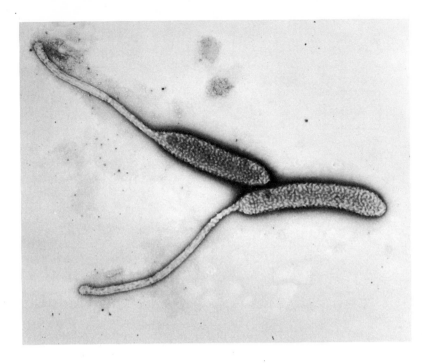

Fig. 3.3. *Caulobacter.* A negatively stained preparation showing the stalked appearance of these bacteria. (Courtesy G. J. Hageage.)

Reproduction in *Caulobacter* is by transverse fission and occurs as follows: The stalk-end of the organism permanently attaches to a surface and the opposite end enlarges into a daughter cell as a cross wall is laid down. At the same time a single flagellum develops at the distal end of the daughter cell. This uniflagellate swarmer (as in *Hyphomicrobium*) separates and swims away from the attached cell. The swarmer then develops its own stalk at the flagellated end, losing the flagellum in the process (Fig. 3.4). *Caulobacter* stalks do not contain a separate nucleus, unlike the budding bacteria. In addition, there is an invariant polarity shown in that the flagellum is always formed at the end opposite the stalk. "Rosette" formations are seen in *Caulobacter,* the attachment being at the holdfast end so that the organisms are oriented in a common mass of material in elegant arrays of from two to more than one hundred cells. Occasionally *Caulobacter* may attach itself to other living microorganisms *(ectocommensalism),* but there

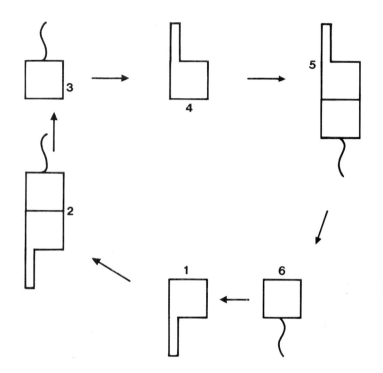

Fig. 3.4. Diagram depicting the mode of reproduction in caulobacters. A stalked sessile organism (1) divides (2) forming a uniflagellated swarmer (3); a stalk develops at the flagellated end, the flagellum being discarded (4), and the process is repeated with cell division and flagella development occurring at the opposite end (5,6).

is no evidence that they are parasites. *Caulobacter* has been isolated from the fresh waters of rivers, ponds, wells, sea water, soil, and even from millipeds.

Some fresh water and soil bacteria, described as *prosthecate,* * produce stalk-like or hypha-like extensions bounded by the cell wall. This term may also be applied to caulobacters and hyphomicrobia. In *Ancalomicrobium,* they are up to 3μm in length, and although these bacteria reproduce by budding, the appendages themselves do not bud, nor are holdfasts produced at their ends. *Prosthecomicrobium* forms many blunt (1 μm or less) appendages, but it reproduces by binary fission.

---

*Prosthecae are smaller than the organism bearing them, and these semirigid structures are usually bounded by peptidoglycan.

Although similar to both *Caulobacter* and hyphomicrobia, *Ancalomicrobium* and *Prosthecomicrobium* differ from the former in that they produce numerous prosthecae per cell, and from the latter in that buds are not formed at the end of these appendages. Flagella are not observed at any time in the development of *Ancalomicrobium*, but *Prosthecomicrobium* may possess a single flagellum. These bacteria may be viewed as forming multiple nonspecialized prosthecae, and it has been suggested that their appendages would retard settling rates, or increase their surface area in nutrient-deficient, aquatic environments. Gas vacuoles present in some prosthecate organisms may help localize them at an optimal depth in their environment.

3.3 | Ribbon-Forming Bacteria

Twisted ribbons some 200 μm in length, on which colloidal ferric oxide hydrate precipitates, are excreted from the concave sides of bean-shaped cells of *Gallionella* (Fig. 3.5A). The ribbons anchor them to solid surfaces in environments containing ferrous iron, but they

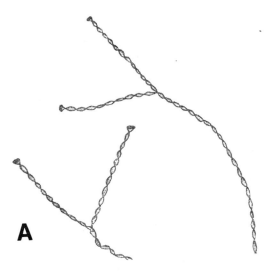

A

Fig. 3.5. *A, Gallionella ferruginea.* Sketch showing twisted and dichotomously branching ribbons on top of which single bean-shaped bacteria are positioned. (From N. Cholodny, *Die Eisenbakterien. Beitrag zu einer Monographie, Pflanzenforschung* **4** (1926).

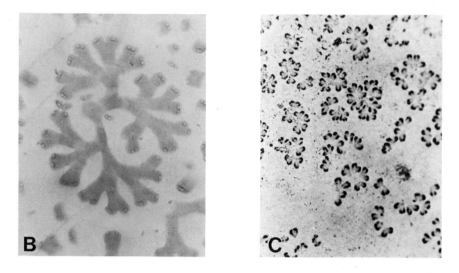

Fig. 3.5 *(Cont'd). B, Nevskia ramosa.* These bacteria are situated at the tips of short, branched ribbons of gum, and form rosette-like aggregations. (From H-D. Babenzein, Zur Biologie von *Nevskia ramosa, Z. Allg. Mikrobiol.* **7,** 89 (1967). Courtesy H-D. Babenzein.)

C, Rosette-forming colonies of *Nevskia ramosa* fixed in formaldehyde and stained with toluidine blue. (From H-D. Babenzein, Zur Biologie von *Nevskia ramosa, Z. Allg. Mikrobiol.* **7,** 89 (1967). Courtesy H-D. Babenzein.)

are not part of the living organism, and appear to be matrices of organic material upon which $Fe(OH)_3$ is deposited, and in any event, they are not stalks in the sense that word is used for hyphomicrobia and caulobacters. It is the ribbon rather than the shape of the organism that is notable here, especially the eccentric spinning of the material comprising it. Some budding bacteria also have been reported to deposit $Fe(OH)_3$ on their surfaces.

Nevskia ramosa is an aquatic bacterium that elaborates short, thick ribbons of gum in which there is no $Fe(OH)_3$. The rod-shaped organisms deposit the gum from one side of the cell at a right angle to their long axis, and form aggregations with the cells oriented to the outside (Fig. 3.5B and C). There is a developmental cycle in which flagellated forms secrete a matrix of gum, losing their flagella in the process. During subsequent divisions, each cell eccentrically excretes gum, forming an array of branched, short, broad ribbons.

3.4 | Fruiting Bacteria

Although the myxobacteria are small, flexible, unicellular rod-shaped organisms,* their so-called fruiting forms may have an intricate developmental pattern leading to formation of individual resting cells called *myxospores* or *microcysts* (Fig. 2.33). These bacteria move by "gliding" (contact motility) rather than by means of flagella and they leave a track of slime behind them as they progress. This mode of motility is also characteristic of blue-green algae and related apochlorotic prokaryotes, such as *Beggiatoa* and *Vitreoscilla* (Fig. 3.6). Myxobacteria remain in the vegetative state as long as nutritional conditions are adequate. However, a deficiency of certain amino acids leads them to aggregate into mounds or heaps of cells. These aggregation centers are conversion sites of vegetative forms into nonmotile spherical myxospores. The fruiting bodies of *Myxococcus* developed at these centers may be white, orange, red, or yellow as a result of the production of carotenoid pigments (Fig. 3.7). The formation of these pigments

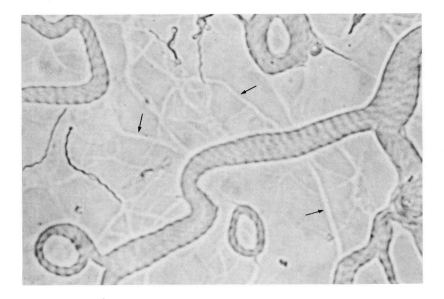

Fig. 3.6. Slime tracks on trails (arrows) formed on an agar surface by gliding chains of *Spirulina albida*. (Courtesy G. J. Hageage.)

---

*There are no known spherical or spiral forms of myxobacteria.

is accelerated by light, and one hypothesis is that they serve as protection against photooxidative damage. Myxospores are spherical bodies that have highly refractile walls and upon germination the vegetative cell simply breaks out of its container. Resting forms are mildly resistant to heat and irradiation but not nearly so much as endospores.

Branched, lobate formations of exquisite beauty are developed by *Chondromyces,* where the fruiting bodies may be red, pink, violet, or yellow, and measure more than 0.5 mm in diameter (Figs. 3.8 and 3.9A and B). The ends of the branches of *Chondromyces* fruiting bodies bear macrocysts 50μm or more in diameter which are attached by pedicles, the main upright portion and branches being made up of hardened slime. Each cyst contains thousands of individual organisms similar in structure to the vegetative cells but somewhat shortened. The large cysts burst upon germination and the vegetative cells then glide out. Myxobacteria such as *Cytophaga,* a cellulolytic form, remain permanently in a vegetative condition, and *Sporocytophaga* develops resting bodies similar to those of *Myxococcus,* but does not form a fruiting body. The ability of myxobacteria to develop fruiting bodies in nature

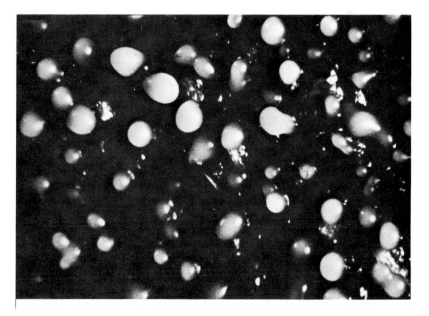

Fig. 3.7. Fruiting bodies of *Myxococcus fulvus.* (From H. D. McCurdy, Studies on the taxonomy of the *Myxobacteriales. Canad. J. Microbiol.* **15,** 1453 (1969). Courtesy H. D. McCurdy and permission of the National Research Council of Canada.)

Fig. 3.8. *Chondromyces apiculatus* fruiting body on dung. (Courtesy H. Kühlwein.)

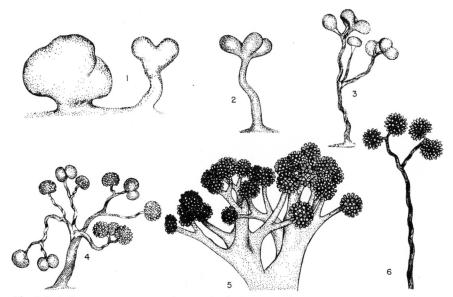

Fig. 3.9. *A,* Successive stages in fruiting body formation by *Chondromyces crocatus.* (Drawings are not all drawn to the same scale.)

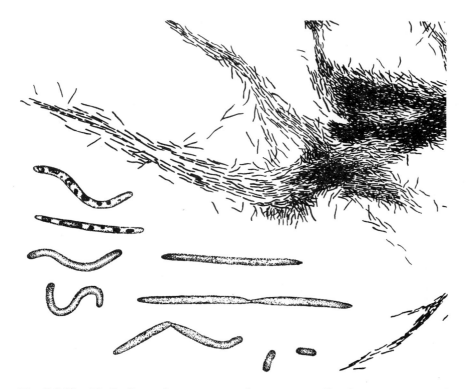

Fig. 3.9 *(Cont'd)*. *B*, General appearance of vegetative cells of *C. aurantiacus* and characteristic swarming phase. (From R. Thaxter, On the *Myxobacteriaceae,* a new order of *Schizomycetes, Bot. Gaz.* **17,** 389 (1892). By permission of the University of Chicago Press.)

is associated with their faculty of lysing and growing on other bacteria. Nonfruiting myxobacteria occur in soils where they are involved in the degradation of insoluble polymers such as cellulose and other polysaccharides, but fruiting forms also occur in the soil and they grow well on animal dung which contains a large bacterial population. One aquatic fruiting myxobacter, *Chondrococcus columnaris,* is pathogenic for fish.

The fruiting myxobacteria exemplify a complex developmental cycle involving alternate individual and colonial cooperative phases, often coupled with alterations in the gross structure of the cells themselves, and they are justly receiving close study by developmental biologists. There is also an organismal analogy between fruiting myxobacteria and eukaryotic cellular slime molds (myxomycetes) that is of interest to evolutionary biologists.

3.5 | Net-Forming Bacteria

The photosynthetic green sulfur bacterium, *Pelodictyon clathratiforme*, is a large (4–7 μm long) rod-shaped organism that forms three-dimensional nets that may extend over 100 μm in diameter (Fig. 3.10A and B). These nets result from the development of branching points leading to the formation of Y-shaped cells. If two adjacent organisms

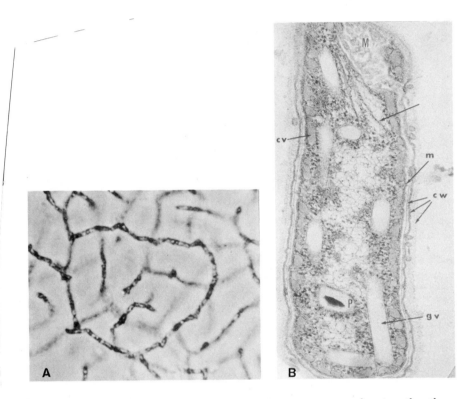

Fig. 3.10 *A, Pelodictyon clathratiforme* strain 1831 preparation showing the three-dimensional character of the net.

B. Pelodictyon clathratiforme strain 2730. Longitudinal section showing complex cell wall (cw), a large mesosome (M), the electron-opaque *Chlorobium* vesicle (cv) limited by a nonunit membrane about 3 nm thick. The gas vacuoles (gv) appear electron transparent, and they are also limited by a nonunit membrane about 2 nm thick. A collapsed gas vacuole is shown by the arrow, and there is a partly volatilized polyphosphate granule (P). (From N. Pfennig and G. Cohen-Bazire, Some properties of the green bacterium *Pelodictyon clathratiforme*, *Arch. Mikrobiol.* **59**, 226 (1967). *A* and *B* courtesy N. Pfennig and permission of Berlin-Heidelberg-New York: Springer Verlag.)

simultaneously branch, and do not separate, a ring structure is formed which enlarges with subsequent cell divisions. Branching always appears to occur simultaneously at the contiguous poles of two organisms. Development of nets, then, results from the ability of *Pelodictyon* to divide both by binary and ternary fission. Nets of bacteria may serve to trap plankton which may then be gathered by organisms higher up in the food chain. The organisms contain numerous gas vacuoles, and have a complex internal structure including involuted mesosomes and a fibrillar organelle of unknown function. Structurally, the green photosynthetic bacteria are the most highly differentiated prokaryotic forms thus far discovered.

3.6   Thiothrix, Leucothrix, and Some "Gliders"

These prokaryotes are characterized by development of unbranched, multicellular filaments termed *trichomes,* in which individual organisms are maintained by a common outer cell wall. Furthermore, all show gliding motility but only when in contact with a solid surface, and this is accompanied by active flexing of the cells and the secretion of a slime track. This form of motility is slow (10–15 μm per minute) and stately, but the mechanism responsible for it is unknown although numerous theories have been proposed. A number of filament-forming bacteria have structural features and movements similar to those of the blue-green algae, and they may be colorless, nonphotosynthetic counterparts of them (leucophytes).

Thiothrix and *Leucothrix* are filamentous organisms that form characteristic rosette arrangements. The former oxidizes hydrogen sulfide and is possibly chemolithotrophic (Section 3.16), whereas the latter utilizes organic compounds. *Leucothrix* is a marine organism, often found associated with seaweeds, that forms slightly tapered filaments, the broader end possessing a holdfast. The filaments may attain a length of 5 mm, and individual cells divide throughout it, not only at the tip (Fig. 3.11A, B, and C). When environmental conditions become unfavorable, rounded cell forms called *gonidia* develop, and these are released separately from the filament tip. Although the entire *Leucothrix* filament does not glide, individual gonidia show gliding motion and generally adhere to a surface, after which a holdfast is elaborated and a new filament develops by subsequent cell divisions.

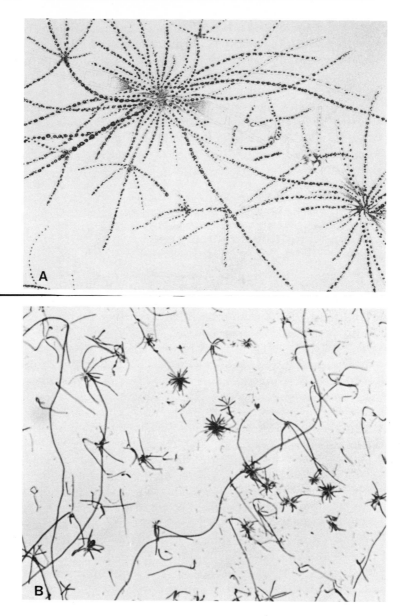

Fig. 3.11. *A,* Habit sketch of *Thiothrix* as seen in nature. (From M. Miyoshi, Studien über die Schwefelrasenbildung und der Thermen von Yumoto bei Nikko, *J. Coll. Sci. Imp. Univ. Tokyo* **10,** 141 (1898).

B, Leucothrix, developing rosettes, filaments, and gonidia. (Courtesy R. L. Harold.)

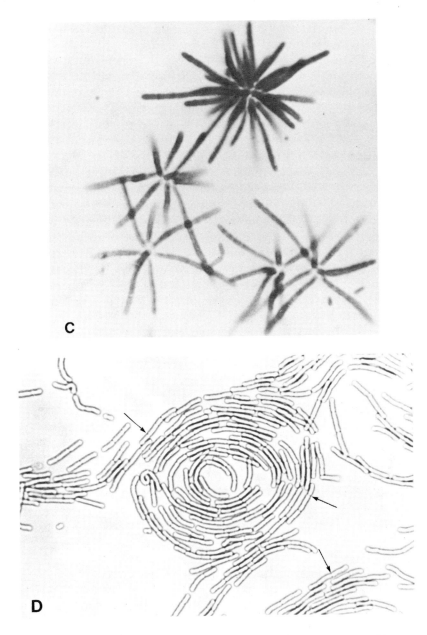

Fig. 3.11 *(Cont'd). C, Leucothrix,* developing rosettes. (Courtesy R. L. Harold.)
D, Vitreoscilla stained to show the cell wall and showing characteristic circular arrangements of the organisms on an agar surface. (Courtesy G. J. Hageage.)

Association of gonidia results in the formation of rosettes of filaments, the cells of which ultimately return to gonidial form (Fig. 3.12). An interesting feature of *Leucothrix* filaments is the occurrence of knots. These appear when the medium is nutritionally rich and filament growth is rapid. One side of a filament develops more rapidly than the other, a loop is formed and, frequently, the filament end will pass through the loop forming a knot. The knots tighten and the adjacent cells form a bulb which is eventually released from the filament, but the fate of these bulbs is not known.

Thiothrix has been observed in sulfur springs and it, too, forms rosettes and occasionally knots. Cells comprising the filaments may be filled with sulfur droplets under certain environmental conditions. As with *Leucothrix,* the filaments are not motile. *Beggiatoa,* an aquatic organism found in sulfide-containing muds and waters, shows gliding motility. Individual cells also may contain sulfur droplets (Fig. 2.36A). Rosettes are not formed, and filaments do not taper at the end. This organism oxidizes hydrogen sulfide to sulfur, which is then stored intracellularly,

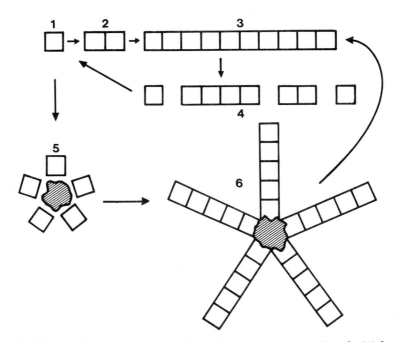

Fig. 3.12. Diagram depicting the mode of reproduction in *Leucothrix.* Gonidia (1) develop into filaments (2, 3) which may either again become gonidia (4) or develop into rosettes (5, 6) attached by holdfast material (hatched center), which then may form free filaments.

as do other sulfur bacteria such as *Chromatium* and *Thiospirillum*. *Vitreoscilla,* a gliding, filamentous organism (Fig. 3.11D), may be distinguished from *Beggiatoa* mainly on the basis of intracellular sulfur deposition in the latter. Gliding *Vitreoscilla* filaments deposit slime tracks on agar surfaces which are easily observed with a microscope, but the chemical composition of the slime, or its precise role in gliding motility, as in others of this kind, has not been established (Fig. 3.6). *Simonsiella* and *Alysiella* are gliding organisms found in rabbit and sheep saliva. They both form multicellular filaments and are interesting in being "ribbon-like," that is, they are 3–4 μm wide, but only 1 μm or so thick. The filaments show gliding motility only when the broad part is in contact with a solid surface, and if oriented on the narrow side no movement is observed.

3.7 │ **Sphaerotilus and Leptothrix**

Sphaerotilus and *Leptothrix* are rod-shaped aquatic organisms that form distinctive *sheaths*. As the cells divide to form long chains, they are pushed beyond the ends of the sheath and they then synthesize additional sheath material around themselves (Fig. 3.13A). The old sheath material may serve as a "primer" for the deposition of this new material.

In environments rich in nutrients, great masses of sheathed cells are formed, and these become slimy nuisances in sewage treatment plants and in polluted streams (Fig. 3.13B).* In nutrient-poor situations, individual motile cells leave the sheath, and such swarmers, each equipped with a fascicle of flagella containing fifteen to twenty individual flagella inserted subpolarly, migrate to a new environment where, if conditions are favorable, they settle down and begin the process of forming a new sheathed filament. The sheath is a protein-polysaccharide complex, which may be considered as either a linear capsule or a kind of excess wall material. The sheaths maintain their structural integrity after the organisms leave them, and they appear as empty transparent tubes when viewed microscopically. Under certain circumstances, oxides of iron or manganese precipitate on the outside of sheaths giving them a yellow or brown coloration. *Leptothrix* is considered to be a form of *Sphaerotilis* showing such depositions.

---

*It has been estimated that the pollution of the Willamette and Columbia Rivers by waste sulfite liquor is such that in the production of 100 tons of paper pulp by this method, the yield of *Sphaerotilus* is about 70 tons.

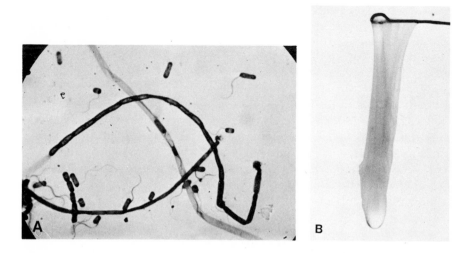

Fig. 3.13. *A, Sphaerotilus natans* showing sheaths, some motile organisms still in the sheaths and free motile swarmers. (From T. Y. Kingma-Boltjes, Function and arrangement of bacterial flagella, *J. Pathol. Bacteriol.* **60,** 275 (1948). Courtesy T. Y. Kingma-Boltjes and permission of Oliver and Boyd, Publishers.)

B, Slimy mass formed by cohering sheaths of *Sphaerotilus natans.* (From E. G. Mulder and W. L. van Veen, Investigations on the *Sphaerotilus-Leptothrix* group, *Antonie van Leeuwenhoek,* **29,** 121 (1963). Courtesy W. L. van Veen and permission of Swets and Zeitlinger.)

3.8 | Filamentous Branching Bacteria

Prokaryotes called actinomycetes develop extensive tubular (about 1 μm diameter) protoplasmic filaments (*hyphae*) that superficially resemble fungal mycelia (Fig. 3.14A and B). The features of these organisms leave no doubt as to their prokaryotic nature and include structural similarity of flagella, lack of sterols, hyphal cross-sectional dimensions, nuclear appearance, chemoautotrophy (some kinds), and cell wall chemical composition. Actinomycetes are prevalent in soil probably because of their ability to utilize a variety of complex substrates as nutrients. In *Nocardia* (aerobic) and *Actinomyces* (anaerobic), the hyphal strands eventually fragment into elements resembling bacterial rods and cocci, and these may then develop a fresh hyphal system. The arrangement of numerous branching protoplasmic tubes

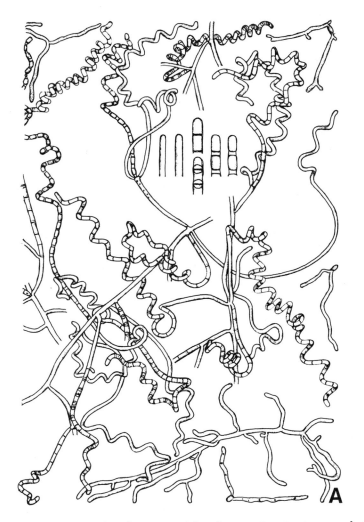

Fig. 3.14. *A, Actinomyces scabies* showing aerial and vegetative structures, and stages in the development of this organism (now known as *Streptomyces scabies*).

provides actinomycetes with surface and subsurface contact with the substrates utilized for growth. In *Streptomyces,* specialized and characteristic *sporophores* produce reproductive elements at their tips called *conidia.* These are "dissemination spores" and are not the equivalent of bacterial endospores (Fig. 3.15A and B; 3.16A and B). Each species in this group forms conidia, varying in surface structure and color,

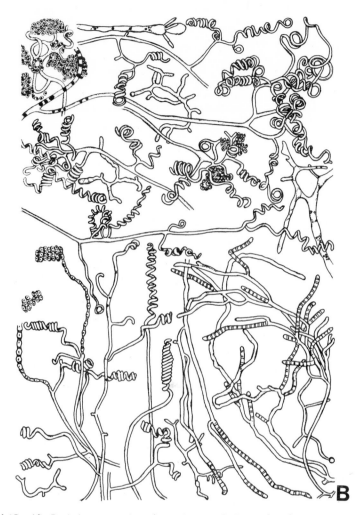

Fig. 3.14 *(Cont'd)*. *B, Actinomyces griseus* (now *S. griseus*). Same details as in 3.14*A*. These are original sketches made by C. Drechsler. (From C. Drechsler, Morphology of the genus *Actinomyces* II, *Bot. Gaz.* **67,** 147 (1919). By permission of the University of Chicago Press.)

by development of cross walls in the multinucleate aerial filament. These are eventually released for aerial distribution, and though they possess some resistance to drying, they are much weaker than endospores in their general capacity to resist adverse environmental conditions.

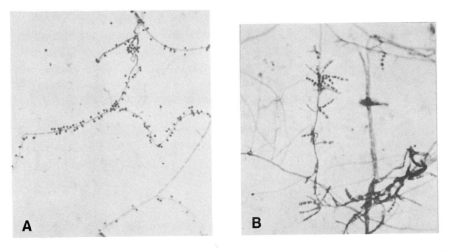

Fig. 3.15. *A, B,* Growth forms of actinomycete hyphae and conidia as revealed by the slide contact technique. (From S. A. Waksman, *The Actinomycetes,* Vol. I, Nature, Occurrence, and Activities. Williams and Wilkins, Baltimore, 1959. Courtesy of S. A. Waksman and permission of Williams and Wilkins.)

Fig. 3.16. *A, Streptoverticillium* sp., showing verticillate chains of conidia.

B, Actinomadura dassonvillei, showing long festoons of conidia. (Courtesy M. P. Lechevalier and H. Lechevalier.)

In *Actinoplanes* some interesting, mainly aquatic, forms produce motile sporangiospores contained within a sporangial envelope located at the tip of a reproductive hypha. When the sporangium ruptures, the motile sporangiospores are released and swim to a new location. After becoming attached to a solid surface, the sporangiospores lose their flagella and begin a new hyphal development. Neither flagellation nor the formation of sporangia is common among actinomycetes and the former characteristic seems definitely related to an aquatic mode of existence. It appears that the "sessile form–motile form" alternation is frequently employed as a mode of disseminating aquatic organisms to fresh environments, and similar behavior is seen in hyphomicrobia, caulobacters, and *Sphaerotilus.*

An obligately thermophilic group of actinomycetes known as *Thermoactinomyces* develops in materials such as compost heaps and damp haystacks. The temperature in these environments is gradually raised by the microbial thermogenic reactions brought about initially by the activities of mesophilic organisms. These actinomycetes and some thermophilic bacilli gain ascendance as the temperature rises and *Thermoactinomyces* may grow at temperatures up to 68°C. A large number of antibiotic substances are produced by actinomycetes, the majority by *Streptomyces,* and they are discussed in Section 4.18.

3.9 | Cell Wall-less Bacteria

Permanently cell wall-less bacteria are known as mycoplasmas. They occur in soil, sewage, insects, man, and vertebrate animals, and apparently survive in nature by having made certain modifications of the fragile cell membrane. The membrane contains sterols and as the *Mycoplasma* are unable to synthesize these compounds they must be supplied preformed from the environment.* Peptidoglycan cell wall components such as N-acetylglucosamine and diaminopimelic acid are absent from these organisms. *Mycoplasma* show a high degree of morphological plasticity or pleomorphism. Blebs, filaments, and coccoid forms appear in cultures (Fig. 3.17A and B). The latter range from 0.3 to 0.9 μm in diameter and may contain the least amount of DNA per cell of any known microorganism (about 10^{-15}g per nucleus).

Mycoplasma are the only bacteria that require sterols, and, in fact, no bacteria are known that synthesize them. *Acholeplasma* is a genus which does not require sterols although it is a mycoplasma.

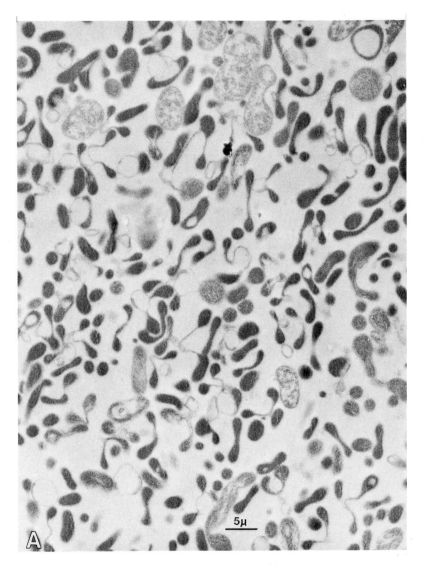

Fig. 3.17. *A, Mycoplasma felis* thin section, showing lobular extensions connected together by a circular membrane. (From E. S. Boatman, Three dimensional morphology, ultra-structure and replication of *Mycoplasma felis, J. Bacteriol.* **101,** 266 (1970). Courtesy E. S. Boatman and permission of the American Society for Microbiology.)

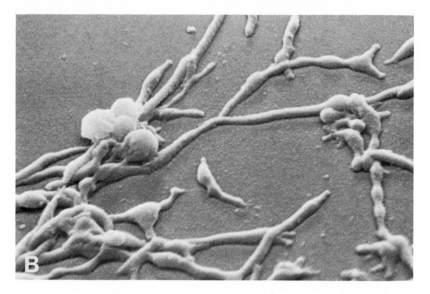

Fig. 3.17 *(Cont'd). B,* Scanning electron micrograph of *Mycoplasma pneumoniae* showing both filaments and blebs. (From G. Biberfeld, Ultrastructural features of *Mycoplasma pneumoniae, J. Bacteriol.* **102,** 855 (1970). Courtesy G. Biberfeld and permission of the American Society for Microbiology.)

Most myocoplasmas reproduce by binary fission; some may develop filamentous elements from spherical bodies. Blebs develop at the filament tips that are ultimately freed to repeat the reproductive cycle.

Some cell wall-less bacteria, called *"L-forms"* (Section 4.6), develop on exposure to antibiotics, such as penicillin, that interfere with the development of a normal cell wall. Such forms are unstable and their media must be buffered against osmotic shock. L-forms revert to the original bacterial form when the inciting agent is removed, and although some remain permanently in that state ("stable L-forms") there is no evidence of any relationship to mycoplasmas.

3.10 | Flexible, Helical Bacteria

Spirochetes are flexible, helically twisted organisms that inhabit a variety of environments (Fig. 3.18A, B, and C). In general, they are long, thin forms that swim by means of an *axial filament,* strands of

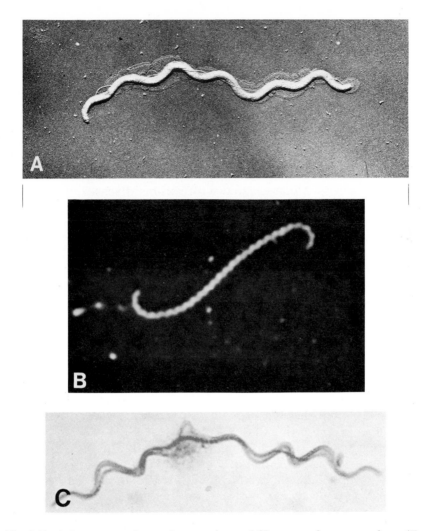

Fig. 3.18. *A, Treponema zuelzerae,* showing the axial filament and outer envelope. (From M. A. Bharier, F. A. Eiserling, and S. C. Rittenberg, Electron microscopic observations on the structures of *Treponema zuelzerae* and its axial filaments, *J. Bacteriol.* **105,** 413 (1971). Courtesy S. C. Rittenberg and permission of the American Society for Microbiology.)

B. Leptospira icterohaemorrhagiae. Dark-field photograph showing "hooks" at each end. (Courtesy T. Y. Kingma-Boltjes.)

C, Cristispira. (From H. Noguchi, The spirochaetes, in E. O. Jordan and I. S. Falk (eds.), *The Newer Knowledge of Bacteriology and Immunology,* University of Chicago Press, Chicago, 1928. By permission of the University of Chicago Press.)

flagella-like fibrils, originating from subpolar sites on the organism, and loosely wound around the helical body. The number of fibrils in a filament vary from greater than 100 in *Cristispira* to 2 in *Leptospira* and *Spirochaeta*. Axial filaments are usually enclosed between a membrane and the organism itself, but there is no satisfactory explanation as to how this arrangement enables spirochetes to move about.

Spirochetes range in length from 500 μm *(Spirochaeta)* to 20 μm *(Leptospira, Borrelia,* and *Treponema)*. The last three are also very thin, perhaps never more than 0.5 μm wide, and they are best observed using dark-field microscopy.

Some spirochetes are associated with serious diseases of man and animals, such as syphilis and yaws *(Treponema),* leptospirosis *(Leptospira),* and relapsing fever *(Borrelia)*. Spirochetes in the genus *Cristispira* have been reported only in bivalves (oysters and clams), where they reside in a part of the digestive system called the "crystalline style." The function these bacteria have in this habitat is unknown. *Treponema pallidum* is the etiological agent of syphilis in man, where it may be seen in fluid exudates from syphilitic chancres.

3.11 | Blue-Green Algae

While blue-green algae are prokaryotes they have certain structural and aggregational complexities not usually seen in bacteria; nevertheless, it is probable that bacteria and blue-green algae arose from a common ancestor, and that they are closely related.

Blue-green algae are a morphologically diverse group of photosynthetic prokaryotes, so-called because of the color imparted to some by a water-soluble pigment-protein complex. Masses of organisms possessing only C-phycocyanin will have a blue-green color. However, if C-phycocyanin is absent or occurs together with another phycobiliprotein, C-phycoerythrin, which is red, the organisms in mass may appear red, bright yellow, pale violet, purple, or black. In many respects the *Myxophyceae* (commonly called *Cyanophyta*) share many morphological features with the eubacteria. They are obligate photoautotrophs (Section 3.19) having a photosynthetic mechanism employing chlorophyll *a* (which all photosynthetic bacteria lack) and a prokaryotic "ground plan" (Figs. 2.24 and 3.19B).

Blue-green algae are cosmopolitan and share the same environments bacteria do, providing there is light. They have been found in the Antarctic, in deserts, in salt lakes, and many other locales. In solar-salt

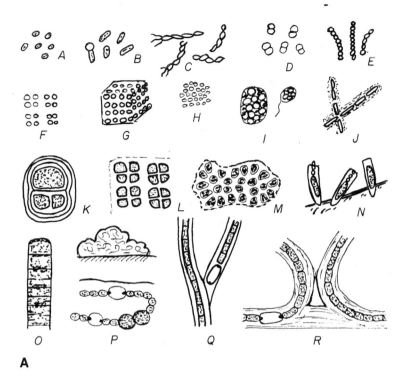

A

Fig. 3.19. *A,* Bacteria *(A–J)* and *Cyanophyta (K–R), A,* cocci; *B,* bacilli—one cell with an endospore; *C,* spirilla; *D,* diplococci; *E,* streptococci; *F,* tetracocci; *G,* sarcina; *H,* staphylococci, *I,* a sulphur bacterium—*Achromatium,* vegative and motile cell; *J,* iron bacteria—*Leptothrix,* with iron deposited in the mucilage sheath; *K, Chroococcus; L, Merisompedia; M, Microcystis* with gas vacuoles; *N, Chamaesiphon; O, Oscillatoria,* note the new cross walls growing inwards as annuli; *P,* mucilaginous colony of *Nostoc* and one filament containing two heterocysts and two akinetes; *Q, Tolypothrix,* showing branching beneath a heterocyst; *R, Scytonema.* (Courtesy F. E. Round, *Introduction to the Lower Plants,* Plenum Press, New York, 1969.)

ponds, *Microcoleus chthonoplastes* and *Lyngbya aestuarii* form layers between the deposited salt and the black underlying sulfide muds. Blue-green algae also are important fresh-water and marine inhabitants (about 20% of the blue-green algae are marine forms) and are both *lithophytic* (growing on rocks) and *epiphytic* (growing on other plants). Rock-colonizing *(saxicolous)* algae select habitats according to the degree of hardness and shading, rather than chemical composition. Tree trunks and rocks are colonized by all sorts of blue-green algae including *Nostoc, Phormidium, Scytonema,* and *Stigonema.*

Some blue-green algae *(endolithic)* may actively tunnel their way into calcareous materials such as rock, mollusk shells, and coral. When

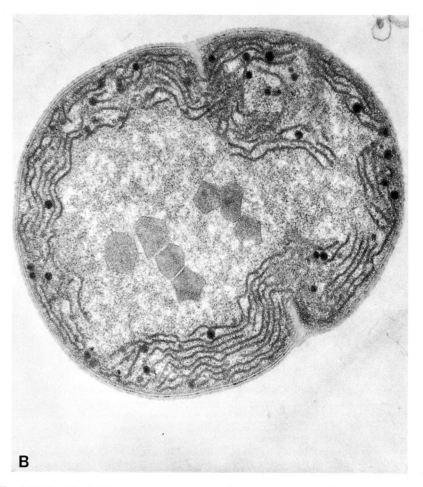

B

Fig. 3.19 *(Cont'd). B, Nostoc muscorum* showing characteristic internal structures of blue-green algae. (Courtesy R. B. Wildman.)

extensive growth occurs the rock becomes porous and soft. Along the limestone and dolomitic cliffs of the Mediterranean Sea, filamentous lime-boring algae such as *Kyrtuthrix, Hyella, Dalmatella, Solentia,* and *Hormathonema* occupy areas in the upper littoral region. The process of rock penetration by algae may be favored by their excretion of oxalic acid, but other theories have been suggested. Extensive calcareous deposits may be precipitated by fresh water *Myxophyceae.* Some fresh water blue-green algae produce "blooms" under certain conditions of nutrient excess, with the result that a large body of water may be colored due to their massive proliferation.

Although the simplest organizational form of blue-green algae *(Chroococcales)* is unicellular, they occur for the most part in aggregates called *palmelloid* colonial forms, that is, organisms embedded in a mucilagenous covering or mass. Some are roughly spherically-shaped organisms, as *Chroococcus, Gleocapsa,* and *Aphanocapsa,* whereas others are roughly rod-shaped, as *Gleothece, Synechococcus,* and *Aphanothece.* All blue-green algae are characterized by the development of pigmented mucilagenous coverings that are often quite extensive. *Chroococcus* and *Gleocapsa* show division in three planes reminiscent of *Sarcina,* and *Merismopedia* forms tablets of cells similar to those in *Lampropedia.* In like manner similarities are seen in the modes of division and aggregation between *Aphanocapsa* and *Micrococcus; Spirulina* and *Spirillum; Phormidium* and *Bacillus;* and *Nostoc* and *Streptococcus* (Fig. 3.19A).

Morphologically, blue-green algae show prokaryotic characteristics. They do not possess flagella, but when motile they move by gliding over a solid substratum in the same manner as gliding bacteria (Section 3.6). Although blue-green algae reproduce primarily by binary fission they show other reproductive modes. Endospores are widespread among blue-green algae not producing hormogonia (see below). The former develop by successive divisions of the protoplast and are a direct means of propagation. The naked endospores are released from the cell by rupturing, or detachment of a "lid," or by gelatinization of the cell wall. They do not have a resting period and these structures therefore have a different function in bacteria than in blue-green algae.

The thallus form of blue-green algae exists as a filament or *trichome,* and the cells are bound together in a common sheath in some species. In such filamentous forms fragmentation of segments occurs at the ends and these glide away from the main filament. These fragments are called *hormogonia* and possess no marked differentiation from the other cells in the trichome. *Akinetes* are resting spores that enable blue-green algae to resist adverse conditions such as freezing or drying. They are usually stuffed with granular food-reserves, possess a thick, pigmented outer layer, and functionally, but not structurally, are most like bacterial endospores. Some blue-green algae may develop *heterocysts,* that is, thick-walled vacuolated cells that are much larger than the normal cells. The walls are perforated so that there is a continuity between neighboring cells in a trichome. Heterocysts serve as a point where trichome fragmentation into hormogonia may occur; they may secrete growth-stimulating substances, and they may be sites of atmospheric nitrogen fixation.

3.12 | Hydrogen Oxidation

The ability of bacteria to colonize and thrive in environments adverse to other living forms has been mentioned, and it seems obvious that they are the most adaptable of all organisms. The important role of bacteria in recycling elements (Chapter 6), and in the decomposition of nearly all naturally occurring substances, attests to their indispensability for the continuation of life on earth. Certain bacteria specialize in metabolizing refractory materials. The biochemical aspects of some of the transformations will be taken up in Chapter 5, but it may be noted here that physiological diversity in bacteria is less restricted than morphological diversity. An example of versatility in the former is that of *Pseudomonas putida,* which is able to utilize 77 different carbon sources including 6 carbohydrates, 10 saturated fatty acids, 3 unsaturated fatty acids, 10 alcohols, 17 amino acids, 9 amides, and 7 amines.

The oxidation of molecular hydrogen (H_2) to water is carried out by some aerobic micrococci, actinomycetes, and particularly by a group of polar flagellated rods, namely, *Hydrogenomonas.* These "hydrogen bacteria" occur in soil and water, where they utilize various organic carbon and nitrogen compounds; consequently, they are not strict chemolithotrophs. Under anaerobic conditions, a number of kinds of bacteria evolve hydrogen by fermentative mechanisms, and this gas may serve as an energy source for *Hydrogenomonas* under appropriate conditions. Hydrogen-oxidizing bacteria may be isolated from soil or water by making enrichment cultures in a mineral medium containing an atmosphere of hydrogen, oxygen, and carbon dioxide.

3.13 | Carbon Monoxide Oxidation

In 1903 Beijerinck found that carbon monoxide (CO) was oxidized to carbon dioxide by certain bacteria, and he described one of these as *Bacillus oligocarbophilus* (now *Carboxydomonas oligocarbophila*). Some pseudomonads also are able to utilize this gas as an energy source. Carbon monoxide lies outside the main carbon cycle and the question arises as to what mechanisms produce it in nature. Volcanic activities may be a source of this gas, but certainly the main factors involved in its production are the gasoline engine and heavy industrialization of modern civilization. Since carbon monoxide is a cytochrome inhibitor and, hence, toxic to aerobes, the mechanism of its oxidation by bacteria, although unknown, is of physiological interest.

3.14 | Methane Oxidation

Methane (CH_4) occurs in nature in widely diverse environments such as the rumens of herbivores, the gases associated with petroleum deposits, and fermenting vegetation. Methane arises as a product of anaerobic fermentations carried out by *Methanobacterium, Methanobacillus, Methanococcus,* and *Methanosarcina.* It is the simplest hydrocarbon and although many organisms are known that oxidize the higher members of this series, ethane, propane, and butane, for example, only *Methanomonas* has been shown to oxidize methane and it is relatively inactive with other gaseous hydrocarbons. These aerobic, polar flagellated bacteria occur in well-drained soils, paddy fields, pond mud, barnyard manure, and sewage, and are able to utilize this gas as an energy source. Under appropriate conditions *Methanomonas* oxidizes methane to methyl alcohol, and ultimately to carbon dioxide and water. Methane-utilizing bacteria may be enriched and isolated using first a liquid and then an agar-solidified medium consisting of mineral salts and a nitrogen source such as nitrate under an atmosphere of methane and air (1 : 1). After 6 to 10 days at 30°C, a pink film or pellicle may develop on the surface of the medium that consists or gram-negative, motile, rod-shaped bacteria. This material may then be streaked on agar plates of the same medium and, after incubation, colonies may be examined for the presence of gram-negative rods.

Ethane (C_2H_6) is associated mainly with petroleum deposits, and ethane-oxidizing bacteria have been used for detection of such areas by examining soil samples for their presence. These bacteria are preferred in oil-prospecting tests over methane-oxidizing forms, since methane may be of more recent biological origin and, therefore, not indicative of petroleum deposits. Aside from methane and ethylene, gaseous hydrocarbon metabolic products have not been reported.

3.15 | Nonsymbiotic Nitrogen Fixation

Nonsymbiotic nitrogen (N_2) fixation by free-living bacteria is carried on aerobically mainly by *Azotobacter* and anaerobically by *Clostridium. Achromobacter, Aerobacter, Pseudomonas, Bacillus polymyxa, Methanobacillus omelianskii, Chlorobium, Chromatium, Rhodomicrobium, Rhodopseudomonas,* and *Rhodospirillum* also have been shown to fix nitrogen nonsymbiotically. This process is not of great importance when contrasted with the symbiotic one, except in tropical or aquatic regions where blue-green algae may be significant in this role.

Azotobacters are generally motile soil organisms requiring an organic carbon source, but they grow in media devoid of nitrogen because of their nitrogen-fixing ability. Some species are quite large (4–7 μm in diameter) and "yeast-like" in appearance and many form a gummy polysaccharide capsule. They may form cysts that serve as resting bodies (Section 2.10), and although these are more resistant to adverse environmental conditions than the vegetative form of *Azotobacter,* they, like actinomycete conidia, and the myxospores of myxobacteria, are not nearly so durable as endospores. *Azotobacter* does not grow below pH 6 and therefore is not present in acid soils, but *Beijerinckia,* a tropically distributed azotobacter, is reported to grow at pH 4.5 in the soil often found in such regions. It has been discovered that minute amounts of molybdenum (1 part in 10^{10}) are needed for nitrogen fixation by these bacteria. Another interesting property of *Azotobacter* is that it has one of the highest respiratory rates of any living organism thus far investigated, its oxygen uptake per gram wet weight per hour being reported between 500 and 1,000 cm^3. *Azotobacter* may be enriched and isolated using liquid and solid media containing mannitol, calcium malate, or sodium propionate as carbon sources, the media not containing added nitrogen sources. After incubation at 25° to 30° C, a surface film develops which may contain large coccoid (from 4 to 7 μm) organisms. These may be isolated by streaking the surface of an agar-solidified medium of the same composition with the enrichment material.

A number of clostridia of the butyric acid-producing group fix nitrogen anaerobically. These bacteria ferment starch, pectin, and other substrates likely to be found in the soil, and the amount of nitrogen fixed is related to the carbohydrate available.

3.16 | Sulfur Oxidation

Sulfur (S), one of the least water-soluble substances known, is oxidized by aerobic, polar flagellated (1–3 μm long by 0.5–0.75 μm wide) bacteria of the genus *Thiobacillus.* These soil and water forms* are of par-

---

*Sulfur oxidizing bacteria have been found in oil brines, coal water, mine wastes, peat soil, concrete, and building stone.

ticular interest in that some are authentic chemolithotrophs incapable of chemoorganotrophic growth, among them are *T. thiooxidans, T. thioparus,* and *T. ferrooxidans.* Most thiobacilli also oxidize other sulfur compounds such as sulfides, thiosulfates, and sulfites. How these bacteria detect elemental sulfur is unknown, but its subsequent oxidation by *T. thiooxidans* often leads to the production of 0.1 N solutions of sulfuric acid (pH 1.0). It also is not known how these bacteria resist this very low pH and continue to function, but their membranes must be impermeable to hydrogen ions. Cytoplasm from *T. thiooxidans* immediately coagulates when exposed to acid solutions of the concentrations it produces, so that unlike halophiles in high salt concentrations (Section 3.20), the continued function of thiobacilli depends on keeping hydrogen ions outside the cell. Sulfur crystals appear eroded after growth of thiobacilli on them, and the problem of how elemental sulfur is metabolized is an intriguing one.

Sulfur resulting from the oxidation of sulfide is deposited intracellularly in the filamentous *Beggiatoa, Leucothrix,* and *Thiothrix* (Fig. 2.36A), as well as in most of the photosynthetic purple sulfur bacteria, but green sulfur bacteria *(Chlorobacteriaceae)* deposit it outside of their cells. The intracellularly deposited sulfur, most likely membrane-bound or enclosed, results when sulfides are present in excess of those required for immediate use. The physical nature of the sulfur stored in droplets is not known, but it must be in a biologically available form. Sulfur isolated from bacteria filled with sulfur droplets is not necessarily in the same form as it occurs in the cells, and very likely it is not (Fig. 2.36B). The process of dissolving out sulfur and recrystallizing the purified product could alter the crystalline structure, and conclusions on its nature must be drawn from studies on the droplets *in situ* (Section 2.11). Upon depletion of external sulfur sources, stored internal sulfur is oxidized to sulfates and starved organisms lack sulfur deposits. Sulfur-oxidizing bacteria may be isolated from water and soil samples incubated at 25° C in an inorganic salt medium containing powdered sulfur ("sulfur flowers") and a source of CO_2 as calcium carbonate. Such enrichment cultures may take from 1 to 3 weeks to develop and they must be examined periodically for evidences of bacterial development. Solid media consisting of silica gel (sodium silicate plus hydrochloric acid) rather than agar, and containing added sulfur in the form of a polysulfide solution, are inoculated with enrichment material. Colonies of sulfur-oxidizing bacteria are detected by clearing of the finely precipitated sulfur around them.

One of the thiobacilli, *T. ferrooxidans,* oxidizes ferrous iron and is involved in the development of acid coal mine waters where metallic sulfides, as iron and copper pyrites, are present. The ferrous iron is solubilized by the acid and oxidized to the ferric form and this precipitates as $Fe(OH)_3$. Such waters are rusty orange or brown in color and do not support any form of aquatic life; consequently, they represent a serious form of environmental pollution.

3.17 | Sulfate Reduction

Bacteria able to reduce sulfates (SO_4^{2-}) to sulfides were first isolated in 1895 by Beijerinck, the great Dutch bacteriologist. Most sulfate-reducing bacteria are limited to anaerobes of the genus *Desulfovibrio,* polar flagellated vibrios occurring in aquatic environments throughout the world. The Black Sea is so named because of vast sulfide production from sulfate-reducing bacteria which form black metal sulfides, generally iron or copper. The gondolas of Venice are painted black to mask the discoloration that would occur from the activities of sulfate-reducing bacteria in the canals of that city. These bacteria thrive under a variety of environmental conditions and thermophilic, halophilic, and fresh water forms have been described. It has been reported that *Desulfovibrio* can form as much as 10 g of sulfide per liter during active multiplication. Sulfides may be produced by anaerobic bacteria metabolizing sulfur-bearing amino acids, so that sulfate-reducing bacteria do not represent the only biological source of H_2S. Sulfate-reducing bacteria may also be involved in the anaerobic precipitation of calcium carbonate by their reducing action on the calcium sulfate dissolved in marine and fresh waters (Section 6.11).

3.18 | Luminescent Bacteria

A number of organisms have the interesting property of emitting visible light. *Bioluminescent* systems are found in organisms of diverse phylogenetic origin, including certain bacteria, fungi, dinoflagellates (*Noctiluca, Gonyaulax*), coelenterates (the jellyfish *Aequorea* and *Halistaura*), annelids (the fireworm *Odontosyllis*), gastropods (the limpet *Latia*), crustacea (the ostracod *Cyprindina*), and insects (fireflies), among others. In many cases, as with the firefly, bioluminescence is a property resulting solely from the activities of the animal itself. However, some marine forms possess specialized organs containing symbiotic luminous bacteria that are responsible for light emission (Fig. 3.20).* The glowing secretions extruded by the fish *Malacocephalus* contain luminous bacteria growing symbiotically in specialized luminous glands. Since bacterial luminescence is continuous and not susceptible to neural control, those forms in which luminescence depends on symbiotic bacteria frequently are found to have a retractible shutter for the organ harboring them. Such organs may serve fish as lures to attract other fish, or in species recognition, or as warning devices.

The striking result of the growth of luminous bacteria, which occurs readily on dead marine fish, permitted their recognition early in the development of bacteriology. After being kept half-covered in sea water for several days at 10° C, a squid or fish often will be found to emit a continuous brilliant blue-green glow (490 nm) in the dark. However, despite an early interest in them, luminous bacteria remain poorly described. The best known species are several gram-negative,

Fig. 3.20. The fish *Photoblepharon,* in which light is supplied by symbiotic luminescent bacteria. These bacteria ("bacteroids") are located in a depression beneath each eye. A black fold of tissue may be drawn over this organ, thereby excluding the light. The lateral line system and gill openings are reflective but not luminous. (From Y. Haneda, Light production in the luminous fishes *Photoblepharon* and *Anomalops* from the Banda Islands, *Science* **173,** 143 (1971). Courtesy of Y. Haneda and permission of the American Association for the Advancement of Science.)

---

*The light organs of fish such as *Photoblepharon* and *Anamalops* contain luminous bacteroids, that is, small rod, oval, and branched organisms of from 2 to 3 μm long and about 1 μm wide. This relationship between the fish and their bacteria is an example of a mutualistic ectosymbiosis (Section 4.18).

rod-shaped marine bacteria of the genera *Photobacterium* or *Vibrio*.* As in many of the various physiological specialists noted previously, they are polar-flagellated organisms similar in form to pseudomonads. Luminescence is linked to biological reactions requiring oxygen, and the phenomenon has been used as a sensitive test for the presence of oxygen (Section 5.11).

3.19 | Photosynthetic Bacteria

Habitats characterized by adequate illumination, low oxygen, and suitable hydrogen donors may support the development of photosynthetic bacteria. In nature one finds algae growing in the surface layers of aquatic environments, usually in a region of high light intensity. Algal photosynthesis results in oxygen evolution so that highly aerobic conditions exist in their presence. Photosynthetic bacteria occur in aquatic environments, often containing hydrogen sulfide, under layers of blue-green or green algae. Since only those wave lengths of light not absorbed by the algae layer are available for bacterial photosynthesis, it is remarkable that the light absorption spectra of algae and photosynthetic bacteria are found to be complementary, that is, the former utilize the mid-regions, and the latter the ends of the photosynthetic spectrum. The evolutionary significance of this has been considered and it is believed that the older, bacterial process was modified after the development of oxygen-evolving photosynthetic processes in such a way that the pigment systems did not directly compete for radiant energy with later-developing, oxygen-tolerant forms.

Photosynthetic bacteria may be characterized on the basis of their pigment systems (Fig. 3.21A and B) and as to whether they accumulate sulfur droplets inside or outside their cells. Accordingly, green sulfur bacteria, purple sulfur bacteria, and purple and brown nonsulfur bacteria are recognized. The green sulfur bacteria have a photosynthetic apparatus whose fine structure and bacteriochlorophylls (*c* and *d*) are not found in other photosynthetic bacteria. Organisms in this group comprise both motile and nonmotile rods and cocci and they deposit sulfur extracellularly. Two well-known green sulfur bacteria, *Chlorobium limicola* and *C. thiosulfatophilum*, are short or slightly curved rods that are obligate anaerobes capable of utilizing hydrogen sulfide,

---

Lucibacterium harveyi is also an interesting luminescent organism in that, although it is peritrichous, the polar flagellum is much thicker (sheathed) than the lateral flagella.

Representative carotenoids of photosynthetic bacteria

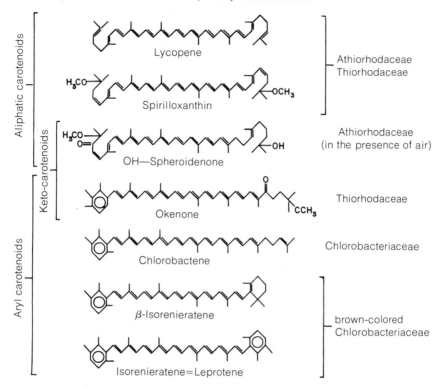

Fig. 3.21. *A,* Carotenoid pigments of photosynthetic bacteria.

thiosulfate, or, rarely, organic compounds as electron donors in the photosynthetic process (Section 5.12).

Purple sulfur bacteria have bacteriochlorophylls *a* and *b,* and they, too, are motile or nonmotile, obligately anaerobic rods and cocci, but, although they utilize the same electron donors as the green sulfur bacteria, most forms accumulate sulfur intracellularly. These organisms proliferate only in shallow layers of water, perhaps because there is strong absorption of the near infra-red wave lengths by water and, consequently, deep layers would not be adequate in regions of maximum absorption by bacteriochlorophylls *a* and *b* (820 nm and 1,025 nm, respectively). Purple sulfur bacteria may be delineated on the basis of gross morphology and structure as follows: polarly flagellated organisms without gas vacuoles such as the spiral-shaped *Thiospirillum*

Relationship between chlorophyll-a and the bacteriochlorophylls

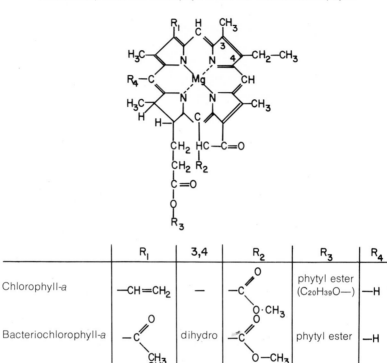

	R_1	3,4	R_2	R_3	R_4
Chlorophyll-a	—CH=CH₂	—	—C (O, O·CH₃)	phytyl ester (C₂₀H₃₉O—)	—H
Bacteriochlorophyll-a	—C (O, CH₃)	dihydro	—C (O, O—CH₃)	phytyl ester	—H
Bacteriochlorophyll-c	—C—CH₃ (OH, H)	—	—H	farnesyl ester (C₁₅H₂₅O—)	—CH₃
Bacteriochlorophyll-d	—C—CH₃ (OH, H)	—	—H	farnesyl ester	—H
Bacteriochlorophyll-b					

Fig. 3.21 *(Cont'd). B,* Relationship between chlorophyll *a* and bacteriochlorophylls. *(A* and *B* from N. Pfennig, Photosynthetic Bacteria, *Ann. Rev. Microbiol.* **21,** 285 (1965). Courtesy N. Pfennig and permission of Annual Reviews, Inc.) Bacteriochlorophyll *b* has been found to have the same R_1, R_2, R_3, and R_4 groups as bacteriochlorophyll *a,* but 3, 4 is not known. (H. Brockman and J. Kleber, Bacteriochlorophyll B, *Tetrahed. Lett.* **25,** 2195 (1970).)

jenense, the short rod-shaped *Chromatium okenii, C. weissii,* and *C. warmingii,* and the spherical-shaped *Thiocystis violacea* and *Thiothece;* and permanently nonmotile organisms also without gas vacuoles, such as the spherical-shaped *Thiococcus* and *Thiocapsa.* Then there are forms that contain gas vacuoles, namely, the polar-flagellated, spherical *Lamprocystis roseopersicina,* and the permanently nonmotile spherical *Rhodothece,* and the rod-shaped *Amoebobacter bacillosus* and *Thiodictyon elegans.*

Rarely, massive proliferations of green or purple sulfur bacteria, also called "blooms" (Section 3.11), occur under favorable conditions, and multiplication may proceed to the point where large areas of a pond or lake are colored green or purple, but most blooms are caused by various algae in fresh water and by dinoflagellates in salt water. The purple nonsulfur bacteria comprise motile rods and spirals, some of which multiply by budding (Section 3.1), and they utilize organic compounds as electron donors when growing photosynthetically and anaerobically, but they may grow aerobically in the dark. Hydrogen sulfide is toxic for these bacteria and, consequently, sulfur deposition is not observed in this group. Common forms are the rod-shaped *Rhodopseudomonas,* the spiral *Rhodospirillum,* and the budding *Rhodomicrobium.*

Purple nonsulfur and sulfur bacteria may be observed in masses colored yellow, orange, brown, red, purple, pink, or violet depending upon conditions such as cell density, culture, age, and environmental conditions. These colors are caused by the carotenoid pigments, which function in absorption of photosynthetically effective radiation at wave lengths not absorbed well by the bacteriochlorophylls. The purple sulfur and nonsulfur bacteria contain lycopene and spirilloxanthin (aliphatic carotenoids) (Fig. 3.21A) and the former group also has okenone, a keto-carotenoid, while the green sulfur bacteria have chlorobactene, an aryl carotenoid. Several other carotenoids are known to occur in photosynthetic bacteria, but none found in the green sulfur bacteria occur in the purple bacteria.

Motile photosynthetic bacteria of the purple sulfur and nonsulfur groups show two kinds of tactic responses, namely, *phototaxis* and *aerotaxis* (Section 2.5). These organisms respond to decreases in light intensity, that is, they are positively photophobotactic, but they are not able to "sense" the direction of a light source. If the intensity of light suddenly decreases, these bacteria respond by immediately reversing the direction of translational motility—a phenomenon known as "Schreckbewegung" or "shock reaction." On the other hand, no reaction is shown to increased light intensity. This remarkable behavior involves both bacteriochlorophylls and carotenoids and is likely related to a membrane potential change linked to photo-dependent reactions. The mechanism by which flagellar activity and photosynthesis is coupled has not yet been satisfactorily explained, but it is remarkable that phototactic bacteria are able to adjust their positions to regions of maximum absorption of their bacteriochlorophylls if a spectrum is focused on a suspension of them (Fig. 2.9C). Photosynthetic bacteria also are negatively aerotactic and they swim away from unfavorable oxygen concentrations, thereby maintaining themselves in areas of optimum hydrogen sulfide concentrations when these are present. There is a negative correlation between aerotaxis and phototaxis in *Rhodospirillum rubrum;* in the presence of air they are not phototactic, and in the absence of light they are positively aerotactic.

Some nonmotile species such as *Rhodothece* and *Thiodictyon* may adjust their vertical positions by means of gas vacuoles contained in them. Gas vacuoles occur in blue-green algae, green and purple sulfur bacteria, and in some halophilic and prosthecate forms. In photosynthetic bacteria a single vacuole is composed of a number of hollow cylinders, 75–100 nm in diameter, which possess pointed ends, and are called *gas vesicles* (Fig. 3.10B). The vesicles are enclosed by a thin nonunit* membrane not connected with the cytoplasmic membrane. In the green sulfur bacteria the vesicles are long and arranged in parallel bundles; in the purple sulfur bacteria, they are shorter and irregularly deployed. Each vesicle is a unit structure and the gas vacuole is a complex of individual vesicles. The nature of the gas, or gases, in the vacuoles is unknown but experiments on some blue-green algae vacuoles suggest that it is nitrogen.

The bacterial photosynthetic apparatus is characterized by a complex intracytoplasmic membrane system termed a *thylakoid.* In *Rhodospirillum rubrum, Chromatium okenii,* and other purple sulfur and nonsulfur bacteria, the membrane system consists of vesicles or tubules invaginating the cell from the cytoplasmic membrane. *Thiococcus* has a tubular membrane system composed of groups of parallel straight tubes, whereas vesicular and lamellar membranes are seen in *Thiocapsa.* Some rhodospirilla have short lamellar membrane systems near the periphery of the cytoplasmic membranes. The green bacteria *Chlorobium limicola, C. thiosulfatophilum, C. phaseobacteroides,* and *Pelodictyon clathratiforme* have distinctive photosynthetic organelles in that they are composed of

---

*A nonunit membrane is one that is less than 6 nm wide and appears in cross-section as a single, electron-dense layer.

oblong vesicles (30–50 nm wide by 100–150 nm long) which are located inside the cytoplasmic membrane around the entire cell. These structures are called *chlorobium vesicles* and each is bounded by a thin nonunit membrane (3 nm thick) underlying the cytoplasmic membrane but distinct from it. Some of the vesicles have a fibrillar structure and the photopigments are associated with it.

3.20 | Extreme Halophiles

In many extreme environments only one or a very few different kinds of organisms may be found. In saline and thermal environments vascular plants and vertebrate animals are absent, although there may be a rich flora of microorganisms. In general, as conditions become more prohibitive, macroorganisms of structural complexity tend to be fewer in number, to a point where only microorganisms flourish. For example, only bacteria and algae are found in the Dead Sea. On the other hand, a more complex biota in an environment is more stable and severe fluctuations are not to be expected.

Bacteria described as "extreme halophiles" require a minimum salt (NaCl) concentration of from 12% to 20% for multiplication and have an optimum of between 20% and 30%. In addition, there is a requirement for Mg^{2+}. There are two forms, the rods called "halobacteria," and the cocci termed "halococci"; both occur in regions of high salinity such as solar salt ponds and marshes. The halococci have well-defined cell walls and maintain their structural integrity even in regions of low salt concentration. The halobacteria, on the other hand, have a minimal cell wall and readily lyse at salt concentrations below from 8% to 10%. The walls of halobacteria are devoid of N-acetyl muramic acid and these bacteria do not appear to synthesize this compound. Some halobacteria have gas vacuoles in them which may function as a floatation organelle.*

Halobacteria are pseudomonas-like organisms of the genus *Halobacterium,* and the halococci occur in arrangements similar to *Micrococcus* or *Sarcina*. Extreme halophiles are sometimes called "red halophiles" because in a mass their carotenoid pigments impart a red or pink color to their surroundings. It has been suggested that these pigments

---

*The gas vacuoles of green and purple bacteria and the halobacteria have identical ultrastructural features.

protect the bacteria from photochemical damage by the strong sunlight likely to be encountered in shallow salt brines.

The salt concentration within extreme halophiles is equal to that of their external medium, and some of the halophile enzymes are adapted to function optimally at 2 M NaCl; indeed, upon removal of salt they become irreversibly inactivated. It has been shown that the ribosomes of *Halobacterium* are unusual in that a high potassium concentration is required for stability. Besides sodium and chloride, potassium is therefore a main component of the internal ionic content in extreme halophiles and in both *H. salinarium* and *Sarcina morrhuae* the ratio (molal concentration) of cell potassium to medium potassium is 100:1 or thereabouts. Furthermore, the ratio of the cell sodium plus potassium to the medium sodium is about 1:1.

The effect of salt on the multiplication of nonhalophilic bacteria is variable. Obligately anaerobic spore-formers are sensitive to it and are inhibited by around 5%, but pseudomonads and coliforms are relatively tolerant and aerobic spore-formers are reported to grow in concentrations as high as 20%. In the Dry Valley region of McMurdo Sound, Antarctica, saline soils contain bacteria, but no halobacteria or strict psychrophiles. The bacteria present have tolerances correlated with incubation temperatures, for example, the total number of colony-forming units per gram of soil is greater in 2.6 M salt media at 15°C than at 5°C or 2°C. It appears that the limited microflora found in such prohibitive areas is the result of a combination of low maximum summer temperatures and high soil salinity. Salt tolerance declines markedly as the temperature is lowered.

References

Dworkin, M. Biology of the Myxobacteria, *Ann. Rev. Microbiol.* **20**, 75 (1966).

Echlin, P., and I. Morris, The Relationship between Blue-Green Algae and Bacteria, *Biol. Rev.* **40**, 143 (1965).

Harold, R., and R. Y. Stanier, The Genera *Leucothrix* and *Thiothrix, Bacteriol. Rev.* **19**, 49 (1955).

Hayflick, L. (ed.), Biology of the Mycoplasma, *Ann. N.Y. Acad. Sci.* **143**, 1 (1967).

Hayflick, L. (ed.), *The Mycoplasmatales and the L-phase of Bacteria,* Appleton-Century-Crofts, New York, 1969.

Hirsch, P., and S. F. Conti, Biology of Budding Bacteria. I. Enrichment, Isolation and Morphology of *Hyphomicrobium* spp. *Arch. Mikrobiol.* **48**, 339 (1964).

Holm-Hansen, O., Ecology, Physiology, and Biochemistry of Blue-Green Algae, *Ann. Rev. Microbiol.* **22,** 47 (1968).

van Iterson, W., *Gallionella ferruginea Ehrenberg in a New Light,* North Holland Publishers, Amsterdam, 1958.

Jensen, H. L., The Azobacteriaceae, *Bacteriol. Rev.* **18,** 195 (1954).

Kushner, D. J., Halophilic Bacteria, *Adv. Appl. Microbiol.* **10,** 73 (1968).

Lechevalier, H. A., and M. P. Lechevalier, Biology of Actinomycetes, *Ann. Rev. Microbiol.* **21,** 71 (1967).

Mulder, E. G., and W. L. van Veen, Investigations on the *Sphaerotilus-Leptothrix* Group, *Antonie van Leeuwenhoek* **29,** 121 (1963).

Pfennig, N., Photosynthetic Bacteria, *Ann. Rev. Microbiol.* **21,** 285 (1967).

Pfennig, N., and G. Cohen-Bazire, Some Properties of the Green Bacterium *Pelodictyon clathratiforme, Arch. Mikrobiol.* **59,** 226 (1967).

Poindexter, J. S., Biological Properties and Classification of the *Caulobacter* Group, *Bacteriol. Rev.* **28,** 231 (1964).

Pringsheim, E. G., Iron Bacteria, *Biol. Rev.* **24,** 200 (1949).

Shilo, M., Morphological and Physiological Aspects of the Interaction of *Bdellovibrio* with Host Bacteria, in *Current Topics in Microbiology and Immunology* (Springer-Verlag, Berlin), **50,** 174 (1969).

Staley, J. T., *Prosthecomicrobium* and *Ancalomicrobium:* New Prosthecate Freshwater Bacteria, *J. Bacteriol.* **95,** 1921 (1968).

Starkey, R. L., Relations of Microorganisms to Transformations of Sulfur in Soils, *Soil Sci.* **70,** 55 (1950).

Starr, M. P., and N. L. Baigent, Parasitic Interaction of *Bdellovibrio bacteriovorus* with Other Bacteria, *J. Bacteriol.* **91,** 2006 (1966).

Starr, M. P., and V. B. D. Skerman, Bacterial Diversity: The Natural History of Selected Morphologically Unusual Bacteria, *Ann. Rev. Microbiol.* **19,** 407 (1965).

Steed, P. D. M., *Simonsiella* fam. nov. with Characterization of *Simonsiella crassa* and *Alysiella filiformis, J. Gen. Microbiol.* **29,** 615 (1963).

Vishniac, W., and M. Santer, The Thiobacilli, *Bacteriol. Rev.* **21,** 195 (1957).

four | Populations and Habitats

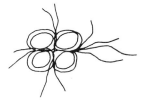

"Take the space which is occupied by the seas of this world, about two-thirds of the terrestrial surface, say with a mean depth of one mile, the collective contents of which would be 929 million cubic miles; by continual progression of multiplication the bacteria which spring from one germ would in less than five days fill the whole world's seas completely full."

Ferdinand Cohn (1866)

4.1 | Bacteria as Open Systems

Bacteria are open systems and they selectively take in substances in solution (osmotrophically), "process" these in various ways, and then return all or part of them to the environment. When the rates of such activities are constant, the system is said to be in a *steady state*. It should be stressed that this is a dynamic condition and not one of stasis. As an organism increases the complexity of its own organization, it reduces that of its environment. In other words, a living bacterium accumulates *negentropy*, by producing and maintaining a less probable state of matter in itself. When bacteria are in an isolated system the total energy content remains constant, hence they obey the First Law of Thermodynamics. What about the Second Law? This states that a system in isolation tends toward greater randomness. *Entropy* is a measure of randomness in a closed system, and it increases as the system tends toward greater randomness. Reactions in closed systems proceed as long as reactants differ in free energy content, but eventually there is no free energy difference and a condition of equilibrium is reached, and a state of maximum entropy or randomness is achieved. Bacteria growing in a closed system appear to decrease their entropy and achieve an increasing degree of nonrandomness; do they, then, not obey the Second Law of thermodynamics? In fact they do, for the following reasons: The highly improbable state of matter in bacteria and other living beings is maintained because of the free energy taken from the environment, and the randomness of the latter is therefore continually increasing during bacterial multiplication and growth. Bacteria are nonequilibrium systems and their environments do not reach the randomness of a closed system as long as a constant supply of energy is available to maintain a steady state. With these points in mind we will turn to some characteristics of bacterial multiplication and growth.

4.2 | Multiplication and Growth

Bacterial multiplication and bacterial growth are two different phenomena, but the words "multiplication" and "growth" are not always clearly defined or used with precision by bacteriologists. Multiplication is a discontinuous phenomenon resulting in an increase in numbers per unit volume, whereas growth is a continuous process increasing the mass of an individual and in which there is a proportional addition of all cell components. In multicellular organisms growth may be accompanied by cell multiplication, but in unicellular forms growth of the individual is limited (Section 2.2), and after a certain size is attained, and following division of the nuclear material, binary fission occurs, thereby resulting in two new individuals. In this process a cross-wall develops centripetally from cell periphery to center, or a constriction develops that separates the organism into two new "daughter" cells (Fig. 2.21). In a rapidly multiplying culture, individual bacteria under specified conditions may divide in as little time as 15 minutes, and this period would be designated as the *generation time*. Generation time varies not only among different bacteria but even in the same organism cultured under different environmental conditions. *Pseudomonas* and related bacteria have about the shortest generation times, closely followed by thermophilic sporogenous bacilli. The rhizobia, phytomonads, and mycobacteria appear to have the longest generation times (Tables 4.1 and 4.2). Little work has been done in studying generation times under natural conditions because of technical limitations. It has been assumed, but not proven, that an absolute limit exists for a minimum generation time, that is, no matter how "complete" the medium might be, an interval is required for cell division which is not amenable to further reduction. Mammalian cells divide about once every 10 hours at 37° C, irrespective of whether they are *in vivo* or *in vitro,* malignant or nonmalignant.

Table 4.1.   Some Representative Generation Times of Bacteria*

Organism	Generation time (min) at optimum temperature
Pseudomonas natriegens	10
Bacillus stearothermophilus	11
Bacillus coagulans	13
Bacillus thermophilus	16
Escherichia coli	17
Bacillus subtilis	26
Staphylococcus aureus	27
Clostridium botulinum	35
Pseudomonas fluorescens	35
Rhizobium meliloti	75
Vibrio marinus	81

*Compiled from various sources. Under natural conditions the generation time of a given organism is much longer than the figures given here. Those times cited above were obtained *in vitro.* This also applies to the data given in Table 4.2. (See T. D. Brock, Microbial Growth Rates in Nature, *Bacteriol. Rev.* **35,** 39 (1971).)

Table 4.2.   Relationship between Generation Time and Growth Temperatures of Psychrophiles, Mesophiles, and Thermophiles*

Organism	Temperature optimum (°C)	Generation time (min)
Vibrio marinus	15	81
Pseudomonas fluorescens	25	35
Pseudomonas natriegens	37	10
Escherichia coli	40	21
Bacillus subtilis	40	26
Bacillus thermophilus	55	16
Bacillus coagulans	70	14

*Compiled from various sources.

The reasons for differences between division times of various kinds of cells are unknown, but it appears a selective advantage for bacteria to reproduce as rapidly as possible.

Growth of an individual bacterium is not often measured by bacteriologists. An understanding of the meaning of growth is prerequisite for a discussion of multiplication. The incorporation of new material into an organism is an ordered process, that is, its rate is governed by physicochemical relationships between the organism and its *milieu*. Limitations on the bulk attained by an organism are referable to its genetic complement and environmental factors, but even in "pure" cultures there is at any given time variation observable in the dimensions of the largest organisms. It has been found experimentally that maximum dimensions of individual bacteria are attained at incubation temperatures below the optimum and, in fact, near the minimum (Section 4.9). Biologists have long known that the same species of animals residing in warm or tropic areas are generally smaller than those living in temperate or cool zones.

Regulation of cell size in bacteria may be related to the ratio of surface area to volume (Section 2.2). In a spherical bacterium, the surface area increases as the radius squared, but the volume increases as the radius cubed. Stated another way, the larger a spherical organism becomes, the less is the ratio of surface area to volume; consequently, less of it is in direct contact with the environment. If the maximum size attainable is related to cellular transport and diffusion processes, there may be a critical ratio at which these no longer serve the organism efficiently enough for continued enlargement. In cylindrical (rod-shaped) bacteria the length rather than the width or radius increases; therefore surface area and volume remain directly proportional. But these organisms also divide, so additional factors govern the size maxima achieved. In some *Escherichia coli* strains, designated as *lon⁻*, selective interference with DNA synthesis in the absence of inhibition of other cellular activities leads to continued cell elongation without cross-wall formation or division. This also happens with organisms grown under high hydrostatic pressure. Separation of division from cell growth (elongation) may be achieved by thymine starvation of thymine-requiring strains, or by ultraviolet or nalidixic acid treatment. Elongation is genetically controlled (by the *lon⁺* gene), and in *lon⁺* cultures no length increase is observed when DNA is selectively inhibited. Thus, in *lon⁺* bacteria, there is an obligatory coordination between DNA synthesis, cell-wall and "cytoplasmic" synthesis, and elongation. In *lon⁻* bacteria, this coordinate development is less stringent, and it is only cross-wall formation which is prevented when DNA synthesis is selectively halted, thus permitting the development of long, serpentine forms. Another point to note is that the ratio of surface area to volume is greatest immediately after division, and it is at this time that the individual is smallest, yet the growth rate is not necessarily fastest.

A bacterial population increases by cell multiplication, and this is accomplished most commonly by binary fission. Each new generation, therefore, results from a doubling of the population, assuming that all of the organisms divide, which, in fact, they do not. It is possible to express the total numbers after n number of cell divisions as exponents of 2. Thus, if multiplication is observed starting from a single bacterium, the total numbers during the first four generations would be as follows: $2^0 = 1$; $2^1 = 2$; $2^2 = 4$; $2^3 = 8$; and $2^4 = 16$. When the logarithms of the population to the base 2 (ordinate) are plotted from time 0 (abcissa), simple inspection of the ordinate reveals the number of generations elapsed, since each unit on the ordinate represents one generation. Before going into details about the characteristics of bacterial multiplication as observed in laboratory cultures, the methods used for estimating viable bacterial numbers will be reviewed.

4.3 | Estimating Bacterial Numbers

Most frequently the so-called *plating method* is used to estimate the number of viable bacteria in a given population. In this operation one counts the bacterial colonies, developing from a suitably diluted sample, that appear in a petri dish of an agar-solidified medium, incubated at a definite temperature and for a specified period. Oxygen-sensitive anaerobic bacteria cannot be enumerated in this way, and, instead, special "roll-tube" techniques are employed. In this case, the agar medium usually contains a reducing agent such as thioglycolate, and redox indicator such as methylene blue. The melted but cooled agar medium is inoculated with the sample to be counted, and while anaerobic conditions are maintained by passing a stream of oxygen-free nitrogen into the tube, the tube is sealed with a sterile rubber stopper and then rapidly rolled on its side so that the agar forms a thin, even film on the entire inside of the tube as it solidifies. Such roll tubes are incubated and the colonies that develop are counted as in the petri dish method. The calculation of the number of colony-forming units (CFU) per milliliter or gram is easily made, since the quantity of diluted sample contained in the petri dish or tube is known. One multiplies the total number of colonies appearing by the dilution factor necessary to give the result on a unit basis. The result of this operation is described as an "estimate" for the following reasons: (1) not all the bacteria in a sample give rise to individual colonies, since a colony may arise from either the multiplication of a single organism or a group of adherent cells; (2) not all the bacteria in a sample will give rise to a colony or colonies because of nutritional or physicochemical inadequacies of the medium, or nonoptimal incubation times or temperatures; (3) since bacteria are unlikely to be homogeneously distributed in the original sample, or in dilutions thereof, no matter how much mixing is attempted, a perfectly representative portion is unobtainable; (4) technical errors in measuring and pipetting; in variations in the actual volume of diluent present in dilution blanks (9 ml or 99 ml are ordinarily used in decimal dilutions) and in their composition and temperature; in the temperature at which the molten agar is poured; in the time and temperature of incubation; and in the colony counting procedure itself, will all give rise to quantitative variations even if other difficulties with the method could be completely eliminated. In order to avoid exposing sensitive bacteria to the heat of molten, but cooled (not less than 45° C to remain fluid), agar, "spread plates" may be used. Here the agar is poured into petri dishes and allowed to cool and solidify. One-tenth milliliter of a sample is placed onto the surface of these prepared dishes and spread over it with a sterile bent glass rod. The plates are incubated and colonies counted as usual. Useful results with this method are not obtained if the bacteria to be counted have a tendency to form spreading colonies and overgrow one another, or if they do not grow well in the presence of atmospheric oxygen. The aim is that the number of CFU's to be counted on any given petri plate be between 30 and 300. This means that a given population may have to be reduced through a series of decimal dilutions. If the bacterial population in a sample is 10^6 per milliliter, for example, a 10^4 dilution must be made to secure a theoretical colony count of 10^2 per plate. Too low a dilution leads to an overcrowded plate with reduced colony size due to competition for nutrients and other factors, and too high a dilution leads to large statistical variations. Furthermore, bacteria in various physiological conditions will react differently, yielding so-called dilution effects. An actively growing culture, on one hand, and a dormant culture as for example in dried material, on the other, will not necessarily be affected or respond in the same manner. Dilution fluids composed of buffered materials are probably better than distilled or deionized water (not synonymous), and the temperature at which these are used ought to be as close to that of the original sample as is possible. Although statistical control methods have been employed to minimize fluctuations inherent in the plating method of estimating bacterial populations, they cannot be completely eliminated and indeterminacy exists in counting bacteria.

Many methods for estimating viable numbers of bacteria have appeared and disappeared since Koch developed the pour plate method in the late 1870's. At best, present-day techniques are expensive, cumbersome, time-consuming, and error-prone, and although many ingenious efforts have been made to improve them by introducing devices such as electronic colony counters, automated shakers and dilution makers, cellulose filters (to concentrate bacteria in very sparsely populated samples), miniaturization of components used, and the like, the number of viable bacteria in a sample still cannot be counted with the same degree of precision that may be attained in counting the coins in one's pocket. Nonetheless, it is the best method available, and perhaps serves bacteriologists in need of quantitative data well enough. It is surprising that it does so, in that what is involved is the problem of enumerating, as precisely as possible, a population of individual living colloidal-sized particles. A good deal of the *corpus* of bacteriology ultimately rests on results of the plating method; for example, the dry weight and net weight, amount of protein, and enzyme activity per bacterium are all based on results of a quantitative estimation of a population obtained by plate counts.

In addition to estimates of the number of viable bacteria in a given population, the total number of organisms per unit volume may be arrived at by microscopic or electronic means. In the former a *direct microscopic count* is made of individuals and clumps of bacteria using specially ruled counting chambers (for example, the Petroff-Hausser Chamber) and known dilutions of sample. One also may spread 0.01 ml of a sample over a 1 cm² area on a glass slide, and then stain the fixed bacteria with a dye such as methylene blue. Since the area of a microscope objective field can be determined, a factor is calculated which, when multiplied by the average number of organisms observed in a given number of fields examined, yields a rough approximation of the total numbers, both viable and nonviable, per unit volume.

Electronic cell counters are devices which accurately enumerate particles in a given volume and yield immediate quantitative results. When the particles or bacteria pass through an aperture across which an electric current flows, the resistance momentarily increases markedly, and this is seen as a recorded "pulse." The number of pulses per unit volume is electronically recorded and calculated, but no differentiation between bacteria and particulate debris is possible. Neither of these direct enumeration methods differentiates viable from nonviable organisms.

4.4 | The Rise and Decline of Bacterial Populations

The rise and decline of bacterial populations was intensively studied by bacteriologists during the first quarter of the 20th century. Essentially, these investigations were made using pure cultures grown in liquid laboratory-concocted media using the plating method or some variation of it. Results of these studies led to the concept of "growth curves," "culture cycles," or "multiplication curves," and, in general, nearly all depicted the viable bacterial numbers per unit volume plotted against time in the form of an "S-shaped" curve. Most work was done using liquid media because of the limitations of solid media, such as slower diffusion of nutrients and oxygen into, and waste products from, the bacteria. It should be kept in mind, however, that in nature bacterial multiplication generally occurs in liquids or in films associated with surfaces or particulate matter and in the presence of a mixed microflora of competitors in environments varying considerably from the sterilized "broths" usually employed. In any case, techniques for studying bacterial multiplication in developing colonies on solid media are nonexistent, so the question of this approach to bacterial multiplication processes is not much discussed, nor would it likely lead to more accurate results.

Multiplication curves (Fig. 4.1) have been mathematically analyzed and the biological significance of various segments exhaustively debated. The results of these cerebrations have been to divide the curves into a number of phases, as many as eight having been proposed, and while all will be dealt with here, some bacteriologists divide such curves into as few as four phases. (1) A *stationary phase* in which no multiplication is detected, and (2) a phase of *increasing rate of multiplication* are sometimes combined into what is termed the *lag phase*. The

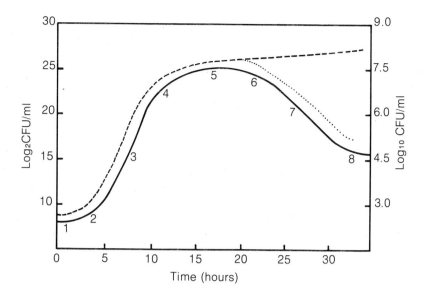

Fig. 4.1. Idealized multiplication curve of a bacterial culture. Solid line (———) represents logarithms of the CFU/ml as determined by the plating method; dashed line (- - - -) represents organisms present per milliliter as determined by direct microscopic count, including those which are nonviable and nonlysing; dotted line (. . . .) also represents total count, but the organisms are lysing. Numbers refer to phases discussed in the text.

time a culture remains in the lag phase varies greatly, and depends not only upon previous cultural history but also on the physicochemical environment into which it is placed. If a sample of an actively multiplying pure culture of an organism is transferred into identical fresh medium, the cells do not exhibit a lag phase but continue to multiply at the same rate as the organisms in the parent culture. If, however, bacteria not rapidly multiplying for some reason are transferred to a fresh medium, there is, in fact, a considerable time lapse before they increase in numbers per unit volume. Lag is a function of the previous history of the culture and reflects to some degree a paucity of cellular ribosomes. It has been found that bacteria in the lag phase may grow larger (increase in mass) instead of dividing, and the total nitrogen content increases linearly with time. The lag phase time decreases with increased numbers in the inoculum, and it has been suggested that some kind of "cell division factor" needs to be synthesized before active multiplication begins. Lag-phase bacteria seem better able to resist laboratory-imposed adverse environmental conditions than do those at a somewhat later culture age. In particular they are more resistant to heating, chilling, and harmful chemicals. One explanation for the greater resistance of lag-phase organisms involves cell-wall synthesis. In rapidly multiplying organisms there is probably greater activity of lytic enzymes involved in wall biosynthesis processes (Section 2.6), and this would serve to weaken the wall if synthetic reactions were halted by inimical environmental conditions without a parallel cessation of lytic enzyme activity.

(3) An *exponential* phase follows the phase of increasing rate of multiplication and here the bacteria divide at a constant rate. If the logarithms of the total numbers of bacteria per unit volume are plotted against time during this phase, a straight line with a positive slope will be obtained. A *growth rate constant (k)* may be calculated which is a measure of the rate of increase per organism. In simple terms,

$$k = \frac{2.3}{t} \log_{10} \frac{n'_{t2}}{n_{t1}}$$

where t is the time of the experiment, and n_{t1} and n_{t2} are the estimated number of bacteria at the beginning and end of the experiment. As values of k increase, so does the number of divisions per unit time.

The *number of generations (g)* during a given period is estimated from the following equation:

$$g = \log_2 n_{t2} - \log_2 n_{t1}.$$

where n_{t2} and n_{t1} are as before, and the assumption is that *all* the bacteria divide. Since the *generation time (gt)* is defined as the time required for the population to double it follows that:

$$gt = \frac{t}{g},$$

where t is the total time of the experiment and g the number of generations. As the generation time decreases, the dry weight and number of nuclei per organism and the amount of protein synthesized per minute per nucleus increase. Generation times vary greatly both as to minima and duration, depending upon the organism and cultural conditions (Table 4.1), but during the exponential phase this time is constant and minimal. In the early part of this phase the bacteria are more susceptible to adverse environmental stresses, and easily succumb to heating, chilling, and certain chemicals. Bacteria in this condition have been described as being in a state of "physiological youth," and they show a number of interesting characteristics, among which are a strong basophilism to dyes and maximum cell water content.

The exponential phase continues until some factor becomes limiting; this might be lack of a specific or general nutrient, accumulation of a toxic and inhibitory product, or the depletion of oxygen (in the case of obligate aerobes). It has been estimated that if a bacterium of the dimensions of *Escherichia coli* divided every 20 min, it would produce a quantity of cells greater than the earth's mass in three days. Obviously, the exponential phase can be maintained longer if nutrients are renewed, waste products are removed, and forced aeration is employed. However, if a required factor is supplied in limiting quantities, an estimate of total numbers of bacteria formed per unit weight of the factor may be calculated. It is known that considerable bacterial multiplication occurs in extremely dilute media, and it is apparent that the large excesses of nutrients provided by the usual artificial laboratory media are rarely encountered in nature. The maximum division rate, or minimum generation time, is achieved by many bacteria at concentrations of a few milligrams per milliliter of compounds serving as carbon or nitrogen sources. Some minimum generation times that appear to be close to a limiting value have been recorded; these include (in minutes): *Pseudomonas natriegens*—9.6 and *Bacillus stearothermophilus*—10.8 (Table 4.1).

(4) Eventually, a *retardation phase* follows in which the multiplication rate decreases and the generation time increases. This phase culminates in a (5) *maximum stationary phase* in which the total number of viable organisms remains constant with time. If the logarithms of numbers of bacteria are plotted against time, a straight line of zero slope results. It is at this point that the maximum viable numbers per unit volume for a given medium are reached; however, the maximum cell yield per milliliter depends on a number of factors in addition to the concentration of substrate available. Using only the plating method, one cannot determine whether all multiplication ceases and the bacteria neither die nor reproduce further, or whether multiplication of a fraction of the population exactly balances those bacteria no longer able to give rise to visible colonies. It has been observed that the total count may continue to rise during this stage, although the viable count remains constant, and this has been interpreted to mean that the numbers failing to reproduce are just balanced by those able to reproduce, the nonviable forms remaining "visible" as deter-

mined by staining or electronic counting. There are some situations when an inhibitory, but not lethal, factor results in the total count and the viable count remaining static, that is, there is neither multiplication nor death. As mentioned previously, many factors influence the magnitude of the maximum numbers per unit volume achieved, but generally this is about 10^{10} per milliliter under the best possible conditions. It has been observed that in some cases, the bacteria in a culture may be removed by centrifugation, and further limited multiplication of the same organism is possible in this "staled" or used medium, thus suggesting that toxic factors and lack of nutrients are not necessarily the reason for the state of affairs assumed to exist at the maximum stationary phase. This phase may persist for some time or only briefly, again depending on circumstances of culture and organism.

Continued sampling of a culture over time will show ever-declining numbers of viable, colony-forming bacteria, and this part of the culture cycle has been described as consisting of: (6) an *accelerating decline phase* in which the viable numbers per unit volume decrease with time; (7) an *exponential death phase* wherein the logarithms of the estimated numbers plotted against time yield a straight line of negative slope; and (8) a *dormancy phase* where ever fewer viable bacteria remain, but in which some survive for an indeterminate time. Following the maximum stationary phase the numbers of bacteria observable microscopically may or may not remain fairly constant, depending upon whether they produce autolytic (self-dissolving) enzymes. If they do, the organisms rapidly lose their morphological identity and disappear; if they do not, they may remain observable for some time, even though not viable.

The concepts of age and death used in reference to bacteria are different from those applied to multicellular organisms. The word "age" refers to the time at which an inoculum is placed into a sterile glass-enclosed, laboratory-concocted medium; thus, 18 hours following this event, the culture is referred to as an "18-hour culture." "Young" cultures are generally those that are from 12 to 18 hours old, although for certain experiments, 2- or 3-hour cultures are required. The individual bacteria comprising an 18-hour culture are, of course, not necessarily that old, since immediately following fission the bacteria are again 0 hours old. An 18-hour culture therefore consists of organisms of all ages from 0 hours old to those that have been dormant since inoculation into the medium, that is, 18 hours old.

Dictionary definitions of death state that it is a cessation of all vital functions in animals and plants without capability of resuscitation. The concept of death in bacteria is not, however, based upon broad physiological considerations, but is narrowly referred to a single character, namely, the ability to reproduce. A viable bacterium is defined as one capable of multiplying; a dead bacterium cannot multiply. Ultimately, all the cells of complex multicellular organisms cease to multiply and eventually die at various times following obvious physiological death, but readily apprehended signs of death are not so numerous in bacteria. The diagnosis of death has been referred to dependence on results obtained from technical laboratory manipulations, that is, whether a colony can be formed on a solid medium, or whether visible turbidity is produced in a liquid medium, or whether an obvious increase in numbers may be observed microscopically. It does not matter that the "dead" bacteria may still be able to produce fermentative changes, or that they may continue to be motile, or that they appear "normal" when examined microscopically; bacterial death is permanent loss of ability to reproduce, and organisms sustaining such a loss therefore are technically dead. It is evident then that bacterial death is a function of the culture medium employed, and it is possible to conclude that an organism is "dead" when inoculated into a single medium, giving negative results for the tests described above, but "not dead" if, for example, the last of a series of fifty culture media tried enables it to multiply.

Another aspect of bacterial death is that in determining the effect of many lethal agents, both physical and chemical, on a population, it is not the dead that are counted but rather the living, that is, reproducing organisms. *Survivor curves* are constructed in which the logarithms of surviving numbers per unit volume are plotted against time of exposure to the lethal agent (Fig. 4.2). In general, these curves are found to be exponential, that is, a straight line of negative slope, but sigmoid and other nonexponential responses have been reported. The "exponential way of death" among bacteria has convinced many bacteriologists to interpret this in terms of a *monomolecular reaction*, that is, the percentage dying at any given instant is a constant fraction of the survivors. In fact, the rate of decrease can be calculated in terms of a death rate constant (k) as follows:

$$k = \frac{1}{t} \log_{10} \frac{n_1}{n_2}$$

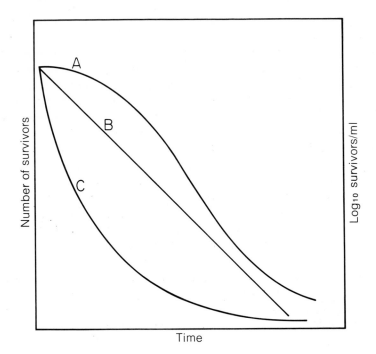

Fig. 4.2. Survivor curves of numbers of mustard seeds exposed to HgCl₂ (*A*) and bacteria exposed to the same agent (*C*). If the log₁₀ per milliliter of bacterial survivors is plotted against exposure time, a straight line (*B*) results.

where t is the time interval between two observations, and n_1 and n_2 the viable bacteria at the beginning and end of the observation time. The value of k varies greatly, depending upon the conditions involved, for example, the presence of water, organic matter, pH, prior history of the organism, kind of organism, and the like.

This peculiar order of death in bacteria is not seen in multicellular organisms, where Gaussian curves are the rule; that is, there is a distribution range from extreme susceptibility to extreme resistance, with the majority of the population being grouped around some average value. It has been suggested that the monomolecular order of death among bacteria is due to inactivation of one limiting molecule per organism. It is, of course, possible that bacteria may actually be more uniform than multicellular organisms since the diploid condition is not involved. Implicit in the argument is the assumption that all the bacteria in a given population are of equal resistance. If there is a range of resistance in bacteria similar to that observed in other organisms, it may be that the techniques employed for investigating bacterial death need to be reexamined. Perhaps the laboratory mode of application of lethal agents to bacteria is such that the most susceptible organisms, which die most rapidly, are indistinguishable from the somewhat more resistant because of technical sampling difficulties during the first few seconds or fractions of a second of exposure, and that these two groups therefore are taken together.

Another interesting aspect of multiplication studies is seen in the method whereby populations may be kept in the exponential phase indefinitely by use of an apparatus known as a chemostat. In the simplest form of this device, which may be made quite unnecessarily elaborate, a nutrient medium is supplied that is limiting in an essential growth factor. For each unit volume that flows into the culture vessel one unit volume is allowed to flow out: consequently, the total volume remains constant. Additionally, the rate of flow is maintained so that the total number of bacteria (or the population density) is the same at any given time. Since the multiplication of the culture depends on the limiting factor available, the rate is controlled by regulating amounts supplied and it is possible, for example, to reach a condition where the bacteria divide only once during the period required to renew the volume of medium in the culture vessel. The chemostat is an example of an open system in which the conditions in the culture vessel are in a constant and steady state. This is in contrast to closed culture vessels where conditions are constantly changing, that is, they are time-dependent. Studies on the influence of medium components on growth rates, and population densities, as well as on the nature of end products formed, may best be performed using continuous culture techniques employing a chemostat.

If a culture of exponentially growing bacteria is taken from its optimum temperature (say 37° C) and cooled to 25° C, and held there for 15 minutes, then returned to 37° C, this temperature shift will induce a large part of the population to divide synchronously. A majority of the bacteria will divide soon after being returned to the higher temperature, and division for two or three generations thereafter will remain, at least partially, synchronized (Fig. 4.3).

In addition to single or multishift, temperature-induced synchrony, other methods have been employed to achieve the same result. For

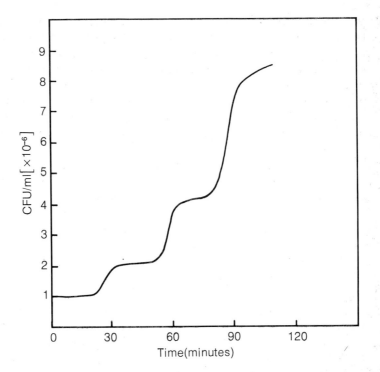

Fig. 4.3. Idealized synchronous multiplication curve showing an approximate doubling of the viable population at intervals equal to the generation time.

example, withholding thymine from thymine-requiring bacteria results in a halt of DNA synthesis and cell division without affecting RNA and protein synthesis. If after 30 min, thymine is supplied, all the bacteria resume DNA synthesis, and after a lag, divide synchronously. Another method takes advantage of the cyclic changes in size that occur in bacteria after they divide. If a bacterial culture is subjected to density gradient centrifugation in a sucrose solution, the bacteria are found distributed according to their densities, and one may recover concentrated suspensions of particular ranges. Filtration techniques also are useful to this end, and if a suspension of around 10^8 organisms per milliliter is placed on a filter paper column, the smaller cells pass through and are thereby separated from the larger organisms. Reincubation of the small cells at 37° C is followed by a lag of about 30 min, and then by a 10-min period in which 50–60% of the organisms divide synchronously. Such synchronously dividing bacteria are useful in studies on induced enzyme synthesis, bacteriophage reproduction, and nuclear division processes. Thus, synchrony may be achieved by two essentially different techniques, that is, by size selection or by physical or chemical interference with nuclear or cellular division. These operations may not yield equivalent results.

The investigation of multiplication of specific bacteria is very difficult in ecological systems, and it is for this reason that most studies have dealt with organisms in isolation as pure cultures. The growth of populations reflects characteristics of the individuals comprising them, and mathematical treatments take these into account as best they can. There is no method whereby the progeny of a given kind of bacterium can be permanently "tagged" in such a way that they are easily identifiable during a long-term observation in a natural environment, but such a technique would be invaluable in ecological experiments (Section 1.3). The interpretation and explanation of events thought to occur at various points on given bacterial multiplication curves require a great deal of faith, and it may be that the practice of reducing bacterial activities into mathematical statements, when all the parameters governing these activities are not known, needs to be reevaluated.

It has been previously noted that loss of some structural features of bacteria, as, for example, pili, flagella, or capsules, may not be a handicap unless limiting conditions of the environment impart survival value to their possessors. In laboratory studies it has been shown that permanent alterations in the *genotype* or genetic complement (mutations) may lead to changes in observable characteristics (*phenotype*). Inability to multiply in a given medium without the addition of a factor because of loss of an enzyme essential in its manufacture is commonly observed, and this is referred to as a change from the *prototrophic* to the *auxotrophic* state. If the enzyme lost by mutation lies in a branched pathway, it may be that multiple requirements are shown. Mutations may also result in other functional changes, such as enhanced resistance to antibiotics or bacteriophage infection, temperature sensitivity (Section 4.9), and the like. In a natural environment those bacteria best equipped to thrive in it will, in fact, do so. If a mutation occurs that in any way places the organism at a disadvantage in competition with others in its microenvironment, it will be rapidly eliminated from the population. One may conclude that the genotype of indigenous bacteria is one which gives them the greatest competitive advantage in a particular situation.

4.5 | Biological Aspects of Dormancy in Bacteria

The ability of bacterial endospores, cysts, or conidia (Section 2.10) to remain in inactive, quiescent, and metabolically dormant states *(cryptobiosis)* for long periods has been noted. These forms must be distinguished from "resting" bacteria which are not multiplying but are in a metabolically active state. Standardized suspensions of the latter type are used in biochemical studies, the bacteria being viewed essentially as if they were tissue preparations derived from multicellular forms.

In the cryptobiotic state resistance to drying is a characteristic feature, but only the bacterial endospore has enhanced powers of surviving high temperatures. However, the utility of this in nature is questionable. Dormancy is reversible and the dormant form constitutes a developmental phase in the life of some bacteria. In them, the form is a characteristic, differentiated element distinct structurally and enzymatically from the vegetative cell; in fact, it is a new cell altogether. What is the biological basis for dormancy? Is it the same for all forms? How did such forms evolve? What is the basis for their activation? To what extent do answers to these questions depend upon the laboratory manipulations applied to them? These are some of the questions biologists are asking themselves concerning dormancy.

It seems that under natural conditions bacterial endospores would survive drying, temperature extremes, and temporary nutritional deficiencies. There probably is some resistance to secondary environmental stresses such as antibiotics or lytic agents produced by other organisms, perhaps including those self-generated. In contrast, they might be deficient in resistance to boiling, toxic chemicals, and various laboratory created conditions. Endospores and cysts are formed on a "one from one" basis, and since no special devices are supplied for their aerial distribution, they favor the survival of the individual. Conidia and myxospores favor survival of the species, first by being produced on a "many from one" basis, and second by being aided in dispersal by association with aerial hyphae or fruiting forms. In addition to serving as agents for survival, it has been suggested that bacterial endospores are "timing devices," ensuring active proliferation in environments that exhibit cyclic variations, some of which are unfavorable to the organism.

There is evidence for the existence of fossil forms of prokaryotes in rocks about 3×10^9 years old (Section 1.1), and in the matter of survival the record appears equally impressive. It has been reported that "modern appearing" pseudomonas-like organisms, as well as bacilli, were recovered from salt deposits 18×10^7 years old. Bacteria have been cultured from sealed cans of food prepared 118 years ago, samples being taken under conditions presumably insuring no contamination from external sources.

Longevity in the dormant state combined with capacity for revival after long periods has continuously attracted the curiosity and envy of mortal man. The mechanisms by which this is accomplished have been assidously sought, but they are not as yet apparent, even though the literature has been amply supplied with biochemical and cytological studies on various points thought important in solving the problem.

Most studies on dormancy in bacteria have been made using endospores and, therefore, more is known concerning this form than any other. The endospore preparations generally employed in such investigations are obtained as follows: Bacteria are grown on a rich medium and allowed to sporulate. The spores and remaining vegetative organisms are collected by centrifugation and washed several times to remove traces of the medium. The preparation is then maintained in water as a dense suspension, or small amounts are added to finely divided sterile soil and dried. This procedure selects forms sporulating under these conditions. Furthermore, such preparations are heterogeneous in that the endospores that comprise them vary both in resistance and in germination ability. Finally, interpretation of results of such experiments on heat resistance of endospores in terms of natural environmental responses would not seem warranted.

The physical structures and biochemical characteristics of endospores and other dormant forms are somehow related to their known properties, but how they are related is unknown. The respiratory rate of bacterial endospores is extremely low, and this also is a characteristic of fungal spores, diapausing insects, and dormant seeds. Cytochromes appear absent from dormant endospores, but cytochrome oxidase and cytochrome c have been reported in myxospores. In *Bacillus* endospores, adenosine deaminase, ribosidase, alanine racemase, pyrophosphatase, and glutamo-transferase are found enriched, but glutamic-transaminase, catalase, acetokinase, glutamine synthetase, and other enzymes are found decreased with respect to amounts present in vegetative cells. It is not easy to assess the significance of these characteristics to the problem of dormancy in bacteria, but obviously, such differences are somehow reflected in the nature of bacterial endospores.

In experiments with *Bacillus* it has been found that the vegetative state may be maintained indefinitely in a chemostat, suggesting that sporulation is induced either by materials accumulating in the growth medium, or by the removal of substances suppressing spore development. The proteins and nucleic acids of endospores exist in a dehydrated state, perhaps developed during the formation of the endospore cortex and synthesis and deposition of calcium dipicolinate (Fig. 4.4 and see below). The heat resistance of mature endospores may be partly explained on the basis of these characteristics.

Dormant endospores are those that do not germinate under supposedly favorable environmental conditions. This state can be overcome by employing a process known as *heat activation*, whereby the spores are heated from 75° to 100° C for 5–60 min before inoculating them into various germination media. It may be that this heat treatment alters spore permeability or structure, destroys germination inhibitors, or releases germination stimulants, but even so, a spontaneous germination does not ensue upon heat activation, and germina-

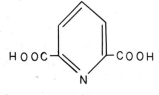

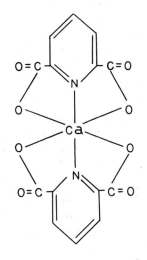

Fig. 4.4. Dipicolinic acid (*left*) and the calcium chelate of dipicolinic acid (*right*).

tion incitants are still required. In the case of endospores of aerobic bacteria, it has been found that in addition to oxygen, a number of chemical compounds may serve as triggering agents for germination; these include inosine, adenosine, L-alanine, and tyrosine. Such agents induce germination but their continual presence is not required for germination, once initiated, to proceed. On the other hand, "germination agents," usually normal metabolites which slowly disappear as germination proceeds, are required throughout the entire process. In addition to these chemical compounds, spores have been found which may be germinated mechanically by abrasion with glass beads, or with surfactant chemicals, both of which may remove permeability barriers. It is difficult to equate heat activation and other laboratory-devised germination methods with processes in nature that also lead to breaking the dormant state. The development of a new vegetative cell from an endospore and its subsequent multiplication is a process that consists of three parts: activation, germination, and outgrowth—each requiring specific conditions for maximum response.

Laboratory evidences for endospore germination are numerous, and among them are loss of spore refractility when observed by phase contrast or dark-field microscopy; decrease in ultraviolet absorption; loss in resistance to heat, as for example to 60° C for 15 min; increased stainability with bacteriological dyes; loss of dipicolinic acid (DPA), and increased metabolic activity. The content of DPA and calcium (perhaps existing as calcium dipicolinate in the endospore) has been found to be correlated with thermoresistance. DPA is believed to be located in the peptidoglycan layer of the spore cortex and its release upon endospore germination is readily understood. The entire process of transforming a dormant endospore into a fully vegetative organism usually takes place in less than 60 min. There is no evidence for sexual processes involving endospores in bacteria. The question whether any selective advantage is conferred upon sporogenous bacteria has yet to be settled.

4.6 | The L-Phase of Bacteria

L-phase variants of bacteria are forms that have had their cell walls altered by structural or metabolic defects, but which retain the potential for reverting back to their original bacterial form. These morphologically aberrant forms were described in 1935 in cultures of *Streptobacillus moniliformis* and given the name "L-form," the "L" being a designation for the Lister Institute of Preventive Medicine. L-phase variants are fragile protoplasmic formations of indefinite morphology (Fig. 4.5A), but they are not mycoplasmas, spheroplasts, or protoplasts. L-phase variants of a given bacterium may show markedly different properties and upon continued cultivation even these may vary. The media employed influence considerably both the morphology and mode of reproduction observed. These variants, when cultivated on a solid medium, produce a usually tiny (0.3–1.0 mm) "fried egg"-appearing colony, characterized by a dark center and a lighter lace-like periphery (Fig. 4.5B). As the cells comprising the colony multiply they penetrate into the agar and do not simply lie on its surface. In liquid media they appear as granular aggregates adhering to the glass sides of the containing vessel or settling to the bottom. Young colonies of *Mycoplasma* (Section 3.9) and some L-phase variants growing on, or in, solid media have a similar appearance when examined microscopically.

Various bacteria spontaneously develop L-phase variants, but it is easy to induce their production *in vitro* in some strains of *Escherichia, Salmonella, Proteus, Shigella,* and many others by the addition of penicillin, cycloserine, bacitracin, vancomycin, and ristocetin, as well as by glycine, antisera, complement, and even salt. These induced forms may be stable or unstable; the former do not revert to a "normal" bacterial morphology when the inducing agent is withdrawn, whereas the latter do. Stable forms possess only a boundary "unit membrane," whereas unstable ones have several outer layers, presumably comprising potential cell wall-forming materials. The protoplasm of L-phase variants is highly vacuolated and small bodies are discernable in the vacuoles, but they are considered nonviable elements.

Involution or *pleomorphic* forms of bacteria have been observed in aging cultures (Fig. 4.6). They are organisms that appear as large blobs of swollen cells, bulging and filamentous rods, and the like, that seem to be on the verge of incipient lysis, but some have normal forms. As the remaining viable bacteria multiply, these pleomorphic forms become normal when subcultured on fresh media. There may be naturally occurring L-phase variants in such pleomorphic cultures.

There is considerable variation in size of these organisms ranging from very small cocci to large bodies approximating the dimensions of a white blood cell (10 μm diameter). No flagellated or spore-forming L-phase variants have been reported. Both the multiplication processes

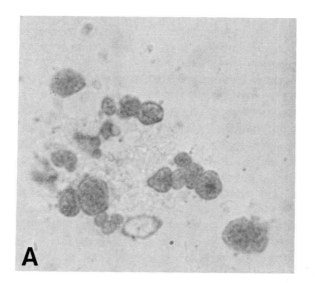

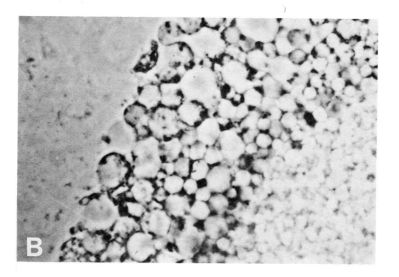

Fig. 4.5. *A*, Large bodies of *Proteus* produced by exposure to penicillin. (Courtesy L. Dienes.)

B, L-form colony of *Neisseria meningitidis*, the periphery showing well-defined and vacuolated L-forms of this organism. (From R. B. Roberts and R. G. Wittler, The L-forms of *Neisseria meningitidis*, *J. Gen. Microbiol.* **44**, 139 (1966). Courtesy R. G. Wittler and permission of the Society for General Microbiology.)

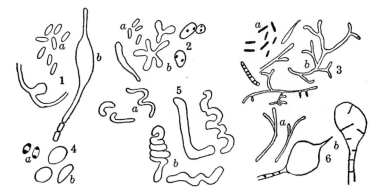

Fig. 4.6. Normal and involution forms of certain bacteria. *a*, Normal; *b*, involution forms. 1, *Acetobacter acetosum;* 2, *Rhizobium leguminosarum;* 3, *Mycobacterium tuberculosis;* 4, *Pasteurella pestis;* 5, *Lactobacillus delbrueckii;* 6, *Actinomyces* sp. (From R. E. and E. D. Buchanan, *Bacteriology,* 5th ed., Macmillan Co., New York, 1951. Courtesy R. E. Buchanan and permission of the Macmillan Co.)

and the function of these pleomorphic forms have been difficult to discern. One hypothesis is that multiplication is essentially similar to that in the bacterial phase, that is, there is a fission process resulting in two daughter cells, the role of the cell wall being substituted for by physical conditions of the environment. Another view is that the multiplication of L-forms is characteristic of them and different from that of bacterial forms. Small elementary bodies (0.5–0.6 μm) are believed to develop and multiply within the large bodies. As these develop, the large bodies disintegrate, freeing the elementary bodies into the environment. The small bodies either multiply directly, or form large bodies, and the cycle then may be repeated. It is likely that L-phase variants multiply by a fission process, but the absence of a boundary cell wall, and the consequent bizarre forms observed, especially as influenced by physical factors in the environment, has led to conjectures of other reproduction modes. Some small bodies are unable to grow because apparently they do not contain DNA. Cell wall-less forms are not the only kind of organisms that produce bodies lacking DNA. Some "mini-cell" mutants of *Escherichia coli* lack some or all of their DNA complement and, therefore, do not continue to multiply. L-phase variants are best cultivated under anaerobic conditions on "soft" agar (0.7–1.2%) medium containing 20% horse serum, the latter serving to detoxify inhibitors in peptones rather than to supply essential nutritional factors.

There are morphological similarities between L-phase variants and mycoplasmas. The latter are similar to the etiological agent of bovine pleuropneumonia, and initially they were called "pleuropneumonia-like organisms" (PPLO). *Mycoplasma* may be pathogens for man and animals, but there are physiological differences between *Mycoplasma* and L-phase variants. The former require cholesterol for growth and L-phase variants do not, and in general, the latter do not require as complex an artificial medium for growth.

L-phase organisms cannot be found after they are inoculated into an animal, but it has been observed that an L-phase induced *in vitro* may be pathogenic for mice upon reversion to the bacterial form *in vivo*. Although the development of an L-phase in organisms is a tempting explanation of the dormancy and latency of certain infections, there is no evidence that they can persist in a host. Antibiotics interfering with cell-wall synthesis, as for example, penicillins and cycloserine, have minimal inhibitory concentrations 10^3 times greater than that when the organisms are in their bacterial form, and in the case of ristocetin, vancomycin, and bacitracin, 10^2 times more is required.

4.7 | Bacterial Distribution

In general, the smaller an organism is, the less space and food it requires, and the faster it reproduces. Furthermore, the absolute abundance on earth of a particular kind of microorganism is related to these parameters in that, roughly, the most numerous organisms are also the smallest. The study of relationships that exist between bacteria and their environments constitutes *ecobiology*. The total of interacting physiochemical and biological factors in a microenvironment is referred to as an *ecosystem*.

One method of determining the kinds of bacteria that will grow under specific environmental conditions involves the technique of *enrichment* or *selective culture,* as developed by the distinguished microbiologists Winogradsky and Beijerinck. Enrichment cultures can be set up to select for multiplication of bacteria under given environmental conditions, such as temperature, osmotic pressure, radiant flux, surface tension, pH, oxidation-reduction potential, or nutritional level. Organisms capable of utilizing specific organic or inorganic compounds are also obtained using this method. The physicochemical conditions selected are imposed upon the bacteria present in a sample of the natural environment such as garden soil, marine mud, or pond water, and after a suitable incubation period only those capable of proliferating under those conditions will reach significant numbers. To obtain bacteria capable of utilizing a specific organic compound as the sole source of carbon and nitrogen, for example, materials from the natural environment are inoculated into a liquid mineral salts mixture containing the compound to be investigated and phosophates, potassium, magnesium, calcium, iron, sulfates, chlorides, and traces of cobalt, zinc, copper, molybdenum, and manganese. After incubation, only those bacteria capable of using the compound will have reproduced to any extent, so that after three or four serial transfers (5–10% by volume of the previous culture transferred to the fresh medium) most adventitious types will be eliminated and the desired kinds will be enriched or "selected." The latter may now be isolated in pure culture by plating on a solidified medium containing the same compound as the sole carbon or nitrogen source. The process of natural selection of organisms best able to thrive in a specified environment results in a population possessing special capabilities. In fact, the elective culture method has revealed the enormous range of physiological and biochemical versatility possessed by bacteria occurring in various environments. In addition, it roughly differentiates a population of bacteria possessing different metabolic capabilities; consequently, it is a useful technique for the bacterial ecologist and numerous variations on the basic method have been reported. Caution must be exercised in interpreting results of enrichment experiments in ecological terms because "physiological artifacts" may be obtained. Bacteria possessing the characteristics observed under laboratory conditions may be enriched by the imposed selective factors but they would be unable to proliferate in nature.

To qualify as a member of an indigenous population, an organism must meet the following specifications: (1) it is observed in all instances in the environment under consideration; (2) it is able to multiply in the environment and metabolize substrates normally present in it; (3) it produces chemical changes known to occur in that environment; and (4) it occurs in sufficient numbers to produce the changes observed.

In addition to facilitating isolation of pure cultures of bacteria possessing special capabilities, enrichment techniques may provide valuable clues to processes and population dynamics in nature brought about by the interaction of a mixed and variable microflora maintained under controlled conditions. Therefore, an experimental model is available for observing environmental parameters affecting the microflora or specific macroenvironments. The effect of detergents, the addition of wastes containing large amounts of fermentable carbohydrates, or thermal pollution, on the microflora of streams are three examples of problems to which the enrichment technique may be applied as a powerful ecological tool.

The morphological and physiological kinds of bacteria, their numbers and range of occurrence, are primary points of investigation in an ecological analysis. A well-defined bacterial flora and other associated microorganisms exist in such habitats as the intestinal tract of man and animals, the oral cavity, the skin, the rumen of some herbivores, and the like. Microbial habitats may allow the development of only a few kinds of organisms, or of many different types. Within a macroenvironment there may be numerous *microenvironments,* the characteristics of which may support proliferation of but one kind of bacterium; in this case it would be referred to as a *niche* for that particular organism. Microenvironments may have physical dimensions whose volumes may be approximately equal to those of the bacteria in them. Also, there is a molecular environment characterized by the physicochemical characteristics of the surfaces of both bacteria and their surrounding elements. The distribution of organisms within a given microenvironment is not homogeneous, although results of estimates of total numbers (Section 4.3) ignore this. There is a heterogeneity of distribution of bacteria that cannot be detected save by a direct microscopic examination of samples using techniques introducing the least amount of disturbance into the system.

4.8 | Extrinsic and Intrinsic Influences

Factors which influence the distribution of bacteria on earth may be categorized as *intrinsic,* that is, arising from the structure and function of the organisms themselves, and *extrinsic,* arising from their environment.

Intrinsic factors include: (1) resistant devices such as endospores; (2) minute dimensions and weight; (3) motility mechanisms, (4) structural features such as stalks, holdfasts, and filaments; and (5) dispersal aids such as myxospores, conidia, or "swarmer" cell production.

Extrinsic factors include both the gross characteristics of the physical environment as, for example, acid and alkaline thermal springs, salt brines, desert sands, arctic tundra, and ocean floor oozes, as well as the animate environment in the form of other microorganisms, plants, insects, birds, animals and man. Superimposed on both intrinsic and extrinsic factors are specific variables of temperature, pressure, radiation, pH, oxidation-reduction potential, and air and water currents. It is the complex of interactions between all factors that determines the ultimate dispersal of bacteria on earth.

Although resistant structures (endospores, for example) may enable bacteria to endure long periods of desiccation, starvation, and other inimical environmental conditions, many bacteria persist without them; they are not absolutely necessary for survival. Furthermore, organisms do not anticipate that they will be subjected at some later time to adverse conditions; thus resistant bodies are likely to be developmental forms fortuitously possessing a high degree of resistance to drying, heat, or poisonous chemicals. Nonsporeforming bacteria are as widely distributed as those forming spores, so sporogenesis *per se* is not an essential factor in bacterial dispersal.

Endospores of *Bacillus* and *Clostridium* seem to be able to survive desiccation indefinitely. Nevertheless, samples which had been buried 3,000 years were taken in 1922 from King Tutankhaman's tomb and were found to be sterile, an unexpected observation if bacterial endospores are as resistant and long-surviving as was thought. The protoplasm of resistant forms is in a dehydrated state, and a favorable environmental "signal" must be received to bring about germination (Section 4.5). Why a response is made to this signal, while many unfavorable or even lethal factors in the environment are ignored or are without effect, is an intriguing biological problem. The extremely light endospores in dust and soil particles may be airborne by wind currents and randomly deposited over the earth's surface. Most resistant or resting forms are produced by soil organisms rather than by marine or aquatic forms. The conidia of actinomycetes also are resistant to drying, although not heating, and they may survive for years in dry soil. They are looked upon as dissemination aids because large numbers per unit structure are formed.

The fact that individual bacteria weigh very little (about 10^{-11} to 10^{-15} g dry weight) means that they may remain suspended almost indefinitely in the air, and that gravitational force does not influence their distribution. The ability of motile bacteria to move about in a liquid milieu aids in their distribution. The tactic motile responses of bacteria have been described in Section 2.5, and it may be that motile organisms possess advantages over their nonmotile counterparts. Aside from the survival value of motility, movement facilitates complete distribution over favorable environments.

Structural features such as stalks in *Caulobacter,* or "ribbons" in *Gallionella,* enable organisms to attach themselves to surfaces in beneficial environments, thereby ensuring that they will remain there (Sections 3.2 and 3.3). The sudden development of unfavorable environmental conditions would be disadvantageous to firmly attached organisms, and generally, their developmental cycle is such that the progeny in the form of a motile "swarmer" is freed to swim away to a new location. Fruiting ability of some myxobacteria is light dependent, which probably aids in their being placed in a position favorable for dissemination (Section 3.4).

Variability in extrinsic factors imposed upon environments determines their ultimate suitability for a given bacterium, for example, so excellent a medium for the multiplication of bacteria as sewage, if kept frozen or boiling, would be useless. In extreme environmental conditions, bacteria appear first and disappear last; this attests to their high degree of adaptability. It has been said that bacteria are found in nature wherever any other form of life is found, and in some environments where no other life can exist.

It is difficult to state exactly that any given environmental factor is the most important one, rather, there are a number of factors, all of which may vary greatly without becoming critically limiting for a given bacterium. One factor is not necessarily more important than another, and the interplay and modifying influences of each on the others make such determinations extremely difficult.

4.9 | Temperature Effects

Temperature is a very important factor affecting the development of bacteria. They have been reported to grow from below 0° C to slightly above 100° C, but any given kind has a much more restricted range. Bacteria are probably at the same temperature as that of their surroundings (*poikilothermal* organisms), but this has not been proven. The concept of an *optimum* temperature implies that there is a point or limited range on the thermometric scale at which the multiplication rate is highest, or the generation time is minimal. There are, consequently, two other points, namely, the *minimum* temperature below which, and the *maximum,* above which, no multiplication occurs (Fig. 4.7). Although bacteria may not multiply below or above these points, they are not necessarily killed. The *range* of an organism is the thermometric distance between minimum and maximum; a greater temperature spread exists between the minimum and the optimum than between the optimum and the maximum. Temperature optima appear stable and do not show much change over long periods of cultivation.

Psychrophilic bacteria are those able to multiply at low temperatures, perhaps even below 0° C, and a distinction has been made between *obligate* and *facultative* types (Fig. 4.7). *Pseudomonas fluorescens* has been reported to grow at −10° C, and *Achromobacter* at −5° C in supercooled media. In some frozen materials liquid water pockets may exist, when dissolved solutes have reduced the freezing point to as low as −20° C, and slow bacterial proliferation may occur in such zones.

In Antarctica at 77° south latitude, 12 miles inland from McMurdo Sound in the Wright Valley, there is a 28-acre brine puddle named

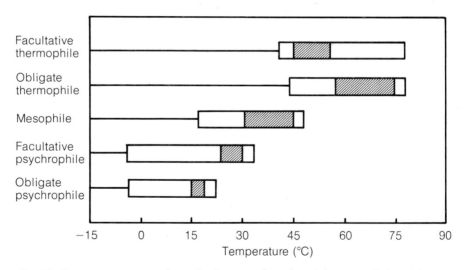

Fig. 4.7. Temperature ranges for multiplication of psychrophilic, mesophilic, and thermophilic bacteria. Hatched portions indicate approximate optima.

Don Juan Pond that is only about 6 inches deep and consists of a saturated solution of calcium chloride. The pond water would not freeze unless the temperature reached −48° C and both bacteria and molds are found there. It may be that some bacterial activity is possible wherever water exists in the liquid state irrespective of the temperature. The freezing poing of cytoplasm is about −1° C, but even at −10°C to −15°C, where ice is present in the external environment, the cell interior remains unfrozen and in a supercooled state. The cell membrane prevents the formation of ice in the supercooled interior and it is thought that cells are not, and do not contain, effective nucleators of supercooled water. Most psychrophilic organisms do not grow much above 25° C and the optimum for most of them is 15 to 18° C. Perhaps they would more properly be described as "cold-tolerant" or *psychroduric* rather than "cold-loving" or psychrophilic. Most psychrophilic bacteria have been found to be gram-negative rods of the genus *Pseudomonas,* others, less frequently observed, include *Achromobacter, Proteus, Flavobacterium, Serratia,* and *Alcaligenes,* and rarely the gram-positive bacteria *Micrococcus, Streptococcus,* and *Lactobacillus.* The number of psychrophiles is smaller in cultivated and greater in uncultivated soils. Various fresh water and marine habitats show numbers varying from 16%–76% of the total bacterial flora. It may be that the psychrophilic population is larger and more generally distributed than is the mesophilic population. Freezing is not uniformly lethal to bacteria. Although they slowly die if held in the frozen state over extended periods, many survive in a metabolically dormant condition and grow readily when thawed. The existence of life is not precluded by the lowest temperatures reached on earth. Bacteria survive much better if maintained at temperatures well below freezing rather than near freezing, i.e., viability is greater at −20° C than at −2° C. The rate of freezing affects survival in that slow freezing kills more organisms than quick freezing. Death by freezing occurs by membrane damage due to structural changes resulting in disorganization of ordered sequences of enzyme reactions. The major part of the world's ocean water never rises much above 5° C, and most marine bacteria are, in fact, facultatively psychrophilic.

Thermophilic bacteria have optima above 50° C, and a few as high as 70° C (Table 4.2). Most thermophilic bacteria are members of the genus *Bacillus,* but there are reports of others, including *Sarcina, Streptococcus, Clostridium,* and *Staphylococcus.* Only a few authentic gram-negative thermophiles are known. Some thermophiles are incapable of multiplication at 37° C, and these are termed obligate thermophiles, or *stenothermophiles,* whereas others grow at 55° C and 37° C and are described as facultative thermophiles or *eurythermophiles.* In contrast to freezing, heating is lethal to all bacteria not in a sporulated state, and boiling at 100°C generally insures no survivors. Resistance to heating varies, and other environmental conditions influence the time required for heat sterilization, as does the kind of organism involved.

There are two general explanations for bacterial thermophily and they are not mutually exclusive. One proposes that the thermophile proteins, both structural and enzymatic, have a greater heat-stability than those of mesophiles. Heat-stable enzymes of thermophiles include malate dehydrogenase, amylase, ATPase, and aldolase, but since results are generally derived from experiments on crude cell extracts, it is not known whether protective factors are present in them. The flagella of thermophilic bacteria have greater heat stability than those of non-thermophiles, and acid-disaggregated flagella of *Bacillus stearothermophilus* are reported to reaggregate into flagellar filaments at 58° C. The other explanation proposes that even if thermophile proteins are inactivated at the temperature optimum for multiplication, rapid resynthesis brings about immediate repair or replacement. As the temperature of a fluid increases the amount of dissolved gases in it decreases, but there have been no investigations on the relationship between gas requirements and multiplication in thermophiles.

Bacteria that have temperature optima around 35° C are called *mesophilic* organisms. Many of those associated with warm-blooded animals and man are included in this group. It is assumed that enzyme systems of bacteria work best at their temperature optima, and above or below this point they are either more rapidly inactivated or operate less efficiently. Bacteria have many optima in the totality of their life processes, but it is believed that the temperature at which multiplication is most rapid is the temperature at which most, if not all, enzyme systems are operating optimally.

Some bacterial mutants have been described as *temperature-sensitive.* One kind is unable to grow in the temperature range of the parent, for example, and does not grow at 37° C, while the parent strain can. This behavior may result from an alteration in the structure of an essential protein that leads to heat lability at the higher temperature. It has been found that certain specific nutritional requirements charac-

It has been found that certain specific nutritional requirements characterize these mutants, and that failure to multiply is due to inability to synthesize essential factors which must therefore be supplied preformed in the medium. If they cannot be supplied exogenously these mutants are *conditionally lethal,* that is, they will die if incubated at the restrictive temperature. *Cold-sensitive* mutants are reported to require additional nutritional factors in order to grow below 20° C, whereas the parent strain easily grows under these conditions.

Actively metabolizing bacteria produce heat, and striking temperature increases have been observed in fermenting masses of green hay, dead leaves, and manure piles. Bacterial *thermogenesis,* therefore, may serve as a means of controlling the kinds of organisms such environments will favor.

4.10 | Pressure Effects

Environmental pressure effects on bacteria are mainly of two kinds, *hydrostatic* and *osmotic.* Bacteria have been isolated from muds and oozes taken from ocean trenches over 10^4 m deep. Pressure at such depths is over 10^3 atm of about 7 tons per square inch. The deepest body of fresh water, Lake Baikal in Siberia, is 1,721 m and the pressure there is less than 200 atm. Few land-dwelling organisms are able to stand the continuous application of such force. Most surface organisms are *barophobic* and do not grow at 600 atm, but some experiments on terrestrial bacteria show that short, but not long (48-hour), exposures to 2×10^3 atm are not particularly harmful. It may be inferred that some marine organisms growing in regions under great pressure do not require such a condition, since at 1 atm large numbers may be cultivated. Sudden decompression of bacteria under pressure may be harmful, so removal of samples for bacteriological study from great depths must be done slowly. The solubility of gases, enzyme volume, diffusion rates, and other physiological parameters are affected by pressure, but specific effects on bacteria in natural environments have yet to be investigated thoroughly. If bacteria grow at pressures of 10^3 atm some interesting questions concerning them arise. If the pressure is equalized how is this done, and under what conditions does the metabolic machinery operate? If the pressure is not equalized, and this seems unlikely, what prevents the cells from being crushed? If there are authentic *barophilic* (pressure-loving) bacteria, what adaptive features do they possess?

The presence of solutes, particularly salts, increases the osmotic pressure of aqueous environments; thus the pressure of sea water can be calculated from its salinity or chlorinity. It has been determined that its freezing point, Δ, is equal to -0.0966 (Cl)-0.0000052 (Cl)^3, and when Δ is $-1.86°$ C, the osmotic pressure is 22.4 atm. The osmotic pressure of sea water is approximately 12.08 Δ at 0° C; at a concentration of 3.5% total salts, it is approximately 23 atm.

In concentrated *(hypertonic)* solutions where the external osmotic pressure is greater than that of the cell interior, water tends to leave the cell, thus inducing cytoplasmic shrinkage and withdrawal of the cytoplasmic membrane from the cell wall (Section 2.7). This phenomenon *(plasmolysis)* is generally reversible and nonlethal. If plasmolysis is sufficiently severe, motile bacteria cease moving, but upon deplasmolysis motility is resumed. Many marine bacteria are affected by slight changes in osmotic pressure and they are referred to as *stenohalinic;* on the other hand, some organisms seem unaffected by wide fluctuations in osmotic pressure and these are the *euryhalinic* types.

The osmotic pressure in many nonmarine environments is probably low, perhaps *hypotonic* (osmotic pressure outside the organism less than inside), and the necessity for a rigid cell wall in bacteria is evident. In regions of high osmotic pressure, such as salt ponds and marshes, extreme halophilic bacteria are found to have made internal adjustments (Section 3.20).

4.11 | Radiant Energy Effects

Most bacteria grow better in the dark than in light, but photosynthetic organisms require light. These bacteria grow photosynthetically as anaerobes, and possess specific chlorophylls and carotenoid pigments capable of absorbing the energy in radiations of specific wave lengths. The absorption maxima for bacteriochlorophylls are 820 nm for *a;* 1,025 nm for *b;* 660 nm for *c;* and 650 nm for *d* (Section 3.19). Bacterial photosynthesis differs from that of plants in many ways, and importantly, free oxygen is not evolved in the process (Section 3.11).

In open sea water the majority of bacteria occur in the top 75 m and in the sediments. It is in this upper *(euphotic)* zone that from 10–100% of the total incident solar radiation falls. Virtually all lethal radiation (around 260 nm) is absorbed by the top meter of water. The depth to which solar radiation extends is dependent on water transparency, atmospheric conditions, latitude, and incident angle. All photosynthetic processes occur in this zone, but 95% of the sea is lightless (the *aphotic* zone).

Ultraviolet radiations (around 260 nm) are lethal to vegetative bacteria, but less so for endospores, and it is known that the primary effect is damage to DNA. The purine and pyrimidine bases absorb extensively at these wave lengths, and genetic effects (mutations) are frequently observed among irradiated survivors. Radiant energy below a wave length of 1,100 nm may induce photochemical reactions that ultimately kill organisms. In the presence of oxygen, photooxidative changes may occur that damage sensitive enzyme systems. Bacteria possessing carotenoid pigments in their membranes may be protected since these absorb the light before such photooxidations are induced.

Bacteria suspended in the atmosphere, either floating freely or attached to dust particles, will be exposed to greater ultraviolet irradiation than those in water or soil. Wet organisms are affected more than dry, and nonpigmented more than pigmented. It has been noted that many nonsporing, airborne bacteria form pigments, perhaps for the reason stated above.

In certain stratified lakes water circulation is confined to the upper layers, the depths being nearly static and anaerobic, and photosynthetic bacteria may gather in layers just below the aerobic-anaerobic interface. Carbon dioxide and hydrogen sulfide generated from bottom sediment microbial activities diffuse upward, and, together with water-filtered light from above, supply the necessary conditions for multiplication of these bacteria. Carotenoid pigments are important since their absorption maxima lie at the wave lengths available (450–550 nm) and the organisms in such zones are richly supplied with them.

Some bacteria may produce colored pigments when grown on solid or in liquid media. The formation of these products depends upon many cultural conditions including the presence of oxygen, temperature, trace factors, and nutritional conditions, and in general, they do not develop anaerobically. Pigments may be released into the medium or they may be contained within the cells; they have been isolated in pure form. Solubility in water or organic solvents such as ether, alcohol, and chloroform has been one basis for classifying them. Carotenoid pigments derive their name from the unsaturated hydrocarbon carotene ($C_{40}H_{56}$). They occur in aliphatic, aromatic, and alicyclic forms that are soluble in ether, chloroform, or carbon disulfide, but not in water. Many bacteria contain pigments of this kind including the photosynthetics, *Sarcina*, *Micrococcus*, and marine and extreme

halophilic forms. Some *Serratia* produce red pigments, called pro-
digiosins, that are pyrrole derivatives possessing antibiotic properties.
A purple water-insoluble pigment named violacein is a tryptophane
derivative and it is formed by *Chromobacterium violaceum;* pseudomonads
develop water-soluble, yellow-green fluorescent pigments of
unknown chemical composition. A number of pigments have been
described, but it is likely that many are products of laboratory cultivation conditions and do not develop under natural conditions.

4.12 | pH Effects

The acidity or alkalinity of the environment is an important factor
governing bacterial proliferation. As with temperature, it has been
shown that bacteria have pH optima, that is, a range in which they
multiply most rapidly, and minima and maxima, below and above
which they do not grow. Very high or very low pH levels are inhibitory
or lethal to all but a few kinds of bacteria. Organisms that multiply
at or tolerate a low pH, for example, between 1.0 and 4.0, are termed
acidophiles or *aciduric;* those that grow at or tolerate a high pH, say
8.0 to 10.0, are *alkalinophiles* or *alkaloduric.* Most bacteria grow best
around neutrality (pH 7.0), with the range being from about pH 6.5
to 7.5 (Fig. 4.8, Table 4.3).

The optimum pH for bacteria is related to their metabolic machinery,
and it has been discovered that pH changes in the environment may

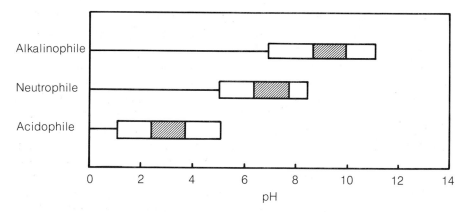

Fig. 4.8. pH ranges for multiplication of acidophilic, neutrophilic, and alkalinophilic
bacteria. Hatched portions indicate approximate optima.

Table 4.3. Minimum, Optimum, and Maximum pH in Multiplication of Various Bacteria*

Organism	pH		
	Minimum	Optimum	Maximum
Escherichia coli	4.4	6.0–7.0	9.0
Proteus vulgaris	4.4	6.0–7.0	8.4
Aerobacter aerogenes	4.4	6.0–7.0	9.0
Pseudomonas aeruginosa	5.6	6.6–7.0	8.0
Erwinia cartovora	5.6	7.1	9.3
Clostridium sporogenes	5.0–5.8	6.0–7.6	8.5–9.0
Nitrosomonas spp.	7.0–7.6	8.0–8.8	9.4
Nitrobacter spp.	6.6	7.6–8.6	10.0
Thiobacillus thiooxidans	1.0	2.0–2.8	4.0–6.0
Lactobacillus acidophilus	4.0–4.6	5.8–6.6	6.8

*Compiled from various sources.

induce compensatory enzymatic changes in the organisms. In *Escherichia coli,* enzymes such as formic, alcohol, and succinic dehydrogenases, as well as catalase, urease, hydrogenase, tryptophanase, and others, are increased so that the total activity remains constant when the pH is altered away from their optima. These enzymes are important for survival and their continued activity (catalase, for example) is mandatory. Other enzymes such as arginine, ornithine, lysine, and histidine decarboxylases are formed in response to increased acidity in the environment, the resulting amines thereby reducing the acidity. Deaminases, such as serine deaminase, form products which tend to reduce the alkalinity of a medium. Changes in medium pH are followed by changes in enzyme activity and content only during active multiplication, and these compensatory alterations serve to maintain the organisms at a constant level of metabolic activity.

A striking example of an acidophilic bacterium is the sulfur-oxidizing *Thiobacillus thiooxidans,* which produces sulfuric acid and is capable of functioning at pH 1.0. This organism and many others can alter the pH of their environments, if not sufficiently buffered, by their metabolic activities. Lactic acid bacteria, for example, can ferment carbohydrates in milk and vegetables, forming lactic acid in the process, thereby rendering them acidic, sometimes as low as pH 3.5. (Resistance to copper ions, ordinarily quite toxic to microorganisms, may be a common feature of acidophilic bacteria. *T. thiooxidans* and *Ferrobacillus ferrooxidans* are resistant to high copper concentrations, perhaps because hydrogen ions form complexes with it at their cell surfaces thus preventing its uptake.) Conversely, bacteria capable of producing ammonia from nitrogenous compounds will make their environment more alkaline.*

Soil pH is important in the development of bacteria; garden soils may be around 6.0 to 7.5, whereas peat and humus may be quite acid (pH 3.0 to 4.5). The pH also affects the soil chemically and in turn the bacterial numbers and activities in it. At high acidity organic phosphorus compounds may be bound by adsorption to clay, whereas at high alkalinity phosphate may be precipitated as heavy metal compounds of iron or manganese even to the point of causing a deficiency of these metals. If soil becomes too acid or alkaline the qualitative and quantitative distribution of bacteria in it becomes disturbed, and may lead to selection of only a few kinds capable of growing under these conditions. The relationship of such changes to alterations of microbial soil processes is evident.

4.13 | Oxygen Relationships

Oxidation-reduction (redox) potential† is a measure of the tendency of a substance to donate electrons (a reducing agent) or accept electrons (an oxidizing agent). The more reduced a material is the greater will be its ability to donate electrons. Well aerated situations will generally have high, and poorly aerated environments will have low, *redox potentials.* Organisms that consume oxygen during their metabolic activities lower the redox potential, and unless oxygen is replaced, anaerobic conditions ensue. Bacteria are classified with respect to their oxygen requirements as *aerobic* or *anaerobic* depending upon whether they multiply only in the presence of oxygen, or only in its absence. The classification of bacteria in these terms seems artificial since there is a gradation of oxygen requirements. A number of bacteria are *facultative,* that is, they may grow under either condition. Aerobic bacteria may be inhibited by lack of free oxygen, but they are able to resume growing once it is again supplied. Pure (100%) oxygen inhibits the growth

---

*In Kenya, Lakes Elementeita and Nakaru are at pH 10 to 11 and their waters contain a high concentration of the blue-green alga *Arthrospira platensis.*

†There is a relationship between the oxidation-reduction potential of a natural environment and the pH. It has been reported that thiobacteria and the "iron bacteria" produce a low pH but function in an oxidation-reduction potential of between ~ 300 and 700 mV. Sulfate reducers and phosynthetic organisms grow at a higher pH, with the oxidation-reduction potential varying from −200 to +500 mV.

of aerobes. There is no satisfactory explanation for this curious behavior, except that the atmosphere contains only about 20% oxygen. The effects of oxygen on obligate anaerobes such as *Clostridium* may be lethal to them when they are in the vegetative state. It has been shown that these organisms may produce hydrogen peroxide which they cannot decompose because they lack catalase. Exposure to air results in the formation of hydrogen peroxide in amounts sufficient to kill the organisms. The presence of oxygen also is a requisite for photooxidative effects in which visible light, absorbed by pigments, causes the inactivation of enzymes. Under anaerobic conditions, photooxidation cannot occur. Numerous anaerobic environments exist in nature and include the intestinal tract of animals, water-logged soils, marine muds and oozes, the depths of the earth, and bodies of water overloaded with organic matter.

Many bacteria exhibit *aerotactic* responses to oxygen, that is, if motile, they will orient themselves in zones interpreted by outside observers to be at their physiological optimum with respect to this gas (Fig. 2.9B). The mechanism by which this response is achieved is not known, but along with phototactic responses it constitutes the most striking and easily observed phenomenon to which the term "bacterial behavior" might be applied.

4.14 | Currents and Surfaces

Winds and currents distribute many kinds of microorganisms over wide areas, and endospores, myxospores, and conidia seem well adapted for aerial distribution. Bacteria resistant to drying, as well as endospores, have been isolated from the upper atmosphere at altitudes of over 16 mi. It has been estimated that an airborne organism 1 μm in diameter would be carried 9,000 mi in a steady 10-mph wind before it settles 100 yd. Generally, airborne bacteria are associated with minute soil particles and aerosols, although they may float about unattached to particulate matter. More airborne organisms occur over land areas than over open seas, and their numbers are roughly proportional to the degree of ground activity and to the nature of the soil surface disturbed. The number of organisms found in the atmosphere decreases with increasing altitude, and it is believed that they are inert and do not multiply while airborne. This need not always be the case, and it has been reported that in airborne dust particles all the ingredients essential for microbial growth are present, concentrated in slowly condensing droplets of water surrounding them. Rain water and snow carry airborne organisms back to earth, and, together with soil surface runoff, constitute a means of distributing bacteria into streams and lakes.

Periphytes are organisms that grow attached to solid surfaces; caulobacters, hyphomicrobia, *Sphaerotilus,* and *Leucothrix,* among others, have a developmental cycle in which surfaces play a role. Some organisms form "holdfasts" to aid them in making a firm attachment to a surface. In environments low in nutrients solid surfaces may concentrate them to some extent by adsorption thus making them favorable for microbial growth. It has been reported that nonperiphytes attach to solid surfaces if the nutrient concentration is low, for example, *Escherichia coli* attaches to glass beads when cultivated in media containing 0.5 to 1.0 mg per liter of organic matter, but when the concentration is 50 to 100 mg per liter, it does not. Extracellular enzymes may be concentrated at organism-surface interfaces where losses by diffusion are minimized. It is commonly observed that microorganisms adhere to glass slides immersed in the sea, but it is not known how much influence the charge on the glass and the adsorption of organic matter has in bringing this about.

4.15 | Extrinsic Biological Factors

Animals, birds, insects, and men are active in the dispersal of bacteria. They may be carried on external surfaces of these agents, or internally, usually through the intestinal tract. Bacteria may be carried under the earth by nematodes, earthworms, and other burrowing forms, so that vertical, as well as horizontal, dispersal is achieved. The association of bacteria with specific living forms may limit their distribution to areas over which these forms range. Next to man, birds travel farthest and, consequently, migratory forms spread their intestinal flora over wide geographical areas, for example, bobolinks fly 4,000–5,000 mi in their migration from Canada and the northern United States to their winter homes in Bolivia, Paraguay, and Argentina. However, a nonmigratory, nonflying species such as the penguin would distribute its flora over a limited area.

Even though certain bacteria appear to occupy specific microenvironments, these seem to be generally distributed throughout the earth, so that only a very few kinds of organisms occur in only a single locale.

4.16 | Habitats

The soil, the sea, thermal springs, desert sands, ocean deeps, and arctic tundras have associated characteristic microflora. The two great ecosystems from which all others are derived are the land and the sea.

Soil is the outer loose material of the earth's surface distinct from bedrock, and it is the site of the dynamic interactions which allow for plant growth and decomposition and recycling of organic matter. Soils have five major components, namely, mineral matter, organic matter, water, gases, and living organisms; the first fluctuates quantitatively less than the second two. Water and gases in soil constitute about half of the soil volume or *pore space,* whereas the living beings constitute less than 1%, but are nonetheless, essential to its fertility. The water in soil is present as a "soil solution" and contains numerous inorganic and organic materials dissolved and suspended in it. Individual soil particles influence the soil structure and range from rocky (diameter >3 mm) to silt (diameter <0.001 mm). It has been calculated that gravel particles (1–2 mm diameter) possess a total surface area of about 11 cm^2/g, whereas clay particles (<0.002 mm diameter) possess 11×10^3 cm^2/g. The effects of soil surface area and hydration are of great importance in determining nutrient availability, water retention, degree of aeration, and other conditions that influence bacterial activities. Pore space depends on soil structure and organic matter content. In clay soils pores are small, in sandy soils they are large, thus the former will retain moisture much better than the latter. The gases in soil are generally not quantitatively the same as in the atmosphere, and usually the carbon dioxide content is greater and the oxygen content somewhat less. The organic matter or soil *humus* is a product of the activities of living organisms, and it is in a continual flux of decomposition and renewal.

Various soil types may be distinguished on the basis of organic matter, water content, and kinds of particles (gravel, sand, clay) comprising them. *Podzols* are soils of temperate humid climates generally poor in organic matter. *Chernozems* show a definite upper layer of organic matter and are typical of subhumid grassland regions. *Latosolic* soils are characteristic of tropical wet areas and they contain organic matter without a definite upper layer or horizon. Arid hot regions have *desert* soils containing little organic matter, and *tundra* soils are those of perpetually cold and frozen regions that also have little organic matter in them. Each soil type and its locale possess numerous characteristics that markedly influence bacterial numbers and kinds. The degree of heating or freezing, and the heat capacity of the soil, the surface color which influences heat absorption, the amount of humidity and ambient air temperature, the soil structure and pore space, the amount of vegetation and organic matter present, the frequency of soil turnover are factors that influence the kinds and numbers of bacteria and other microorganisms found in various soils (Tables 4.4, 4.5, and 4.6).

The soil microflora contains bacteria, actinomycetes, yeasts, fungi, protozoa, and bacteriophages. The total number of bacteria per gram of soil fluctuates from 10^5 to 10^9 depending on soil composition and the determinative method used. Plating techniques using a variety of laboratory media, direct counts on soil using microscopic methods, buried slide techniques, and other contact preparations all yield variable and varying results (Table 4.7). In general, upper soil layers are higher in total numbers of all kinds of microorganisms than deeper layers (Table 4.6). Most soil bacteria are mesophiles but variations in the kinds of mesophiles are numerous. For example, warm soils may show actinomycetes, marshy and water-logged soils yield anaerobes in abundance, dry soils show sporeforming bacteria and actinomycetes, and saline soils reveal halophiles; dry soils give rise to the atmospheric dusts which may be responsible for numerous airborne bacteria. Predominant soil bacteria include representatives of the genera *Pseudomonas, Arthrobacter, Clostridium, Achromobacter, Bacillus, Micrococcus, Flavobacterium,* and various actinomycetes. Morphological characteristics of bacteria in soils taken from under crops show variations depending upon the plant grown (Table 4.4).

The stable microflora of soil has been called its *autochthonous* population, and it is not influenced to any degree by added plant and animal matter. *Arthrobacter* and *Nocardia* are prominent members of this group. The numerically fluctuating population is termed *zymogenous,* and includes bacteria active in the decomposition of organic matter such as cellulose and protein. *Pseudomonas, Bacillus,* and actinomycetes, among others, are included here. The latter are widely distributed in soils (Table 4.5), and may constitute up to 50% of the total microflora, especially in areas of high pH. It has been reported that in certain sections of the Bikini Atoll, 95% of the total microflora consists of actinomycetes, presumably because of the high alkalinity.

The microflora adjacent to plant roots is different from that of the characteristic nonroot soil population. It has been found that plant roots excrete bacterial growth-stimulating substances and are, in turn,

Table 4.4. Incidence of Bacterial Groups in Rhizosphere and Control Soils*

Classification	Tobacco soil Control	Tobacco soil Rhizosphere RH. 211	Tobacco soil Rhizosphere CH. 38	Corn soil Control	Corn soil Rhizosphere	Flax (pot expt.) Control	Flax (pot expt.) Rhizosphere Bison	Flax (pot expt.) Rhizosphere Novelty
Plate count (millions)	94.7	269.2	505.4	28.2	389.2	98.3	439.9	2751.3
Nonpleomorphic	(%)	(%)	(%)	(%)	(%)	(%)	(%)	(%)
Cocci	1.7	1.7	0.0	0.0	0.0	1.9	0.0	0.0
Rods, spore forming	10.0	1.7	0.0	9.6	0.0	5.8	1.0	0.0
Rods, nonspore forming, total	21.6	41.7	40.3	28.8	44.4	11.5	24.2	54.9
Rods, nonspore forming, gram positive	18.3	20.0	20.9	21.2	24.4	5.8	10.1	18.3
Rods, nonspore forming, gram negative	3.3	21.7	19.4	7.5	20.0	5.8	14.1	36.6
Rods, nonspore forming (percent gram positive)	(84.6)	(45.8)	(52.0)	(73.3)	(55.0)	(50.0)	(41.7)	(33.3)
Rods, nonspore forming (percent motile)	(38.5)	(48.0)	(41.7)	(6.7)	(30.0)	(33.3)	(50.0)	(66.7)
Rods, nonspore forming (percent chromogenic)	(0.0)	(32.0)	(28.0)	(13.3)	(53.0)	(0.0)	(41.7)	(26.2)
Pleomorphic								
Total (*Proactinomycetaceae*)	65.0	55.0	58.1	53.9	48.9	80.7	74.5	45.1
Acid-fast, nonmotile, gram positive (*Mycobacterium*)	8.3	0.0	1.6	1.9	2.2	3.8	0.0	0.0
Non-acid-fast, nonmotile (*Corynebacterium*)								
Total	55.0	45.0	45.2	42.3	35.6	57.7	50.5	16.9
Gram positive	43.3	33.3	38.7	40.4	24.4	44.2	42.4	8.5
Gram negative	11.7	11.7	6.5	1.9	11.1	13.5	8.1	8.5
Non-acid-fast, motile (*Mycoplana*)								
Total	1.7	10.0	11.3	9.6	11.1	19.2	24.2	28.2
Gram positive	1.7	1.7	6.5	9.6	8.9	13.5	13.1	5.6
Gram negative	0.0	8.3	4.8	0.0	2.2	5.8	11.1	22.6
Non-acid-fast (percent gram positive)	(81.9)	(63.7)	(80.0)	(96.3)	(71.4)	(75.0)	(74.3)	(35.5)
Non-acid-fast (percent motile)	(2.9)	(18.2)	(20.0)	(18.4)	(23.8)	(25.0)	(32.4)	(61.3)
Non-acid-fast (percent chromogenic)	(2.9)	(24.2)	(31.4)	(0.0)	(19.0)	(10.0)	(24.3)	(22.6)
Non-acid-fast (percent liquefied)	(14.7)	(24.2)	(54.3)	(22.2)	(23.8)	(52.5)	(56.8)	(54.8)
Non-acid-fast (percent reduced NO$_3$)	(47.1)	(30.3)	(42.8)	(22.2)	(18.0)	(50.0)	(14.9)	(41.9)

*From A. G. Lochhead. Quantitative Studies of Soil-Microorganisms, IV. Influence of Plant Growth on the Character of the Flora, *Canad. J. Res.* (C), **18,** 42 (1940). Courtesy of A. G. Lochhead and permission of the National Research Council of Canada.

Table 4.5. Numbers (1000's/g soil) and Percentages of Microorganisms in Different Types of Soil*

Soil	Condition	Total microbes	Total bacteria	Bacterial spores	Actino-mycetes	Fungi	% Bac-teria	% Spores	% Actino-mycetes	% Fungi
Tundra-gley, gley-podzolic	Virgin	2140	2040	13	30	70	95.6	0.7	1.4	2.9
	Cultivated	4847	4750	27	84	13	98.0	0.6	1.6	0.4
Podzol and gley-podzolic	Virgin	1086	970	130	90	26	89.3	12.0	8.1	2.7
	Cultivated	2620	1800	430	790	30	70.7	14.9	28.2	1.1
Chernozem	Virgin	3630	2300	750	1300	30	63.8	21.4	35.4	0.8
	Cultivated	4533	2940	1000	1570	23	64.4	24.5	35.1	0.5
Chestnut	Virgin	3482	2260	690	1200	22	64.8	19.3	34.7	0.6
	Cultivated	6660	4540	1680	2100	20	67.6	23.0	32.0	0.3
Brown soil and serozem	Virgin	4490	2920	770	1550	20	63.4	17.7	36.1	0.5
	Cultivated	7378	4980	1470	2380	18	66.1	19.8	33.6	0.3

*From E. N. Mishustin. The Law of Zonality and the Study of Microbial Associations of the Soil, *Soils and Fert.* **19**, 385 (1956). By permission of Commonwealth Agricultural Bureaux, Harpenden, England.

Table 4.6. Numbers of Various Kinds of Microorganisms as Determined by Plating ($\times 10^3$/g of soil)*

Depth (cm)	Profile	Horizon	Aerobic bacteria Absolute	Aerobic bacteria Relative	Actinomycetes	Bacterial spores Absolute	Bacterial spores Percent	Anaerobic bacteria Absolute	Anaerobic bacteria Percent
3–8		A₁	7800	100	2080			1950	20.2
20–25		A₂	1804	23.1	245			379	17.4
35–40	I	A₂B₁	472	6.1	49			98	17.2
65–75		B₁	10	0.1	0.5			1	8.3
135–145		B₂	1	—	—			0.4	25.5
3–8		A₁	1291	100	191	1097	84.9	156	10.8
20–25	II	A₂	1424	110.3	177	554	38.9	101	6.6
35–40		A₂B₁	276	21.4	123	115	41.7	36	11.5
65–75		B₁	61	4.7	16	16	26.2	7	10.3
3–8		A₁	2922	100	590	1215	41.7	296	9.2
20–25	III	A₂	1264	43.6	671	415	32.8	38	3.0
35–40		A₂B₁	491	16.8	6	50	10.2	4	0.9
65–75		B₁	74	2.5	6	2	3.4	1	1.7
3–8		A₁	16740	100	4920	6300	37.8	784	4.5
20–25		A₂	4840	28.9	960	2630	54.4	556	10.3
		B	2090	12.5	292	207	9.9	60	2.8
		B C	606	3.6	7	105	17.3	100	14.2

*From A. Starc. Mikrobiologische Untersuchungen einiger podsoliger Böden Kroatiens, *Arch. Mikrobiol.* **12,** 329 (1942).

Table 4.7. Bacteria in Soil as Determined by Various Methods ($\times 10^3$/g)

Soil*	Direct microscopic count	Plating method	Liquid media
Podzol	560,000	7,500	500,000
Chernozem	9,000,000	25,000	7,000,000
Garden soil	7,000,000	16,000	6,000,000

*In the plow layer under perennial grasses (adapted from Krasil'nikov).

influenced by their activities, an example of an *ectosymbiotic* association (Section 4.18). This somewhat ill-defined area is known as the *rhizosphere* and, as the bacterial population here approaches 10^9/g of soil, it is an area of great activity and competition.* Many of the bacteria in the rhizosphere produce biotic substances such as vitamins, auxins, and amino acids that stimulate plant growth; however, plant

*Leaf surfaces also support a characteristic microbial flora, and this environment is known as the *phyllosphere*.

growth inhibitors *(phytotoxins)* are produced by certain organisms. The actinomycetes, algae, and protozoa seem not to be influenced by proximity to plant roots.

Antibiotic substances are produced in soil by some bacteria and particularly by *Streptomyces* (Section 4.18). It has been reported that although neomycin and biomycin are active against *Bacillus subtilis* in laboratory media in a concentration of 0.1 μg/g, they are without effect in the soil even at very high (500 μg/g) concentrations. One must be cautious, therefore, in relating the results of laboratory experiments to conditions existing in the soil. Although many soil microorganisms, including bacteria, produce antibiotic substances, their effect on the total soil population is difficult to assess. Many antibiotics are rapidly inactivated in soil either by bacterial activities or by adsorption. Clay adsorbs basic antibiotics such as streptomycin quite rapidly, but not acid or neutral ones; on the other hand, chloramphenicol, penicillin, and chlortetracycline may be decomposed by soil microorganisms. Theoretically, antibiotic production in soil should have ecological significance. However, antibiotic production may not necessarily favor survival of a given soil bacterium if other environmental conditions become limiting; furthermore, specific antibiotic formation in the soil is difficult to detect, and generally is observed only if the soil is sterilized and supplemented with organic materials. Finally, there seems to be no relationship between predominant groups in the soil and sensitivity to antibiotics, and it appears that they are more likely to be sensitive than not when tested in the laboratory. It may well be, then, that production of antibiotics, while a fortuitous discovery for mankind, plays only a small part in the ecology of soil microorganisms. A distinction needs to be made between knowledge obtained from impressing laboratory methods on isolated cultures of bacteria taken from the soil, and bacterial activities as they involve processes occurring under natural conditions; much more is known of the former than the latter at present.

The other great arena of microbial life is the sea. About 72% of the earth's surface is covered by water, and 98% of this is sea water. Inland fresh waters of rivers, lakes, marshes, and reservoirs account for less than 0.2% of the total. It is a curious fact that water is extremely scarce in the solar system and apparently only the earth has free water. An even more interesting statistic is that one-third of the earth's surface is under 3,800 m or more of water at pressures ranging from 380 to 1,100 atm. If the earth were perfectly smooth without continents or ocean trenches, it would be covered uniformly with 12,000 ft of water. The *hydrosphere* (the earth's aqueous envelope) has salt concentrations ranging from less than one part in a million in some mountain lakes to about 3.5% in the oceans. Only 86% of the salt concentration is sodium chloride and the most abundant chemical entities of sea water include (in grams per liter): chlorine, 19.75; sodium, 10.98; sulfate, 2.7; magnesium, 1.3; calcium, 0.4; potassium, 0.4; bicarbonate, 0.15; bromide, 0.07; boric acid, 0.03; and strontium, 0.01. The pH of sea water is between 7.5 and 8.5, and more than 90% of the marine environment is at a temperature of 5° C or lower. There is, in fact, a seasonal variation of less than 10° C, and below depths of 200 m the temperature of the sea is nearly constant.

The organic matter of the sea ranges from <0.4 to 10 mg per liter in the open ocean to over 100 mg per liter in shallow productive areas. Much of the organic matter exists in a colloidal form, and less than 25% is in a dissolved state. This finely divided matter (*leptopel*) consists of chitin, humus, cellulose, lignin, protein, and other complex materials.

Are marine bacteria merely land forms carried out to sea, or are they authentic products of marine environments? If the latter, what specific properties distinguish them from land forms (Fig. 4.9A). It has been noted above that marine bacteria encounter conditions of salinity, temperature, and pressure, in combinations not usually occurring elsewhere on earth. There is great variation in the distribution of organisms in water and probably an even greater variation in their rates of reproduction. In general, the range of bacteria in fresh water lakes is from 10^3 to 10^6 per milliliter, in inland and shallow seas from 10^3 to 10^5 per milliliter, and in the open ocean from 10^1 to 10^3 per liter. The study of the bacterial ecology of the sea is limited by the techniques available, and, as previously mentioned, with regard to soil bacteria, many more can be seen (10–100 ×) microscopically than can be cultivated or isolated in the laboratory. The main genera appear to be *Vibrio*, *Pseudomonas*, *Spirillum*, *Corynebacterium*, *Nocardia*, and *Streptomyces*, with *Micrococcus*, *Sarcina*, and *Bacillus* generally isolated from sediments and shallow environments (Table 4.8). About 80% of the bacteria occurring in the sea are gram-negative and the majority are motile. Bacteria occur in large numbers only in marine sediments. The *littoral* (near the shore) zone, where biological activity is greatest, also shows the greatest diversity in bacterial species. Total bacterial counts in the littoral zone range from 1 to 10^8 (average 10^2 to 10^3) per milliliter of water, whereas in *pelagic* (open water) areas the population may be so sparse as to require examination of 50–1,000

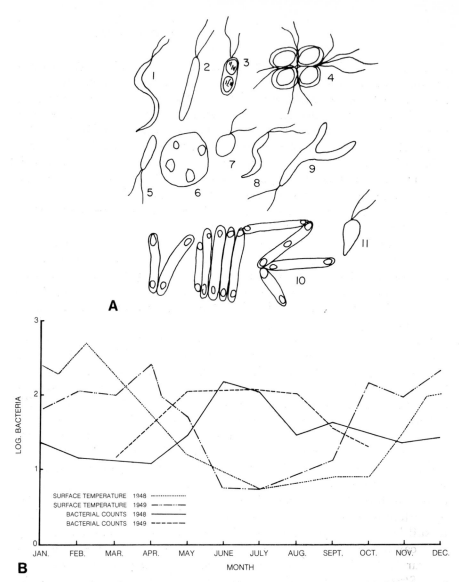

Fig. 4.9. *A,* Marine bacteria. 1, *Spirillum;* 2, *Pseudomonas;* 3, *Pseudomonas*-like form of *Mycoplana* usually seen in cultures; 4, *Sporosarcina ureae;* 5–9, *Pseudomonas-Spirillum,* branched form and "cystite" of *Mycoplana* (all these may be seen in one strain when freshly cultured), 10, *Corynebacterium;* 11, *Vibrio.* (From E. J. Ferguson Wood, *Marine Microbial Ecology.* Reinhold Publishing Co., New York, 1965. Courtesy E. J. Ferguson Wood and permission of the Reinhold Publishing Co.)

B, Viable bacterial counts in the waters of Botany Bay, showing an inverse correlation with temperature. (From E. J. Ferguson Wood, *Marine Microbial Ecology,* Reinhold Publishing Co., New York, 1965. Courtesy of E. J. Ferguson Wood and permission of the Reinhold Publishing Co.)

Table 4.8. Generic Distribution of Bacterial Types in the Mud, Water, and Animal Samples*

Taxonomic group	Mud		Water		Animals	
	Mar-umsco Bar (%)	East-ern Bay (%)	Mar-umsco Bar (%)	East-ern Bay (%)	Mar-umsco Bar (%)	East-ern Bay (%)
Vibrio spp.	37	22	56	17	47	21
Pseudomonas spp.	10	11	18	0	27	9.5
Achromobacter spp.	16	6	13	42	11	16
Corynebacterium spp.	26	22	0	0	1	4
Cytophaga/ Flavobacterium spp.	0	38	6	8	3	32
Micrococcus/ Bacillus spp.	0	0	0	0	0	8
Enterics†	10	0	6	25	7	0
Other‡	0	0	0	8	4	9.5
Total in sample	19	18	16	12	90	73

*Total sample studied = 229. From Lovelace, T. E., H. Tubiash, and R. R. Colwell, Quantitative and Qualitative Commensal Bacterial Flora of *Crassostrea virginica* in Chesapeake Bay, *Proc. Nat. Shellfisheries Assoc.* **58**, 82 (1968). Courtesy of R. R. Colwell and permission of National Shellfisheries Assoc.
†Enterics: *Enterobacter* spp., *Proteus* spp.
‡Other: *Caulobacter* spp., *Saprospira* spp., *Spirillum* spp.

ml of water to detect even one organism. In contrast, bottom sediments may contain up to 10^8 *benthonic* (inhabiting the sea floor) organisms per gram wet weight (Table 4.9), the numbers dropping off as core depth increases.

So far the majority of bacteria which have been isolated from sediments appear to be facultative aerobes, most of which grow better with air than without it. Organisms in bottom deposits decompose complex materials such as lignin, cellulose, and chitin, and in addition are involved in the deposition of manganese and iron, the precipitation of carbonates, the reduction of sulfates, and the formation of gases. Much of the ocean bottom is covered by clay-like sediment. This material is low in lime, and rich in $Fe(OH)_3$ and MnO_2, and it contains volcanic matter, skeletons of diatoms and radiolarians, and calcareous animal remains. Red clay is estimated to cover about 40×10^6 mi^2 of ocean floor. There is also present a calcareous ooze in the form of foraminifera casts that covers around 50×10^6 mi^2, and a diatomaceous ooze that spreads over 12×10^6 mi^2. It has been suggested that the microbial flora of sediments are the same irrespective of their source. The study of microbial life in marine sediments is important since the ocean trenches represent the deepest penetration of life in the biosphere.

Table 4.9.   Microbial Populations of Sediments and Their Functions*

	780 m	505 m	1,322 m
Depth of overlying water	780 m	505 m	1,322 m
Total aerobes (plate count)	930,000	31,000,000	8,800,000
Total anaerobes (oval tube), including facultative	190,000	2,600,000	1,070,000
Ammonification (peptone NH_4)	100,000	1,000,000	1,000,000
Ammonification (nutrose —NH_4)	10,000	1,000,000	100,000
Urea fermentation	100	present	1,000
Gelatin liquefaction	100,000	10,000,000	1,000,000
H_2S from peptone	10,000	1,000,000	100,000
Denitrification (to N_2)	100	1,000	10,000
Nitrate reduction (to NO_2)	100,000	10,000,000	10,000
Nitrogen fixation	0	0	0
Nitrification	0	0	0
Sulphate reduction	1,000	1,000	10,000
Starch hydrolysis	10,000	100,000	1,000
Glucose fermentation	10,000	100,000	10,000
Xylose fermentation	10,000	present	10,000
Cellulose decomposition	1,000	present	1,000
Fat hydrolysis	1,000	present	present
Chitin digestion	100	present	present

*From E. J. Ferguson Wood, *Marine Microbial Ecology,* Reinhold Publishing Co., New York, N.Y., 1965. Courtesy E. J. Ferguson Wood and permission of the Reinhold Publishing Co.

Estuarine waters, those areas where fresh and sea water meet, show large variations in temperature, salinity, pH, and oxidation-reduction potential not seen in the open sea. In such places there may be serious pollution problems of the "world's greatest septic tank" in the form of crude sewage containing pathogenic bacteria and phosphates, oil spillage, industrial wastes, insecticides and toxic chemicals, thermal pollution from atomic reactors, and other refuse from man's activities. Other factors affecting the character of the microflora are tides, prevailing winds, geophysical properties of the adjacent shore, and the currents generated by the moving waters.

Marine bacteria are only slightly halophilic when compared with *Halobacterium* spp., for example. Studies show, in fact, that nonmarine halophiles withstand much higher NaCl concentrations. Marine bacteria generally are inhibited by 10% salt. It has been observed, however, that marine organisms grow well in sea water because of the presence of inorganic ions needed for multiplication. The sodium ion is a distinctive and mandatory requirement for bacteria isolated from sea water and it cannot be substituted for by other monovalent cations such as lithium, cesium, or rubidium. The sodium requirement ranges from 0.24 to 0.3 M for optimal multiplication, that is, it is no less than one-half of the concentration present in natural sea water.

The sodium requirement is stable and the few mutants which have been described as having lost it in fact demonstrate a reduced, but still detectable need for sodium. It is thought, among other reasons, that sodium is required for activation of membrane permeases involved in the transport of substrates into the cell. Insofar as sources of carbon, nitrogen, and energy are concerned, marine forms are indistinguishable from nonmarine forms.

Sea water appears to be bactericidal for some nonmarine bacteria, but autoclaving or the addition of certain organic materials abolishes the toxicity. It may be that heat-labile antibiotics exist in sea water, which are not effective against marine bacteria, or else certain toxic metallic ions (for nonmarine forms) exist in trace amounts, and these may be removed by combining with organic compounds such as cysteine, amino acids, or other chelating agents. The low organic content of sea water is a major factor operating against growth of terrestrial forms.

At least 90% of the marine environment is 5° C or less, and most marine bacteria appear to be capable of multiplication at the lower end of the temperature range. Most of these could be described as "facultatively psychrophilic," that is, they generally grow at 0°C, but many have optima between 20° and 25°C, with a maximum at 30°C (Fig. 4.9B). Insofar as pressure is concerned, nonmarine bacteria are markedly inhibited at hydrostatic pressures over 300 atm (4,500 lb/in²). Marine bacteria occurring at the ocean surface possess pressure sensitivities similar to those of nonmarine forms. Great technical difficulty is encountered in studying barophilic bacteria. For example, up to 18 hours may be required to retrieve a sample taken at a depth of 10^4 m, and there is really no satisfactory method for stabilizing the pressure during this operation (Section 4.10).

Marine bacteria have important roles to play, but chief among these is the regeneration of nutrients, that is, conversion of organic substances to inorganic so that they may be used by plants. The euphotic zone (up to 100 m deep) of the sea receives sufficient sunlight for photosynthesis and it is here that the primary productivity of the sea occurs.

The high temperatures (up to 100° C) of thermal springs limit the active bacterial population mainly to thermophiles. In neutral or alkaline waters bacteria may be found growing at 100° C, but organisms do not actually grow in acid thermal spring waters, but rather along the thermal gradient of cooling water as it flows away. Such accumulations form macroscopic masses at temperatures around 60°–65° C. In general, as the pH of the water decreases, the upper temperature limit at which the bacteria can be observed also decreases. It has been reported that both thermophilic acidophiles and alkalinophiles may be isolated from these hot acid (pH 2) or alkaline (pH 8) environments. Gunflint chert (2×10^9 years old), possibly formed in hot spring deposits, shows fossil organisms resembling modern flexibacteria* commonly seen in thermal waters. It has been speculated that these may be primordial relics of Precambrian organisms.

Many areas of the polar regions are permanently frozen, and although not sterile, they have limited bacterial activity. Psychrophilic forms may be active under certain rare circumstances, but in general, polar areas are locales of scant microbial action. Curiously, thermophilic bacteria have been isolated from permafrost areas, but it is likely that they have been carried there adventitiously.

Essential bacterial activities, such as decomposing waste materials and recycling elements, will proceed very slowly in these frozen areas. It is most important that they do not become excessively fouled through human activities, since the existing flora could not cope with the demands placed upon it. Any environmental pollution would be likely to persist for a long time, and ecologists' warnings concerning this problem must be heeded by potential developers of these areas.

The presence of bacteria is attributed either to their proliferation in the above mentioned habitats or to their chance contamination of them. If the latter, they may not actually grow, but they may either remain dormant, slowly die off, or possibly be carried elsewhere. Some adventitious forms, however, may be able to adapt to their new surrounding, and may eventually colonize it. The interplay between factors, such as competition for nutrients, physicochemical alteration of the environment during multiplication, production of antibiotics, differences in growth rates, and the like, determines which kinds of organisms will appear to be the "real" or indigenous members of a particular habitat, and which the adventitious intruders.

Man's activities play the most important role in altering habitats. The installation of nuclear power-generating plants and the accompanying temperature increase in adjacent waters probably change the kinds and numbers of organisms dwelling there. The eradication

---

*Flexibacteria are heterotrophic, flexible, filamentous, gram-negative bacteria that have gliding motility on solid media. (See R. A. Lewin and D. M. Lounsbery, Isolation, Cultivation, and Characterization of Flexibacteria, *J. Gen. Microbiol.* **58**, 145 (1969).

of natural drainage areas, the indiscriminate disposal of untreated industrial wastes and sewage into rivers and lakes, and the introduction of vast quantities of biologically nondegradable detergents into them have done more to alter the microbial balance of nature than any combination of natural factors possibly could.

Animate nature presents a variety of habitats for bacteria. The gastrointestinal tract, the oral cavity, the respiratory tract, and the outer surfaces of man and animals possess characteristics of temperature, chemistry, and structure that combine to support a "normal" flora of organisms. Improved techniques yield fresh ecological insights; for example, it has been shown that the coliform group of bacteria is quantitatively not the most important component of the flora of the human gut. Using strict anaerobic techniques, *Bacteriodes* and anaerobic "lactobacilli" are found to be present in feces at numbers approaching 10^{10} per gram (wet weight), while the coliforms are found in numbers around 10^8 per gram. Studies on, and new techniques for the investigation of, mixed bacterial populations and their interactions are currently being pursued vigorously in many laboratories, but the interpretation of results is not as "simple" as with single pure culture studies.

4.17 | Ecological Aspects of Chemolithotrophy

Chemolithotrophy is a biological mode of securing energy by dark chemical reactions involving the oxidation of exogenous inorganic substances (Chapter 5). It has been said that the ability to utilize this mode of biological energy generation is more common than the necessity to do so. Chemolithotrophic bacteria may utilize carbon dioxide as a sole source of carbon, and ammonia, nitrite, sulfur, thiosulfate, and hydrogen may serve as oxidizable substrates for various kinds. Clearly then, such bacteria are able to synthesize all of the cell constituents from inorganic compounds—an achievement of extraordinary biochemical virtuosity. The chemolithotrophic bacteria of the soil are not only of great importance but, interestingly, appear to have no competition for their tasks.

Nitrifying bacteria are slow-growing organisms that do not occur in large numbers. They exist in soil as obligate aerobes, but generally cannot be detected and isolated without enrichment. There are two kinds of nitrifying bacteria; one oxidizes ammonia to nitrite, and the other oxidizes nitrite to nitrate, and these two must occur together since soil nitrites are seldom observed. The process of *nitrification* is favored by the presence of calcium and phosphate, and by soil conditions which are not too acid (above pH 6.0), and, of course, a source of ammonia, usually from protein decomposition, must be available. Nitrate does not accumulate in the soil under natural conditions (50 parts per million), but one must distinguish between nitrate production and nitrate accumulation. There are, however, areas where nitrate does accumulate, generally in regions high in nitrogen-containing organic material. Manure piles may contain large amounts of nitrate, but under natural conditions much of it is leached out by rain water.

Nitrifying bacteria comprise a number of genera of authentic chemolithotrophs; among them are: *Nitrosomonas, Nitrosococcus, Nitrosospira, Nitrosocystis,* and *Nitrosogloea,* which oxidize ammonia to nitrite, and *Nitrobacter,* which oxidizes nitrite to nitrate. *Nitrosomonas* is a rod or ovoid organism (1.0 × 1.5 μm) which, when motile, possesses a single polar flagellum, whereas *Nitrosococcus* is a coccus, 1.0–2.0 μm in diameter, and *Nitrosospira* is spiral shaped. *Nitrobacter* comprises motile and nonmotile ovoid rods. The chemical activities of *Nitrobacter* are suppressed by alkaline soil conditions; this may lead to an accumulation of nitrite, with consequent damage to crop plants since it is phytotoxic.

Nitrifying bacteria may be isolated from the soil on a silica-gel medium containing ammonia or nitrite as an oxidizable substrate together with a basal salts mixture at a pH of 8.5.* One method of doing this is to place crumbs of soil onto the surface of such a medium, after first coating it with a paste of carbonate or kaolin. Nitrifiers develop haloes around the colonies as the calcium carbonate is solubilized. Liquid media also have been employed to obtain enrichment cultures of nitrifying bacteria, but pure cultures are impossible to obtain in this manner.

Studies on the soil nitrification process may be made using soil perfusion or percolation devices, wherein an oxygenated nutrient solution is percolated or cycled over a soil sample contained in a glass column. Samples may be removed periodically for qualitative and quantitative chemical analysis, as a guide and measure of the bacterial activity taking place under the conditions imposed. Nitrification is

---

Nitrobacter and *Nitrosomonas* have been reported to grow at pH 13.0 but not 13.4.

one of the most important bacterial processes occurring in the soil, since nitrogen, in a form readily utilized by plants, is produced. Furthermore, as nitric acid is formed the pH drops and this tends to increase the availability of K^+, PO_4^{3-}, Mg^{2+}, and Ca^{2+} to plants (Section 4.12). Nitrification also occurs in marine environments, and this can be demonstrated in samples of bottom deposits inoculated into inorganic media made with sea water.

Nitrification proceeds in the soil in the presence of appreciable quantities of organic matter and obviously it is not toxic to nitrifying bacteria. As stated above, nitrates are easily leached out of soils by rain water, and agricultural practice has tended toward the use of ammonia fertilizers (urea or ammonium sulfate), which plants also utilize, and which do not wash out so readily. The addition of certain specific inhibitors of ammonia oxidation, such as 2-chloro-6-(trichloromethyl) pyridine (commercially known as "N-Serve"), has been found to prevent nitrification for three months in a concentration of 1.0 part per million. Soil nitrification processes apparently may be circumvented without penalty.

Thiobacillus denitrificans is noteworthy in that it is a facultatively anaerobic chemolithotroph capable of oxidizing thiosulfate to sulfate by using nitrate as an electron acceptor, thereby reducing it to gaseous nitrogen. The nitrate is not a nitrogen source for this organism and ammonium salts must be provided to serve this function.

The inorganic compounds of the *sulfur cycle* include sulfur, sulfides, or sulfates, and transformations between the various oxidation states of sulfur are carried out in part by obligately or facultatively chemolithotrophic bacteria. *Beggiatoa* and *Thiothrix* oxidize sulfides, and are able to store sulfur within their cells. This biologically available sulfur can be oxidized to sulfate in the absence of exogenous sulfide. The sources of sulfide are sulfide-containing amino acids.

In the soil, the efficiency of phosphate fertilizers is increased by encouraging the growth of *Thiobacillus thiooxidans*. This is accomplished by adding sulfur to the soil, and the sulfuric acid formed by this organism solubilizes the phosphate and makes it more readily available. The proliferation of this organism may also be used to counteract alkaline conditions in barren so-called "black alkali" soils.

A series of environmental pollution problems are caused by acid bituminous mine water. These mine effluents may have a pH as low as 3.0, and they are not only a serious source of environmental pollution but are exceedingly corrosive to barges, bridge structures, canal locks, pumps, and pipes. It has been estimated that the equivalent of 3×10^6 tons of concentrated sulfuric acid flows into the Ohio River and its tributaries annually from such wastes. An organism involved in producing acid mine water has been identified as *Thiobacillus ferrooxidans*. It not only produces sulfuric acid, but oxidizes ferrous to ferric iron at a pH of around 2.0 to 2.5 and this ultimately forms insoluble rusty brown precipitates of iron oxides and hydroxides. This form of pollution may be minimized by limiting the amount of water entering the mines. Methods of making conditions anaerobic, so that sulfur oxidation can not occur, and of introducing specific bacteriophages for *Thiobacillus ferrooxidans* are still in the exploratory stage.

Concrete sewer pipes are constantly exposed to the hydrogen sulfide of the sewage they carry. Such pipes may be seriously corroded by the sulfuric acid formed during oxidation of hydrogen sulfide by thiobacilli. Hydrogen sulfide itself is a corroding agent for concrete, and it may be formed by sulfate-reducing organisms.

Stone statues are often pitted and corroded chemically by the acid atmosphere of cities, but a particularly graphic example of the effect of thiobacilli on stone is provided by the great temple of Angkor Wat, located in the Cambodian jungle. In the moist, warm climate, with a soil containing sulfides which coat the stones or are absorbed by them, the thiobacilli have slowly obliterated the features of many of the carvings of this magnificent edifice. By way of contrast, the great capitol of red standstone near Agra, India, built by Akbar the Great in 1569, has survived without noticeable deterioration in the hot, dry climate.

4.18 | Bacterial Interactions

Since bacteria do not exist naturally in isolation, they react in various ways with one another, as well as with other living forms, and many such bacterial interactions are distinguished according to the kinds of influences operating. An association of one organism with another form is called a *symbiotic* relationship. In *mutualistic* symbiosis both partners benefit; in *parasitic* symbiosis one partner benefits while the other obtains nothing or is seriously impaired. Furthermore, an organism may remain external to the cells of the *host* (the larger of the symbiotic pair) in *ectosymbiosis,* or multiply within the host's cells in *endosymbiosis.* The latter condition involves a much closer relationship.

An example of a casual mutualistic symbiosis is that of a bacterium attacking an insoluble substrate and producing soluble products that can then be utilized, not only by the bacterium, but also by organisms not able to attack the substrate. The overall result is that a change occurs that neither organism acting in isolation could bring about; hence, such interactions are sources of novel processes. These interactions are important in the development of "food chains" in nature, whereby several different organisms degrade large complex organic molecules into smaller, utilizable molecules. A point of interest here is that the "waste product" of each kind of organism in the chain serves as a nutrient (carbon or nitrogen source) for another kind of organism farther down the chain. In the natural souring of milk, for example, lactic acid-producing streptococci and lactobacilli ferment lactose to lactic acid until they are eventually inhibited by the low pH. Lactic acid-utilizing yeasts and molds may oxidize lactate to carbon dioxide and water, and consequently, raise the pH to a point where the lactic acid bacteria may again attack the remaining lactose. Thus, the waste product of several kinds of lactic acid bacteria becomes a source of carbon and energy for several other kinds of microorganisms in this particular food chain. The net result of such interactions is the recycling of chemical elements from complex molecules to very simple ones (Chapter 6). The interactions occurring among microorganisms in their natural environment are not easily defined by laboratory techniques, and they may involve a number of kinds of organisms, sometimes acting in sequence, sometimes in concert. The analysis of complex interactions awaits the development of innovative techniques for the investigations of such populations. In the laboratory a phenomenon known as *syntrophy* is sometimes observed. In this case, one bacterial culture produces an excess of a factor required for the growth of a second. Substances such as amino acids or vitamins excreted around a colony growing on a solid medium give rise to the development of *satellite colonies* of a second, factor-requiring, organism. An even more complex reaction occurs in the case of "cross-feeding," wherein each of two kinds of bacteria produce a factor the other needs, but cannot synthesize for itself.

Interactions of bacteria with other living organisms, both plant and animal, are sometimes beneficial, and sometimes even absolutely necessary for their survival. The development of mutualistic ecto- and endosymbiotic associations would serve a number of possible functions, among them: (1) as parts of "quasi-organs," as for cellulose digestion in ruminants, and bioluminescent lures or recognition devices in fish; (2) as producers of essential nutritional factors; (3) as niches protected from outside competition; and (4) as innovative catalysts in processes or structures that otherwise would not exist. A well known example of mutualistic endosymbiosis, first demonstrated by Hellriegel and Wilfarth in 1888, involves the relationship between leguminous plants such as clover, alfalfa, and soy beans, and bacteria of the genus *Rhizobium*. These bacteria infect and grow in the root hairs of legumes and the proliferation of tetraploid root cells in the infected hair results in the development of *nodules* or tumor-like processes (Fig. 4.10A).

Although *Rhizobium* is rod-shaped or coccoidal in pure cultures (2–3 × 0.5 μm), when in the nodules it develops swollen, branched, and T-shaped forms known as "bacteroids" (Fig. 4.10B). It is in this condition that nitrogen fixation is believed to be most effectively carried out. The rhizobia apparently stimulate the root hairs to produce a pectinase (polygalacturonase) which allows them to gain entrance, and the root hairs release excretion products that specifically stimulate bacterial growth. At first motile organisms may be observed only within the root hair, but later the bacteria form hyphae-like threads which pass from cell to cell within the root cortex. Such an arrangement is called an "infection thread." This structure has a membrane composed of cellulose and pectin, and is laid down by the plant, rather than the invading rhizobia. It has been found that from 10^6 to 10^9 rhizobia per gram of soil must be present for infection and nodule formation; even so, only a small number of root hairs are invaded. The growth of the nodule is stimulated by the production of indole-acetic acid, which is a known plant auxin, but less than 5% of infected root hairs develop nodules. One other interesting point about the association between rhizobia and leguminous plants is the occurrence of leghemoglobin (a form of hemoglobin found only in such nodules); in fact, the bacteroid membranes are colored pink by it. Leghemoglobin is never found in nonleguminous plants, and apparently it plays some role in nitrogen fixation. Leghemoglobin is not formed by rhizobia or leguminous plants alone, but only when they are growing symbiotically, and the amount in the nodule is correlated with the amount of nitrogen fixed.

This process, then, consists in the bacterial *fixation* of atmospheric nitrogen into combined forms utilizable by the plant, and the latter, in return, supplies the rhizobia with all necessary nutriments, as well as a protected niche free from competition with other soil organisms.

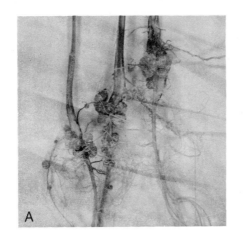

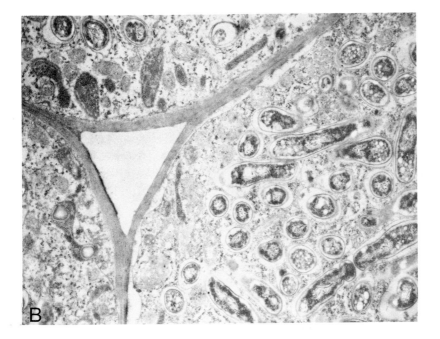

Fig. 4.10. *A,* Soybean root nodules. (Courtesy V. Miller, U.S.D.A.)

B, Development of soybean nodule cells showing the host cell cytoplasm filled with bacteria. Division of bacteria is commencing. (From D. J. Goodchild and F. J. Bergersen, Electron microscopy of the infection and subsequent development of soybean nodule cells, *J. Bacteriol.* **92,** 204, (1966). Courtesy F. J. Bergersen and the American Society for Microbiology.)

Nitrogen is fixed by *Rhizobium* only when no other, more readily us-able, nitrogen source is available. A variety of combinations of specific rhizobia and leguminous plants exist (Table 4.10), and although a single species may be able to infect only one kind of plant, others are able to infect several related kinds belonging to the same so-called *cross-inoculation group*. Nodules develop on the roots of certain non-leguminous plants but the kind of bacteria in them has not been definitely determined, although actinomycetes may be involved. There are no known examples of nitrogen-fixing eukaryotes.

In mutualistic ectosymbiosis between cellulolytic and amylolytic bacteria and herbivorous ruminants, such as cows, goats, sheep, llamas, buffaloes, camels, reindeer, and caribou, the bacteria hydrolyze cellu-lose and other insoluble polysaccharides of foods taken into the rumen to soluble carbohydrates, and these in turn are fermented mainly to acetic, propionic, and butyric acids. These acids are absorbed into

Table 4.10. Estimated Nitrogen Fixed by Legume Bacteria in Different Legumes*

Legume	Nitrogen fixed per acre (lb)
Alfalfa	194
Sweet clover	119
Red clover	114
Ladino clover	179
White clover	103
Alsike clover	119
Crimson clover	94
Lespedezas (annual)	85
Soybeans	58
Peas	72
Cowpeas	90
Winter peas	50
Vetch	80
Beans	40
Peanuts	42
Lupines	151
Velvet beans	67
Sour clover	98
Kudzu	107
Fenugreek	82
Bur clover	78
Lentils	103
Garbanzos	66
Pastures with legumes	106

*From the Farmers' Bulletin 2003, U.S.D.A., 1953.

the animal's blood stream where they are further metabolized; ruminants do not directly depend upon absorption of glucose for energy. In return for serving as a source of hydrolytic enzymes which the ruminant host lacks, the rumen provides its resident microorganisms with a continuous supply of nutrients, together with a constant environment. Ultimately, rumen organisms are carried into the intestinal tract where they are utilized as a source of protein and vitamins. In this relationship the rumen microflora is highly specific and serves as a "fluid organ," that is, an entity which, although composed of a variety of organisms, functions in the rumen to produce a constant array of end products upon which the ruminant animal is absolutely dependent. Since the rumen organisms are harbored in the rumen, but not in the cells or tissues, this association is ectosymbiotic.

Rumen pH ranges from 5.5 to 7.0, and the organ is maintained at 39°–40° C under anaerobic conditions in an atmosphere of carbon dioxide and methane (approximately 3:1). The volatile fatty acids in the rumen do not fall below 40 μmoles per milliliter, that is, absorption into the animal's blood stream does not reduce their total below 0.3%, and generally it lies between 0.6 and 0.9%. The maximum amount of acetic, propionic, and butyric acids (with a ratio of 6:2:1.5) produced in ruminants fed hay occurs around six hours after feeding when there are about 150 μmoles per milliliter of rumen fluid.

The total number of viable organisms cultivable from rumen ingesta samples is always less than that revealed by direct microscopic observation. Total counts of 2×10^{10} per milliliter of rumen contents have been reported in animals fed a grain diet, but the numbers fluctuate with the time elapsed after feeding or drinking. Unlike the microflora of soil, it is the indigenous (autochthonous) rumen flora that increases in number as cellulose and starch are brought into it, and there is no zymogenous flora *per se*. Many rumen bacteria are killed by exposure to oxygen so that cultural methods must attempt to duplicate the rumen environment as much as possible. Little is known about the distribution of bacteria in the rumen, but the whole digesta and solids have larger populations than the liquid portion, probably because many rumen organisms attach to solid feed particles.

The majority of authentic rumen bacteria are gram-negative obligate anaerobes and include the important cellulose-hydrolyzers. Around 5% of the total number of rumen bacteria are cellulolytic. *Bacteroides succinogenes, Butyrivibrio fibrisolvens,* and *Ruminococcus albus,* among others, hydrolyze this insoluble polymer to soluble carbohydrates and then to volatile fatty acids. Other *Bacteroides,* namely *B. amylophilus* and *B. rumincola,* as well as *Streptococcus bovis* and *Succinomonas amylolytica,* hydrolyze starch. Some rumen bacteria are distributed throughout the world, for example, *B. succinogenes* has been found in cattle in the United States, United Kingdom, and Africa. Rumen organisms are passed on from generation to generation through salivary and fecal contamination of the young by mature animals.

"Antagonistic" relationships among bacteria have been demonstrated in the laboratory by growing a large number of bacteria, usually from soil, on an agar plate. As colonies develop, certain ones will show clear spaces around them, whereas others will grow directly to the borders of their neighbors (Fig. 4.11). The clear zones are inter-

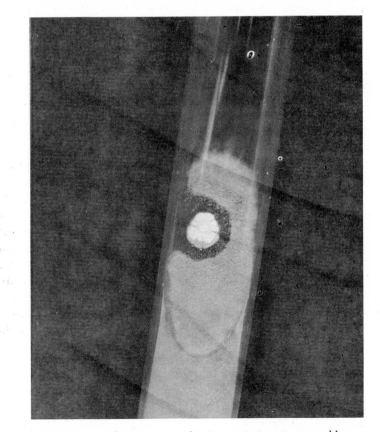

Fig. 4.11. *Streptomyces* sp. showing an antibiotic reaction against a mold.

preted as signifying an antagonistic reaction between the bacteria of one colony and those of its neighboring colonies. Simple competition for nutrients may be seen in the development of small colonies on a very crowded plate rather than the few larger colonies usually seen. Antagonism may occur for various reasons: an antibiotic compound or compounds produced by one kind of organism may inhibit the growth of other organisms; there may be inhibitory waste products, such as acids, formed; there may be competition for a limited nutrient supply. Antibiotics are organic compounds (Fig. 4.12) produced by microorganisms that are inhibitory or lethal to other, generally not related, organisms. They are formed in bulk only under conditions of laboratory cultivation and their purpose in nature, if in fact they are produced in nature, is not clear. They are biosynthetic substances and it seems that in the ordinary nutrient-poor conditions of the soil, synthesis and excretion of such material would be uneconomical. It appears that antibiotic producers are not sensitive to their own antibiotics and this has been thought to be ecologically significant. The isolation, purification, and application of antibiotics in medicine has been spectacularly successful in combating some of man's most serious infectious diseases. The question of antibiotic production in the soil is taken up in Section 4.16.

Although antibiotic substances are effective against a range of organisms, they are selective. Most antibiotics used therapeutically are elaborated by *Streptomyces* and are effective against certain bacterial pathogens. The mode of action of antibiotics has been thoroughly studied and in many cases the cellular process or organelle affected is known (Table 4.11).

A parasitic symbiosis results in the harm or destruction of one organism by another. An interesting case of bacterial parasitism was discovered in 1962, when it was found that an organism named *Bdellovibrio bacteriovorus* was able to attack certain host bacteria for its benefit. Only gram-negative bacteria, such as *Pseudomonas* and *Escherichia,* are attacked, and they must be in an actively metabolizing state; dead organisms are not attacked. *Bdellovibrio* first attaches to the host at the blunt end opposite its single, large, sheathed, polar flagellum. The process causes the host bacterium to cease its motility almost immediately. The host is then converted, after 5–10 minutes, into a spheroplast-like body, and the smaller *Bdellovibrio* (0.3–0.4 μm × 0.8–2.0 μm) penetrates it, but it is separated from the cytoplasmic contents proper by the membrane of the host cell. It therefore lies

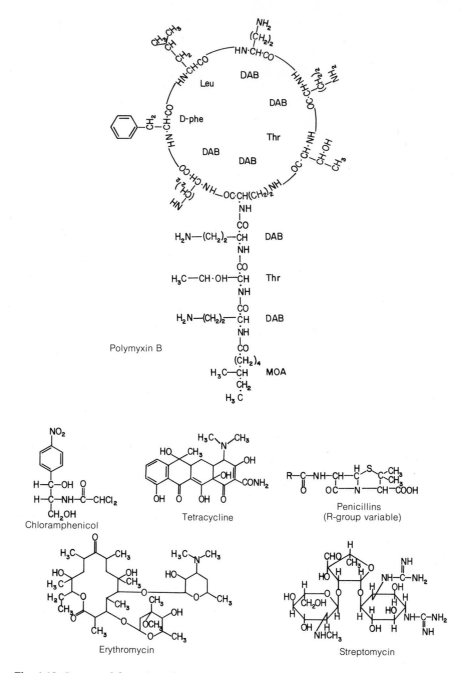

Fig. 4.12. Structural formulas of some common antibiotics.

Table 4.11. Some Antibiotics of Therapeutic Importance

Antibiotic	Chemical species	Producing organism	Affected group/ organisms	Effect on
Penicillin	β-Lactam	*Penicillium* spp.	Gram-positive bacteria	Cell wall synthesis
Gramicidin	Polypeptide	*Bacillus brevis*	Gram-positive bacteria	Membrane function
Bacitracin	Polypeptide	*Bacillus licheniformis*	Gram-positive bacteria	Cell wall synthesis
Polymyxin B	Polypeptide	*Bacillus polymyxa*	Gram-negative bacteria	Membrane function
Chloramphenicol	Aromatic	*Streptomyces venezuelae*	Gram-positive and negative bacteria, rickettsiae	Protein synthesis
Erythromycin	Macrolide	*Streptomyces erythreus*	Gram-positive bacteria	Protein synthesis
Streptomycin	Aminoglycoside	*Streptomyces griseus*	Gram-positive and negative bacteria, *M. tuberculosis*	Protein synthesis
Kanamycin	Aminoglycoside	*Streptomyces kanamyceticus*	Gram-positive bacteria, *M. tuberculosis*	Protein synthesis
Chlortetracycline	Tetracycline	*Streptomyces aureofaciens*	Gram-positive and negative bacteria, rickettsiae	Protein synthesis

between the cell wall and the cytoplasm of the host (Fig. 4.13A and B). There follows a period of elongation and division, and finally the host cell is lysed and the *Bdellovibrio* progeny are released. Single infected *Escherichia coli* host cells yield about six bdellovibrios in from 3.5 to 5 hr. These bacteria, parasitic to other bacteria, are now known to be widely distributed in soil and water, and it is incredible that they remained undetected for so long a time.

Many different kinds of bacteria are parasitized by specific *bacterial viruses* or *bacteriophages*. The infectious particle (*virion*) contains a single kind of nucleic acid, that is, either single or double-stranded DNA, or single-stranded RNA, surrounded by a proteinaceous coat (*capsid*). In the lytic cycle of infection the nucleic acid of the bacteriophage

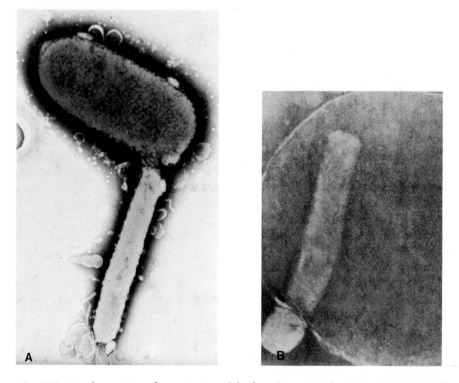

Fig. 4.13. Attachment to and penetration of the host bacterium by *Bdellovibrio. A,* Initial attachment by means of a specialized polar portion of the parasite. *B,* Rounded spheroplast showing final stages of parasite penetration. (From M. Shilo, Morphological and Physiological Aspects of the Interaction of *Bdellovibrio* with Host Bacteria, in *Current Topics in Microbiology and Immunology,* Springer-Verlag, Berlin, **50,** 174 (1969). Courtesy M. Shilo and M. Kessel and permission of Springer-Verlag.)

utilizes the host's metabolic machinery to direct the production of more viruses, and eventually the organism bursts, liberating 100 or more newly synthesized particles. Some viruses specifically attach to receptor sites on or in the cell wall of the infected bacterium (Fig. 4.14A, B, and C) and thereafter, as in the "T" bacteriophages of *Escherichia coli*, the nucleic acid is injected into the organism by means of an attachment organelle or "tail"; the protein coat remains outside. Some filamentous, icosahedrally- and polyhedrally-shaped bacteriophages do not have tails and the mode of nucleic acid entry into host bacteria is not clear. Occasionally, lysis does not follow bacteriophage infection and a condition known as *lysogeny* results. Here the viral genome is carried along with that of the host, but no protein capsids are formed and no virions are liberated. The balance is not completely stable and a lytic cycle may develop, resulting in bacteriophage synthesis and cell lysis.*

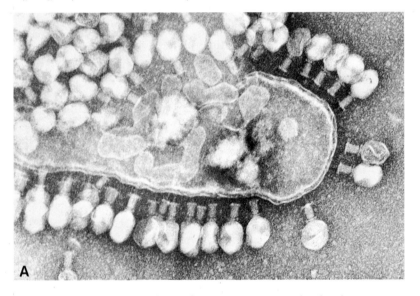

Fig. 4.14. *A,* T₄ bacteriophages adsorbed on *Escherichia coli* B; the sheaths are contracted and their baseplates are 30–40 nm from the cell wall.

---

*Bacteriophages may selectively influence a bacterial population in various ways: (1) directly by eliminating sensitive organisms and replacing the population with resistant mutants of differing potentialities; (2) directly by destroying part of the population, then indirectly by conferring an ability on the survivors to produce and transmit the bacteriophage to the progeny *(lysogeny);* and (3) indirectly by altering the genetic characteristics of the survivors by *transduction,* that is, genetic transfer via the bacteriophage itself.

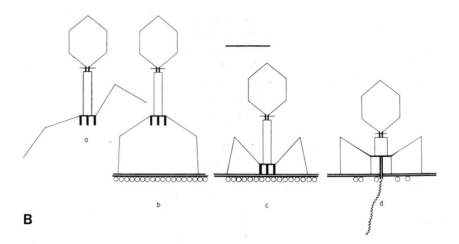

Fig. 4.14 *(Cont'd)*. *B,* Schematic illustration of the major steps in the attachment of T₂ and T₄ bacteriophages to the cell wall of *Escherichia coli* B. *a,* An unattached phage—a few of its long tail fibers and tail pins are shown; *b,* the long tail fibers have attached to the cell wall; the long fibers are situated about 180 nm apart. The baseplate is 100 nm from the cell wall; *c,* the phage has moved closer to the cell wall—its tail pins are in contact with the wall. *d,* The tail sheath has contracted and short tail fibers and long tail fibers link the baseplate to the cell wall. The needle has penetrated the wall and the DNA is being injected. (From L. D. Simon and T. F. Anderson, The Infection of *Escherichia coli* by T₂ and T₄ Bacteriophages as Seen in the Electron Microscope. 1. Attachment and Penetration, *Virology* **32,** 279 (1967). *A* and *B* Courtesy L. D. Simon and permission of Academic Press.

Many of the earliest triumphs of bacteriology centered around eluci-dation of the relationship between diseases of man, animals, and plants, and associated bacteria. In parasitic symbiosis the parasite satisfies all necessary conditions for its continued existence at the expense of another, usually larger organism or host, and thereby substitutes it for a portion of the nonliving environment. If the parasitic process is inefficient and so unbalances the host that it exhibits characteristic symptoms (an *infectious disease*) the organism is called a *pathogen.* A parasitic symbiosis may be permanent or transient depending upon specific conditions of the relationship; for example, if the parasite kills its host there must be a way for it to leave *(portal of exit),* contact a new host *(mode of transmission),* and enter it *(portal of entry),* all of which are characteristic for that particular disease. Pathogenic bacteria generally have a limited range of hosts, and it appears that this mode

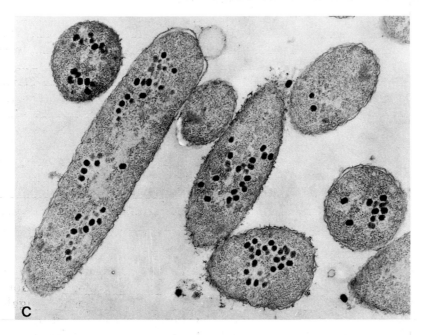

Fig. 4.14 *(Cont'd). C, Escherichia coli* infected with bacteriophage T₂, observed 30 min after infection, showing intracellular proliferation of bacteriophage particles. (Courtesy D. Kellenberger and E. Boy de la Tour.)

of existence is a fortuitous one in the life of parasitic organisms. If parasitism is efficient and does not lead to host derangement the relationship is a benign one. Numerous ectosymbionts of man, animals, and plants have been described. A number of different kinds of bacteria may proliferate on various parts of the host, and they are referred to as its *normal flora*. The ability of a parasite to harm or kill the host is a measure of its *virulence* and imperfect host-parasite relationships sometimes lead to an elimination of nearly all susceptible hosts. Such *epidemics* (among people) or *epizootics* (among animals) may have important ramifications in altering ecological relationships.

The host has defenses against invasion by parasites and the degrees to which these are effective are a measure of its *resistance*. Host defenses include outer static barriers such as the skin, mucous membranes, or waxy leaf coats; dynamic structures developed after invasion, such as walling off infected areas; and inner physiological responses such as *phagocytosis* (ingestion and destruction of bacteria by certain white blood cells such as granulocytes) and the development of antibodies. On the other hand, the virulence of pathogenic parasites is enhanced by factors known as *aggressins* that resist or defeat host defensive mechanisms.

A significant feature of pathogenic, parasitic bacteria is their toxin-forming ability. Toxins are of two kinds: (1) pharmacologically active, specific, potent, heat-labile proteins *(exotoxins)* synthesized by the parasite at localized sites in the host, and which in the form of bacteria-free filtrates cause the disease symptoms; and (2) relatively heat-stable, less specific, less potent protein-lipopolysaccharide complexes *(endotoxins)* that are parts of the cell walls of gram-negative bacteria, and whose production in the host is generalized. Diphtheria and botulism exemplify exotoxin diseases, and typhoid fever and cholera are endotoxin diseases. In both, the toxic materials are liberated after the death and dissolution of the bacteria.

Bacterial toxins have various modes of action ranging from destruction of vital tissues to interference with essential physiological functions. The host may or may not succumb to the pathological actions of toxins and the characteristic features of the host-parasite struggle constitute the disease syndrome. The possible outcomes of disease are death, or recovery with immunity to, or recovery with tolerance to, the parasite. Only a relatively few bacteria are pathogens, and probably the majority of bacteria are free-living *saprophytes* securing all the requirements for existence from inanimate inorganic and organic substances in their environments.

When two kinds of bacteria coexist in the same environment without obvious effect, they are said to be neutral toward one another. This phenomenon is probably quite rare in nature because of the similarity of many of the things required for continued life. It is possible that two widely disparate groups having little or nothing in common might show neutralism, but little is known of this kind of relationship.

4.19 | Some Exobiological Problems

Terrestrial life forms are remarkably resistant to what are generally considered lethal conditions. Tropical tarantula spiders, for example, have been found not to be harmed by exposure to ultraviolet radiation for one month, at levels believed to exist on Mars, or at simulated altitudes of 8 mi, and turtles have lived in atmospheres composed of 90% carbon monoxide. The question of unusual environments is paramount in space biology research programs, but the primary question is not one of survival of life forms, but rather their proliferation. Furthermore, experiments on the adaptation of terrestrial forms to non-tellurian conditions provide little information about the evolution of life elsewhere.

Since the exploration of Mars and Venus is imminent, the question of whether life exists on them has arisen in a less theoretical context. Space bioscience missions are charged with the duty to detect, and if found to describe, extraterrestrial life, and vast sums of money have been, and doubtless will be, invested in this endeavor. The activities, techniques, and hypotheses gathered for this attempt have been subsumed under the appellation "exobiology." The question of whether, in fact, such a science exists is moot, since thus far nothing living or "life-like" has been available for laboratory study and samples from the moon have thus far proven to be sterile. It is assumed that detection devices, whether placed on Mars and Venus during "fly-bys," or applied to samples brought back to earth, as in the case of the moon rocks, will be programmed to make determinations based on life as we know it and, furthermore, will involve a search for microbial rather than "advanced" forms of life. The investigations involve inoculating samples into aqueous media containing carbohydrates, amino acids, peptones, physiological salt solutions, and also into inorganic media of various kinds, in other words, against a background of terrestrial biochemistry and geochemistry. Evidences for life will also be sought by testing for nucleic acids, proteins, lipids, and carbohydrates in samples brought back to earth.

Whether life in the form of bacteria or other microorganisms has come to earth from elsewhere in the universe is an ancient and much-debated question. *Cosmozoe* or *lithopanspermia* are hypothetical life-bearing particles assumed to have been brought here by means of meteorites, and presumably they could have "seeded" the earth in this manner. Even Pasteur examined carbon-containing meteorites, but like other investigators before and after him, he was not able to obtain positive evidence for the presence of bacteria. The suggestion also has been made that *radiopanspermia* might be responsible for the distribution of life throughout the universe. Living, but dormant, spores were supposed to have been wafted through space by means of radiation pressure, and it was calculated that spores 0.15–0.2 μm in diameter would reach Mars in 20 days, Jupiter in 80 days, and Neptune in 15 months after leaving the earth. The spores would probably have to be freely suspended, since association with any particle resulting in a diameter greater than 1.5 μm would cause them to be drawn into the sun. The survival of terrestrial "life as we know it" under conditions of extreme cold, almost absolute vacuum, and exposure to cosmic radiations for long time periods of space travel seems extremely doubtful.

Some physical conditions of space may be duplicated in the laboratory, and a number of experiments using them have been performed on terrestrial organisms. It has been found that *survival* is possible under space-simulated low temperature and high vacuum conditions. But the introduction of ultraviolet radiation imposes a complication in that, for example, at $-190°$ C irradiated bacterial endospores did not survive exposure to a high vacuum after 6 hr.

"Mars jars" have been devised in which physical conditions believed to exist on that planet, insofar as we know them, are imposed on samples of terrestrial organisms in culture media or in natural environments such as soil. Thick-walled glass vessels are filled with a simulated Martian atmosphere of carbon dioxide, nitrogen, and argon, at about 85 mm of mercury pressure, and containing less than 1% water. The jars are subjected to diurnal temperature variations of from $-26°$ C to $-60°$ C and under these conditions it has been found that terrestrial bacteria survive, but do not multiply. Vegetative cells of *Clostridium tetani,* for example, survive two weeks and their spores ten months, but there is no detectable multiplication.

The evidence thus far available supports the conclusion that life on earth has probably arisen here and is not referrable to panspermia. There is no evidence concerning extraterrestrial life but this, of course, excludes life as we do not know it, since obviously, present detection devices would fail to recognize it. One of the stimuli for seeking extraterrestrial life is the discovery that some meteorites have organic compounds in them. Another is that fossil microorganisms are thought to be seen in some of them. The Orgueil (1864)* and Ivuna (1938) carbonaceous chondrite meteorites (Fig. 4.15A and B) have been chemically analyzed and they show the presence of insoluble organic matter identical to sporopollenin, an oxidation polymer of cartenoids. This has been found in spore microfossils several million years old, and it also is present in modern pollen. Its synthesis is believed to require processes occurring only in living organisms. Interestingly, meteorites that have fallen to earth at great distances from one another, and, at

---

*This meteorite is believed to have originated in the belt of asteroids lying beyond Mars, and dates from the time of formation of the solar system about 4.5×10^9 years ago.

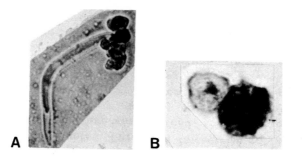

Fig. 4.15. *A,* Possible life-form observed in material from the Orgueil meteorite.
B, Spherical form in contact with a contaminating pollen grain (dark) found in the
Ivuna meteorite. (From P. Tasch, Life-forms in meteorites and the problem of terrestrial
contamination: A study in methodology, *Ann, N.Y. Acad. Sci.* **105,** 927 (1964). Courtesy
P. Tasch and the New York Academy of Sciences.)

different times, show similar formations in them. The question still
remains, however, whether aerial and terrestrial contamination is possible, or whether these bodies are artifacts.

With regard to our solar system, life as we know it appears to be
possible only on Venus and Mars, and even these seem predominantly
hostile. Since bacteria are likely to exist at environmental extremes
when all other forms are absent, it seems reasonable to search for
them in these situations.

4.20 | Gnotobiosis

The environmental condition in which a sterile ("germ-free") animal
is inoculated with a known kind of bacteria is referred to as *gnotobiotic.*
A colony of animals must be established before studies of this sort
may be pursued, but presently germ-free birds, rabbits, guinea pigs,
mice, and rats are available. The point of such work is to determine
the role of specific bacteria in the host's physiology, as for example,
in digestion, antibody formation, tooth decay, and deficiency diseases.
It has been found that animals survive in a germ-free state, but dietary
supplements, such as vitamins, must be provided. Such animals show
abnormal developments of the digestive tract and in the manufacture
of immunoglobulins. It has been shown that tooth decay (dental caries)
is definitely associated with bacteria, particularly lactic acid-forming
streptococci, since high carbohydrate diets fed to germ-free animals

induce no caries, whereas single strains of such streptococci added to the same diet soon result in tooth decay.

Germ-free animals and birds are obtained by caesarian section and by removal from surface-sterilized eggs, and since the tissues of normal healthy animals are sterile it is technically possible to keep the new-born in the same condition by circumventing normal natal processes. Germ-free plants are easily obtained by surface-sterilization of seeds and placing them in a sterile environment.

References

Aaronson, S., *Experimental Microbial Ecology*, Academic Press, New York, 1970.

Alexander, M., *Introduction to Soil Microbiology*, Wiley, New York, 1961.

Alexander, M., *Microbial Ecology*, Wiley, New York, 1971.

Barber, M., and L. P. Garrod, *Antibiotics and Chemotherapy*, Livingston, Edinburgh, 1963.

Brock, T. D., *Principles of Microbial Ecology*, Prentice Hall, Englewood Cliffs, N.J., 1966.

Brock, T. D., Life at High Temperatures, *Science* **158**, 1012 (1967).

Dean, A. C. R., and C. Hinshelwood, *Growth, Function, and Regulation in Bacterial Cells*, Oxford University Press, Oxford, 1966.

Farrell, J., and A. Rose, Temperature Effects on Microorganisms, *Ann. Rev. Microbiol.* **21**, 101 (1967)

Gray, T. R. G.. and D. Parkinson, *The Ecology of Soil Bacteria*, Liverpool University Press, Liverpool, 1968.

Gregory, P. H., *The Microbiology of the Atmosphere*, Interscience, New York, 1961.

Henry, S. M. (ed.), *Symbiosis*, Vol. I: *Associations of Microorganisms, Plants and Marine Organisms*, Academic Press, New York, 1966.

Henry, S. M. (ed.), *Symbiosis*, Vol. II: *Associations of Invertebrates, Ruminants and Other Biota*, Academic Press, New York, 1967.

Heukelekian, H. and A. Heller, Relationship between Food Concentration and Surface for Bacterial Growth, *J. Bacteriol.* **40**, 547 (1940).

Hungate, R. E., *The Rumen and Its Microbes*, Academic Press, New York, 1966.

Katznelson, H., A. G. Lochhead, and M. I. Timonin, Soil Microorganisms and the Rhizosphere, *Botan. Rev.* **14**, 543 (1948).

Krasil'nikov, N. A., *Soil Microorganisms and Higher Plants*, Acad. Sci., U.S.S.R., Moscow, 1958.

Lochhead, A. G., Quantitative Studies of Soil-Microorganisms, IV: Influence of Plant Growth on the Character of the Flora, *Canad. J. Res.* (C) **18**, 42 (1940)

Lovelace, T. E., H. Tubiash, and R. R. Colwell, Quantitative and Qualitative Commensal Bacterial Flora of *Crassostrea virginica* in Chesapeake Bay, *Proc. Nat. Shellfisheries Assoc.* **58,** 82 (1968).

Luckey, T. G., *Germfree Life and Gnotobiology,* Academic Press, New York, 1963.

Mishustin, E. N., The Law of Zonality and the Study of Microbial Associations of the Soil, *Soils and Fert.* **19,** 385 (1956).

Mitchell, R. (ed.) *Water Pollution Microbiology,* Wiley-Interscience, New York, 1971.

Oppenheimer, C. H. (ed.), *Symposium on Marine Microbiology,* C. C. Thomas, Springfield, Ill., 1963.

Rodina, A. G. *Methods in Aquatic Microbiology,* translated, edited, and revised by R. R. Colwell and M. S. Zambruski, University Park Press, Baltimore, 1972.

Shilo, M., Morphological and Physiological Aspects of the Interaction of *Bdellovibrio* with Host Bacteria, In *Current Topics in Microbiology and Immunology,* Springer-Verlag, Berlin, 50, 174 (1969).

Smith, H., and J. Taylor (eds.), *Microbial Behaviour in Vivo and in Vitro,* 14th Symp. Soc. Gen. Microbiol., Cambridge University Press, Cambridge, England, 1964.

Stanbridge, E. Mycoplasmas and Cell Cultures, *Bacteriol Rev.* **35,** 206 (1971).

Starc, A., Mikrobiologische Untersuchungen einiger podsoliger Boden Kroatiens, *Arch. Mikrobiol.* **12,** 329 (1942).

Starkey, R. L., Interrelationships between Microorganisms and Plant Roots in the Rhizosphere, *Bacteriol. Rev.* **22,** 154 (1958).

Stent, G., *The Molecular Biology of Bacterial Viruses,* Freeman, San Francisco, 1963.

Sussman, A. S., and H. O. Halvorsen, *Spores, Their Dormancy and Germination,* Harper and Row, New York, 1966.

Williams, R. E. O., and Spicer, C. C. (eds.) *Microbial Ecology.* 7th Symp. Soc. Gen. Microbiol., Cambridge University Press, Cambridge, England, 1957.

Wood, E. J. Ferguson, *Marine Microbial Ecology,* Reinhold Publishing Co., New York, 1965.

Zimmerman, A. M., *High Pressure Effects on Cellular Processes,* Academic Press, New York, 1970.

ZoBell, C. E., The Effect of Solid Surfaces on Bacteria, *J. Bacteriol.* **46,** 39 (1942).

five | Bacterial Energetics

"Nature which governs the whole will soon change all things thou seest, and out of their substance will make other things, and again other things from the substance of them, in order that the world may be ever new."

Marcus Aurelius (121-180 A.D.)

Bacteria characteristically are associated with natural processes involving chemical change: biodegradation and deterioration of materials, oxidations, fermentations, putrefaction, and decay. One of Pasteur's important contributions was his demonstration that bacteria are causative agents of these alterations. As Pasteur emphasized, these chemical changes are not incidental, but rather they are a direct result of the operation of vital processes. These are largely chemical processes, and all organisms depend upon and function by means of numerous chemical reactions, the sum of which constitutes *metabolism*.* It embraces the flux of materials and energy between the organism and the environment, as well as all the transformations of matter and energy taking place within the organism itself. The chemical reactions

---

*The following abbreviations will be employed in this chapter: NAD and NADH, for the oxidized and reduced forms of nicotinamide adenine dinucleotide, respectively; NADP and NADPH for the oxidized and reduced forms of nicotinamide adenine

involved in the breakdown and utilization of exogenous nutrients, the generation of energy by oxidation reactions, the formation of precursor compounds needed to produce cellular constituents, and the synthesis of new components all are a part of metabolism. An understanding of the functioning bacterium requires an understanding of its metabolism. In a bacterium this consists of several thousand different chemical reactions catalyzed by highly specific catalytic proteins, termed *enzymes*.

An attempt is made in Figure 5.1 to present an overall view of the flow of material through a bacterium. Nutrients, generally organic compounds, are taken into a bacterium and degraded into simpler compounds. Some of these metabolic intermediates are oxidized in order to supply the cell with energy, and others constitute, or are converted into, the compounds needed to synthesize additional cellular material. Energy-yielding, cellular oxidations may involve the utilization of atmospheric molecular oxygen, in which case the process is termed *respiration*. The respiration of organic compounds usually involves complete combustion of the energy substrate to yield carbon dioxide and water as end products. Oxygen acts as the ultimate oxidant, that is, as the acceptor of electrons and protons released by substrate oxidation, and consequently it is reduced to yield water. When oxygen is absent, some other acceptor must be employed. *Fermentation* is the term used to denote anaerobic energy-yielding processes in which organic compounds serve in place of oxygen as the final electron acceptor, and ordinarily a metabolic intermediate produced from the substrate is used for this purpose. In fermentations a considerable portion of the energy substrate is converted to end products at a lesser oxidation state than carbon dioxide. A number of interesting and useful end products are formed in bacterial fermentations, among which are lactic acid, acetic acid, butyric acid, ethanol, isopropanol, acetone, and glycerol.

Many of these processes in the bacterium are similar, if not identical, to those found in other organisms. There is an underlying similarity

---

dinucleotide-3'-phosphate; ATP, adenosine triphosphate; FAD, flavin adenine dinucleotide; FMN, flavin mononucleotide; DNA, deoxyribonucleic acid; UDPG, uridine diphosphoglucose; ETS, electron transport system; E_0, standard electrode potential; ΔG^0, standard free energy change. References to amino acids are to the L-isomers of α-amino acids, unless otherwise specified. The designations for the undissociated form and the carboxylate anion of organic acids, e.g., acetic acid and acetate, will be used interchangeably.

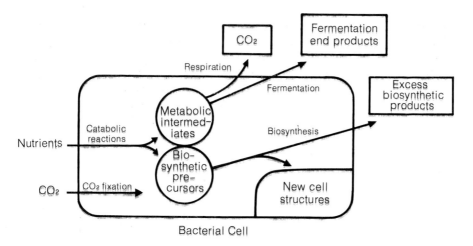

Fig. 5.1. Flow of materials in bacterial metabolism. Nutrient materials are broken down in order to generate energy by oxidation reactions (aerobic respiration and anaerobic fermentations), and are used for formation of new chemical compounds (biosynthesis).

at the molecular level between many of the fundamental biochemical processes in different organisms. If, as we are now convinced, such pathways are essentially the same in most organisms, then it follows that information gained in the study of one organism should generally be applicable to others. The elucidation of metabolic pathways and the mechanisms of enzyme action represent key problems in biochemistry, and bacteria have provided the biochemist with convenient experimental systems with which to investigate these problems. One reason for the utility of bacteria in biochemical studies is the relative ease with which experimental modifications can be made. The content of certain enzymes may be considerably enriched by appropriate manipulations so that they may comprise as much as 10–20% of the total cell protein in some instances. Extensive studies over the last three decades have resulted in a large body of information on the biochemistry of bacteria, albeit mostly concerned with a rather limited number of species.

The concept of the "unity of biochemistry," which is of indisputable validity, has been of great importance in the development of comparative biochemistry, and has provided a unifying principle around which the special biochemistry of particular biological groups is meaningfully organized. This latter point is of interest because there is a possibility that the pervasive and continued emphasis on common, central pathways may obscure the fact that there is also a special biochemistry of the bacteria, including unique processes or reactions which are *not* found in all organisms. The easy interchangeability of data from various organisms, which is so evident in reviews of the technical literature, should not lead one to conclude that all bacteria are the same, or that *Escherichia coli* is merely some sort of "small man," and, indeed, there are many differences between bacterial metabolism and that of higher organisms.

No attempt will be made here to give a systematic and comprehensive account of general biochemistry since excellent treatments of this material are already extant. Attention will be directed to some selected biochemical processes which, it is hoped, will illustrate several points: (1) the importance of bacteria in nature; (2) the variety of those bacterial "life-styles" which have evolved; and (3) some unique features of bacteria which may be absent in other organisms.

A word of caution must be introduced concerning the extrapolation of results from analytical biochemistry back to the intact organism. The chemical events observed *in vitro* on extracts and homogenates are purely scalar, whereas those occurring in the living cell are vectorial. Organizational checks and balances of an intact organism are lost upon its fragmentation and the increase in quantitative data obtained is at the expense of information concerning whole-cell behavior. The usefulness of biochemical studies on bacteria is not questioned, but one ought not be deluded into assuming that a description of cellular events derived from such investigations is necessarily applicable to the living, intact organism.

5.1 | Enzyme Localization

Biologists have long appreciated that cellular structure and function are inseparable and are but two aspects of the same thing. The complex cellular metabolic apparatus cannot be thrown together at random, and clearly a bacterium does not consist merely of a bag of enzymes. There are real, if currently ill-defined, requirements for spatial orientation and compartmentalization of chemical reactions, in order to accomplish the vectorial transfer of materials throughout the cell. In the case of bacteria an understanding of these requirements has only

Table 5.1. Location of Representative Enzymes in the Bacterial Cell

Cellular Location		
Endoenzymes		
Intracellular	Cytoplasmic membrane	Exoenzymes
Glucose-6-P isomerase	NADH oxidase	Proteinases
Adolase	Cytochrome oxidase	Phospholipases
Glyceraldehyde-3-P dehydrogenase	Succinic dehydrogenase	Penicillinase
DNA polymerase	Murein transpeptidase	Amylases
RNA polymerase	Murein polymerase	Cellulase
UDPG synthetase	ATPases	Hyaluronidase
Aminoacyl-adenylate synthetases	Thymidine phosphorylase	Neuraminidase

just begun. Progress has been impeded by their small size which hampers application of histochemical procedures, and by the presence of a strong cell wall, which generally requires rather drastic treatment to rupture. Some promise is offered by the development of new histochemical techniques for use with the electron microscope, and of gentler methods involving removal of the cell wall with muramidase, or use of mutants with defective walls.

The cytoplasmic membrane of bacteria (Section 2.7) is the site of a number of metabolic functions, among which are: (1) transport of materials into and out of the cell; (2) respiration and oxidative phosphorylation; and (3) formation of the cell wall and other surface structures. As indicated in Table 5.1, a number of enzymes concerned with respiration, such as succinic dehydrogenase and cytochrome oxidase, are found in isolated membranes of protoplasts prepared with muramidase. In fact, the entire electron transport chain from NADH to oxygen is located in the membrane. However, the associated respiratory dehydrogenases of the tricarboxylic acid cycle are not always so firmly situated, and sometimes are dissociated during isolation procedures and appear in the cytoplasmic fraction. This is particularly true of isocitric and malic dehydrogenases.

Energy generated by respiratory enzyme systems is needed not only for biosynthetic reactions to occur within the cell, but also to transport materials across the cytoplasmic membrane. Located in the membrane are specific proteins, termed *porters* or *permeases,* which mediate the stereospecific transfer of substances from the external environment into the cell. The genetic and physiological regulation of the porter for lactose and related β-galactosides has been given considerable study, and the porter for inorganic sulfate has been isolated in crystalline form. These proteins exhibit a fine discrimination between closely related chemical structures including diastereoisomers, so that an individual porter may transport the L-isomer but not the D-isomer. Unlike enzymes, transport proteins do not catalyze reactions involving the breaking and formation of covalent bonds; rather they simply permit the translocation of a molecule in space without chemical change.

In many cases it appears that energy must be expended in this transport, inasmuch as it is immediately halted by blockade of the respiratory system by cytochrome inhibitors, such as cyanide, or uncouplers, such as 2,4-dinitrophenol. Energy-dependent transfer of solutes across the cytoplasmic membrane is referred to as *active transport*. Some of the "ATPase" activity in the membrane must be attributable to enzymes associated with this active transport. The materials accumulated by these processes appear to reach intracellular concentrations greatly in excess of those in the external environment. Clearly the transport of a solute from a region of low concentration to one of high concentration would not be a spontaneous process because of thermodynamic considerations. However, it has been asserted that the high intracellular concentrations are more apparent than real, and that actually the intracellular materials are not freely in solution, but bound to proteins or some other macromolecules, thereby giving much lower effective concentrations. Thermodynamically, the chemical activity or effective concentration, and not the absolute concentration, is important in determining the direction of diffusional flows. While it would be valuable to have more information on total versus effective concentrations of dissolved materials inside the cell, the concept of active transport does not rest simply on observations relating to putative concentration differentials, but also on the demonstrable need for metabolic energy for transfer of solutes into the cell.

Although it is thought that the transport proteins themselves do not catalyze chemical changes in the transported substances, the overall process may not always be so straightforward as it would seem. The appearance within an organism of the externally supplied compound in itself cannot be taken to imply that no chemical reactions have occurred. Some of the complexities which may be encountered are illustrated by the example of the entry and incorporation into nucleic acids of exogenously supplied nucleosides. In some cases thymidine present in a culture medium is not incorporated into DNA. This is a result of its rapid cleavage by a membrane-bound enzyme, thymidine phosphorylase:

$$\text{thymidine } PO_4{}^{3-} \longrightarrow \text{thymine} + \text{deoxyribose-1-phosphate} \qquad (5.1)$$

However, when a second nucleoside, such as deoxyadenosine, is present the thymine produced in this reaction can be converted again to thymidine by the addition of a deoxyribosyl residue derived from the deoxyadenosine. In this case the thymidine found within the cell is not the original molecule but a new one produced by phosphorolysis and resynthesis.

Cell wall biosynthesis (Section 2.6) and the formation of other surface structures are intimately associated with the cytoplasmic membrane. The amino sugar precursors of the wall, including N-acetylmuramic acid, are synthesized within the cytoplasm in the form of nucleotide-linked molecules, for example, UDP-glucosamine or UDP-muramic acid, by means of "soluble" enzymes. Polymerization of these precursors to form the murein substance, or peptidoglycan itself, is accomplished by a membrane-bound enzyme. For further cross-linking of the peptide side chains to produce the finished murein sacculus membrane-bound transpeptidases are involved. Certain enzymes may be found bound to the cell wall material, as in the case of muramidases and peptidases, and sometimes enzymes are present in the region between the wall and membrane (*periplastic space*).

Most of the enzymatic reactions carried on by a bacterium take place in the cytoplasmic region. Some enzyme systems, such as those involved in glycolysis and biosynthesis of amino acids and other metabolites, are soluble or at least so loosely bound that internal distribution cannot be definitely fixed, whereas others are associated particularly with identifiable intracellular components. Certain enzymes are bound to DNA and can be recovered from cell-free extracts by precipitating the DNA with streptomycin or protamine; among these are DNA-polymerase, RNA-polymerase and deoxyribonuclease. Possibly because of a need for spatial orientation, bacterial DNA is attached loosely to the cytoplasmic membrane, and can be sedimented along with the membranes of osmotically lysed spheroplasts. There is evidence suggesting that mesosomes are attachment sites for DNA (Fig. 2.22), and it is possible also to demonstrate attachment of newly synthesized messenger RNA to the DNA, as well as the linkage of ribosomes to the mRNA molecules.

Bacteria do not possess the special subcellular organelles for the compartmentalization of cytoplasmic enzymes such as the mitochondria, lysosomes, peroxisomes, and glyoxisomes found in eukaryotes (Section 2.1). Lysosomes are membrane-enclosed sacs containing various hydrolytic enzymes, such as cathepsins, nucleases, and phosophatases, which if free in the cytoplasm would cause cell damage. Peroxisomes, such as those of liver cells, are small particles containing flavoprotein oxidases including uric acid oxidase, D- and L-amino acid oxidase, and catalase. It is believed that localization of hydrogen peroxide-generating flavoprotein oxidases along with catalase, which acts to destroy the peroxide formed, would tend to protect the eukaryotic cell from the deleterious effects of peroxide. There is no similar compartmentalization of catalase and flavoprotein oxidases in bacteria, and it is not clear how the bacterium is able to prevent self-destructive action of its proteolytic enzymes, nucleases, and phosphatases in the absence of protective structures such as the lysosome.

In addition to those enzymes exclusively located within the cytoplasm or cytoplasmic membrane of the bacterium, significant quantities of extracellular enzymes or *exoenzymes* are produced by some bacteria. These are usually hydrolases such as proteinases, nucleases, amylases, and phospholipases. Certain exoenzymes produced by pathogenic bacteria are correlated with virulence and invasiveness, among which are coagulases which cause plasma to clot, plasminogen activators which promote digestion of fibrin clots, hemolysins causing lysis of erythrocytes, and hyaluronidase which is a β-glucosaminidase attacking the tissue-cementing substance, hyaluronic acid. The α-toxin of the gas gangrene organism, *Clostridium perfringens,* is an extracellular phospholipase which attacks the lecithin component of the host cell's cytoplasmic membrane.

There has been a question as to whether exoenzymes are secreted by intact living bacteria, or are simply released as a result of the rupture of the cytoplasmic membrane or the dissolution of the entire organism. Although some so-called exoenzymes may result from lysis, it seems clear that others, such as penicillinase, are in fact released without lysis. It is, of course, a technical problem of extreme difficulty to specify that *no* cell in a population of billions has lysed, but in a number of cases it has been shown that the extracellular enzymes are produced under conditions where lysis is undetectable, that is, there is no release of soluble metabolites or of "cytoplasmic" enzymes. Bacteria do not have a Golgi apparatus or enzyme-filled vesicles seen in eukaryotic secretory cells, and the manner in which bacterial exoenzymes are selectively passed from the cell interior through the membrane and released into the external environment is a matter of conjecture. The cell wall presents a coarse barrier to free passage of exoenzymes, but there is no more difficulty in envisioning the selective release of an enzyme from an intact cell than the insertion of entire enzyme systems, for example, the electron transport system, into specific positions within the fabric of the cytoplasmic membrane.

5.2 | Energy Sources and Carbon Compounds Used by Bacteria

Bacteria require a source of energy and utilizable carbon compounds, although they differ in the manner in which they satisfy these requirements. Thus, organisms may be categorized according to the source of energy used, and, in addition, the type of carbon compounds employed. This permits recognition of a number of different nutritional and metabolic systems, which in many cases are not found in other forms of life. Indeed, the study of bacterial nutrition provides a widening of horizons which may be a necessary corrective for the more narrow outlook engendered by an exclusive preoccupation with the metazoa and metaphyta.

There are two major divisions with respect to the source of bacterial energy: radiant energy and chemical bond energy (see Table 5.2). Bacteria using the former are called *phototrophs* (from the Greek— *photos*, light; and *trophos*, feeding), and those using the latter are called *chemotrophs*.

Phototrophs, which ordinarily utilize CO_2 as a carbon source, face an additional problem of obtaining reducing power for CO_2 fixation.

Table 5.2. Categories of Energy Metabolism Utilized by Bacteria

Type of energy metabolism	Energy source	Source of reducing power	Example
A. Phototrophy			
1. Photolithotrophy	Light	H_2S	*Chromatium*
2. Photoorganotrophy	Light	Organic substrate	*Rhodospirillum*
B. Chemotrophy			
1. Chemolithotrophy	Inorganic substrate	Same as energy substrate	*Nitrosomonas*
2. Chemoorganotrophy	Organic substrate	Same as energy substrate	*Pseudomonas*
3. Mixotrophy	Both inorganic and organic	Same as energy substrate	*Hydrogenomonas*

The assimilation of carbon dioxide requires a reduction step, that is, addition of electrons and protons. Although energy is obtained in the light-dependent reaction, this does not supply the electrons and protons needed for CO_2 assimilation. In the case of the purple and green sulfur bacteria, the inorganic compound hydrogen sulfide is used as the source of reducing power (Section 3.19). Bacteria such as *Chromatium* and *Chlorobium* are termed *photolithotrophs* (from the Greek—*lithos*, stone, signifying use of an inorganic substance photosynthetically). There is another possibility, realized by the purple nonsulfur bacteria, such as *Rhodospirillum*, which is the oxidation of an organic compound to yield reducing power. This is termed *photoorganotrophy*.

In the case of the chemotrophs a further distinction is made as to whether the chemical compounds used for energy are inorganic *(chemolithotrophy)*, organic *(chemoorganotrophy)* or whether both are used simultaneously *(mixotrophy)*. The reality of mixotrophy was doubted, but it now has been verified in several cases.

If one considers the types of carbon compounds used, he finds the same three possibilities: inorganic (CO_2), organic compounds, or both simultaneously. Formerly it was common to designate those organisms exclusively using CO_2 for their carbon source as *autotrophs,* and those using organic compounds as *heterotrophs.* However, the term autotroph has carried with it the connotation of chemolithotrophic or photolithotrophic energy metabolism. It is now recognized that the sources of energy and carbon are separate and distinct entities, and that there is no requirement that a given type of energy metabolism should demand a given carbon source, or *vice versa;* they occur in all possible combinations. If we consider chemotrophs, each of the three categories with respect to energy source may be combined with the three different categories for carbon source, yielding nine distinct types (Table 5.3). There will be limits to the versatility of a single species, but considering the bacteria as a whole, examples can be found of those capable of operating under all nine conditions, with the possible exception of condition 7. The examples given in Table 5.3 are not to be taken in a restrictive sense, in that an organism appearing in one category also may be capable of growing under other conditions as well. Thus *Desulfovibrio* is shown as being able to utilize an inorganic energy source, namely elemental hydrogen, H_2, and an organic source of carbon (condition 2). But under most circumstances these bacteria probably utilize organic compounds for both carbon and energy (condition 5). Similarly it is likely that many bacteria fix

Table 5.3.   Carbon and Energy Sources for Bacteria*

Condition	Energy source	Carbon source	Example
1	Inorganic	CO_2	*Nitrosomonas*
2	Inorganic	Organic	*Desulfovibrio*
3	Inorganic	CO_2 + organic	*Thiobacillus*
4	Organic	CO_2	*P. oxalaticus*
5	Organic	Organic	*Pseudomonas*
6	Organic	CO_2 + organic	*Neisseria*
7	Mixotrophic	CO_2	?
8	Mixotrophic	Organic	*Hydrogenomonas*
9	Mixotrophic	CO_2 + organic	*Hydrogenomonas*

*From S. Rittenberg, The Roles of Exogenous Organic Matter in the Physiology of Chemolithotrophic Bacteria, *Advances in Microbial Physiology* **3:** 159 (1969). Courtesy S. C. Rittenberg and permission of Academic Press, Inc., New York.

CO_2 during growth on organic compounds (condition 6) even if they are capable also of growth in the absence of CO_2 (condition 5). The difficulty with condition 7 is that it demands that the organism use both an organic and inorganic substrate for energy, but not incorporate the organic compound into cell substance, and, while possible, this case has not been verified. Condition 9 has been observed with *Hydrogenomonas,* and when H_2 and an organic substrate are supplied together, both are utilized simultaneously as energy sources, and both CO_2 and the organic substrate are assimilated at the same time.

The sharp dichotomy between energy and carbon sources, while long recognized, is often unappreciated. In the case of a chemolithotroph such as *Nitrosomonas*, which obtains energy from the oxidation of ammonia and all of its carbon from CO_2 (condition 1), the reality of this distinction is clear and inescapable, but even in organisms operating under condition 5 a similar separation may be noted. The lactic acid bacteria, for example, including *Streptococcus* and *Lactobacillus,* obtain energy almost exclusively by fermentation of carbohydrates. These organisms ferment glucose with a very high degree of efficiency, converting 95–98% of the glucose fermented to lactic acid. Almost none of the glucose supplied is assimilated into cell substance, which instead is synthesized from amino acids, purines, pyrimidines, and other precursors which must be supplied in the growth medium.

5.3 | Energy Metabolism in Bacteria

Respiration and fermentation. Most bacteria are nonphotosynthetic organisms in which the provision of energy, which is to say the synthesis of ATP, is accomplished via the oxidation of appropriate chemical compounds supplied by the external environment. Only in the few photosynthetic bacteria, which use radiant energy, is ATP generated by other means.

Some bacteria are restricted to respiratory processes and are unable to grow in the absence of O_2; in consequence these bacteria, such as *Pseudomonas fluorescens,* are obligate aerobes (Section 4.13). Similarly, there are other organisms which exclusively utilize fermentation reactions and are unable to gain energy by means of aerobic respiration. Some of these strictly fermentative bacteria, including many of the *Bacteroides* species of the rumen and human intestinal tract, are extremely sensitive to the presence of O_2. Not only are they unable to initiate growth if free oxygen is present, but often they are killed by exposure to the air for as little as 30 sec. Such bacteria are of necessity strict anaerobes, but the reason for their extreme sensitivity to oxygen is unknown. Other fermentative organisms, such as *Streptococcus* and *Lactobacillus,* are more tolerant of O_2 and grow in the presence of air, although only fermentation reactions are used. A third group of bacteria, which includes *Staphylococcus aureus, Escherichia coli,* and other common bacteria, is able to employ either respiratory or fermentative reactions, depending on the presence of O_2. These facultative organisms show much greater yields of cells per unit of substrate utilized when grown aerobically. This is a consequence of the greater amount of ATP formed by the respiration of the energy substrate as compared with its fermentation. The basis of this is readily apparent from a comparison of the free energy change for the fermentation of glucose to lactic acid with that for the aerobic oxidation of glucose to CO_2:

$$C_6H_{12}O_6 \longrightarrow 2CH_3CHOHCOOH \quad \Delta G^0 = -52 \text{ kcal} \quad (5.2)$$

$$C_6H_{12}O_6 + 6O_2 \longrightarrow 6CO_2 + 6H_2O \quad \Delta G^0 = -686 \text{ kcal} \quad (5.3)$$

Free energy (G) is a measure of the ability to do work, so that free energy can be thought of as useful energy. The change in free energy in a given system during a reaction carried out under "standard conditions" (molar concentrations of reactants, etc.) is designated ΔG^0. Spontaneous reactions proceed with a release or decrease in free energy, and thus the change in free energy (ΔG) is negative in this notation.

Bacterial respiration and the tricarboxylic acid cycle. Bacterial respiration rates generally are high and commonly exceed those for mammalian tissues when considered on a basis of the activity per unit of mass. Q_{O_2} values (microliters of oxygen consumed per hour per milligram of dry bacteria) of 100 are commonplace, and may reach 500 or more with highly active bacteria.

Both inorganic and organic compounds are utilized for energy via respiratory mechanisms. Discussion of the oxidation of inorganic compounds by chemolithotrophic bacteria will be deferred until Section 5.9. A large number of substances may serve as energy sources for chemoorganotrophic bacteria. All of these diverse organic materials are fed into a common terminal respiratory system consisting of the Krebs tricarboxylic acid cycle (or citric acid cycle) linked to an electron transport system:

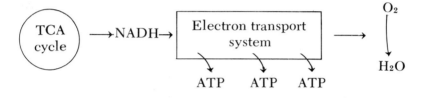

The transfer of electrons (and protons) from the tricarboxylic acid cycle to the electron transport system is effected by a special carrier molecule, nicotinamide adenine dinucleotide, which exists either in the oxidized form (NAD) or the reduced form (NADH). As stated above, these enzyme systems are located in the cytoplasmic membrane, which when ruptured yields small particles. Since the respiratory enzymes are found in this particulate fraction, such preparations have been referred to somewhat loosely as "respiratory particles" or "electron transport particles." The term "electron transport particle" was applied originally to fragmented mitochondria of eukaryotic cells, and carried with it an implication that these derived fragments were individual mitochondrion subunits. As applied to bacterial preparations the term means simply the total comminuted membranes of the bacterium. Although there are some indications that respiratory enzymes may be localized in mesosomes, especially the vesicular mesosomes of *Bacillus*, little is known about the actual arrangement of the respiratory system in the cytoplasmic membrane.

The sequence of reactions in the tricarboxylic acid cycle and the points of entry for some representative compounds which are respired via this system are shown in Figure 5.2. The Krebs cycle operates

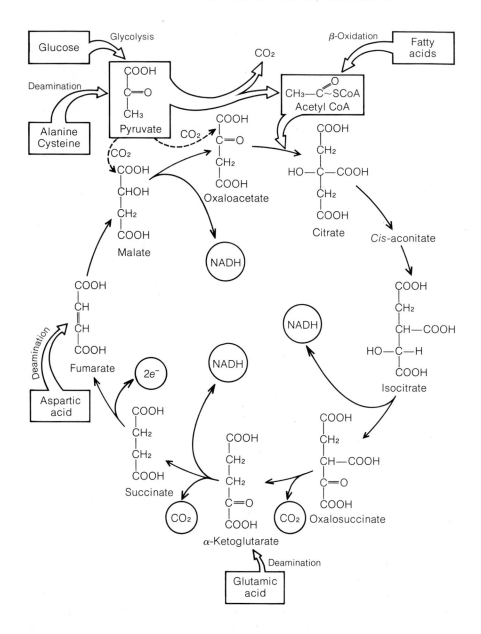

Fig. 5.2. Entry of materials into the tricarboxylic acid cycle. Carbohydrates, amino acids, and fatty acids are metabolized by way of a common respiratory cycle.

in bacteria, as in other organisms, essentially as a mechanism for the aerobic combustion of acetate. The two CO_2 molecules formed in the cycle represent the equivalent of a complete oxidation of one acetate molecule. Any compound which can be dissimilated to yield acetate directly, or to pyruvate, which is convertible to acetate, or to one of the other intermediates of the cycle, such as α-ketoglutarate, can be fed into this cycle and oxidized. Among such compounds are various carbohydrates, amino acids, long chain fatty acids and hydrocarbons, and aromatic compounds.

Three oxidation steps of the cycle are catalyzed by dehydrogenases which use nicotinamide adenine dinucleotide (NAD) as the coenzyme, and these are the isocitrate, α-ketoglutarate and malate dehydrogenases. Oxidation of the substrates of these enzymes is coupled to the reduction of the NAD coenzyme. Although in mammalian mitochondria the isocitrate dehydrogenase is specific for NADP (the 3'-phospho-NAD), both NAD- and NADP-specific isocitrate dehydrogenases are found in bacteria. The nicotinamide coenzymes are the major electron (and hydrogen) carriers of the cell, and mediate a great many oxidation-reduction reactions in addition to those of the tricarboxylic acid cycle. These coenzymes are present in only catalytic amounts and serve simply in an intermediary capacity as links between the oxidation of particular substrates and the reduction of another compound. This coupling between the oxidations and reductions is represented in the following three reactions:

$$\text{malate} \xrightarrow{\text{malate dehydrogenase}} \text{oxaloacetate} + 2e^- + 2H^+ \qquad (5.4)$$

$$NAD + 2e^- + 2H^+ \longrightarrow NADH + H^+ \qquad (5.5)$$

$$NADH + X + H^+ \longrightarrow NAD + XH_2 \qquad (5.6)$$

If the NAD were not regenerated by transfer of the electrons and hydrogens to some acceptor compound (X) in reaction 5.6, then the oxidation of substrate (reaction 5.4) would cease as all of the coenzyme of the cell became reduced (reaction 5.5).

Bacterial electron transport systems. As biological oxidations proceed with release of electrons, and often protons as well, the organism is constrained to make use of some substance as their ultimate acceptor.

Aerobic bacteria are able to employ molecular oxygen for this purpose by linking substrate oxidation to reduction of oxygen in a multienzyme electron transfer system (ETS), consisting of flavoproteins, quinones, cytochromes, and a cytochrome oxidase capable of combining O_2, electrons, and protons to yield water. The flow of electrons in the electron transfer system is presumed to be as follows:

Substrate, NAD

Oxidized, NADH $\longrightarrow$ flavin $\longrightarrow$ quinone $\longrightarrow$ cytochrome b
Product

$O_2 \longleftarrow$ cytochrome $a + a_3 \longleftarrow$ cytochrome c

The reoxidation of NADH to NAD is accomplished by transferring a pair of electrons and a proton to the first component of the ETS, which is a protein utilizing a flavin coenzyme, usually flavin adenine dinucleotide (FAD). The FAD coenzyme is reduced to FADH, and then in turn it is reoxidized by passing the pair of electrons on to the next member of the system.

Essentially, then, aerobic respiration consists of using atmospheric oxygen as an "electron sink" to permit continued substrate oxidation. While useful in itself to some extent, the electron flow to O_2 is of significance primarily because it can be coupled to the generation of ATP. In mitochondria the reoxidation of one molecule of NADH, which involves the passage of two electrons through the ETS, results in the formation of three molecules of ATP. In bacteria the ATP yield per electron pair has not been established with certainty, although one may reason by analogy to the situation in mitochondria. Electron transport usually appears to be partially uncoupled from ATP formation in cell-free preparations from bacteria, which results in rather low P/O ratios, that is, the ratio of orthophosphate esterified into ATP to oxygen consumed. It is thought that in intact organisms the P/O ratios approach those observed in mitochondria.

Although generally analogous to the mitochondrial system, bacterial electron transport systems differ in detail. Some, but not all, bacteria show the presence of the same types of cytochromes found in mammalian mitochondria. Components of the electron transport chain of some representative bacteria are listed in Table 5.4. The cytochromes are enzymes in which an iron-porphyrin (heme) coenzyme is bound to the protein portion. The exact structure of this heme portion differs in the various cytochromes which are designated by letters (see Table

Table 5.4. Components of the Respiratory Chain of Certain Aerobic and Facultatively Anaerobic Bacteria

Bacterium	Cytochrome pigments	Quinone*
Pseudomonas aeruginosa	b, c_{551}, c_{552}, c_{554}, a_1, d	Ubiquinone-45
Mycobacterium phlei	b, c_1, c, a, a_3	Naphthoquinone-45
Nitrobacter agilis	c, a_1, a	?
Haemophilus parainfluenzae	b, c_1, $(c?)$, a_1, a_2, o	Naphthoquinone-35,30,25 (demethyl K_2)
Escherichia coli	b_1, b_{562}, c, c_{552}, a_1, d	Naphthoquinone-45 Ubiquinone-40
Staphylococcus aureus	b_1, a, o	Naphthoquinone-45
Bacillus subtilis	b, c, $a + a_3$	Naphthoquinone-35 (K_2)
Azotobacter vinelandii	b_1, $c_4 + c_5$, a_1, d, o	Ubiquinone-40

*The number following the quinone indicates the chain length of the isoprenoid side chain.

Table 5.5. Characteristics of Some Cytochromes

Type of cytochrome	Approximate absorption maximum (α-band, reduced)	Type of heme	Approx. E_0'	Accepts e^- from	Donates e^- to
b	560 nm	Protoheme	0 v	FAD	Cyt c
c	550 nm	Protoheme	0.25 v	Cyt b	Cyt a
a	605 nm	Heme a	0.30 v	Cyt c	O_2

5.5). In cytochromes of the c-type the protoheme group is covalently bonded to the polypeptide chain, whereas in the other cytochromes the heme groups are more loosely bound and can be removed by treatment with acid and/or organic solvents. The iron in heme groups is readily oxidized and reduced, which enables cytochromes to transfer electrons from one to the other in a series of alternating oxidation and reduction reactions.

Cytochrome components are identified by means of the characteristic absorption spectra of the reduced forms, which are determined primarily by the nature of the heme moiety. Bacterial cytochromes showing absorption spectra identical to the mammalian counterpart do not necessarily exhibit equivalent biological activity. Purified bacterial cytochrome oxidase preparations often will reoxidize reduced bacterial cytochrome c purified from the same bacterium, but not cytochrome c of ox heart mitochondria. In the reciprocal experiment mitochondrial cytochrome oxidase may fail to oxidize reduced bacterial cytochromes of the c-type. The terminal cytochrome oxidase of some bacteria does not show the spectrum typical of mammalian cytochrome

$a + a_3$, and instead may consist of cytochrome a_1, cytochrome a_2 (also called d), or cytochrome o. The latter pigment shows a spectrum more akin to cytochrome b, but is a true oxidase capable of interacting with O_2 and binding carbon monoxide. In addition to differences in absorption maxima, bacterial cytochromes may not show the same standard electrode potentials (E_0') or sensitivity to inhibitors as mammalian cytochromes. Classical inhibitors of mitochondrial electron transport such as antimycin A, and uncoupling agents such as 2,4-dinitrophenol, often are much less effective in bacterial systems.

In some bacteria, such as *Pseudomonas,* the quinone component of the electron transport chain is one of the so-called ubiquinones (also termed coenzyme Q). The ubiquinones consist of a group of closely related 2,3-dimethoxy-5 methylbenzoquinones which have a long isoprenoid side chain in the 6 position. In bacteria this side chain usually contains 7–10 isoprenoid units of 5 carbons each, or a total of 35–50 carbons:

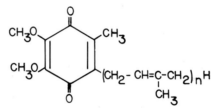

The length of the side chain usually is characteristic of the species, which frequently, if not universally, contains a single isoprenolog. *Pseudomonas denitrificans* contains only ubiquinone-50 (i.e., 10 isoprene units), whereas *P. fluorescens* has been reported to form both ubiquinone-45 and small amounts of ubiquinone-40.

In other organisms, notably gram-positive bacteria such as *Bacillus,* the benzoquinone is replaced by a naphthoquinone related to vitamin K_1 or K_2. In some bacteria, including *Escherichia coli,* both a ubiquinone and naphthoquinone are present. The significance of the use of naphthoquinones to replace or augment ubiquinone is uncertain, although there is apparently some correlation with the gram reaction. Ubiquinones appear to be absent in gram-positive bacteria, whereas in the gram-negatives there may be either ubiquinone alone or in combination with naphthoquinone. It has been suggested that the naphthoquinone component might be used for certain specific ETS chains and the ubiquinone for others. Another suggestion is that the presence of naphthoquinone could be related to the ability to grow as a facultative anaerobe, either fermentatively in the absence of air or by respiratory mechanisms. However, this does not adequately account for the exclusive utilization of naphthoquinones by other facultative organisms, such as *Staphylococcus aureus,* or by strict aerobes such as *Bacillus subtilis.*

Electron transport components in obligate anaerobes. Generally the electron transport system is thought of in connection with aerobic respiration, and it might be supposed that it would be inoperative in the absence of O_2, and completely absent in strictly anaerobic bacteria. The notion that cytochromes are concerned only with aerobic respiration has proven to be erroneous. In certain facultative anaerobes the respiratory chain, or a portion thereof, is formed during anaerobic growth in the presence of nitrate. In such organisms the ETS can operate in the absence of air using nitrate instead of O_2, as described in Section 5.8.

There are certain strict anaerobes, such as *Clostridium,* that appear devoid of cytochromes and respiratory quinones. However, cytochromes or other components of the ETS are found in a number of other obligate anaerobes. The rumen organism, *Vibrio succinogenes,* contains cytochromes *b* and *c,* and another rumen bacterium, *Bacteroides ruminicola,* possesses electron transport components including a cytochrome *b*. The presence of electron transport enzymes in *Bacteroides* might have been inferred from their nutritional requirements, since some species require both heme and vitamin K for growth. The sulfate-reducing bacterium, *Desulfovibrio desulfuricans,* contains a cytochrome c_3, which possesses an unusually low oxidation-reduction potential (Section 5.8). Photosynthetic anaerobes such as *Chromatium* also contain cytochromes and quinones, which are involved in photophosphorylation reactions (Section 5.10).

An interesting problem is presented in the case of *Streptococcus faecalis,* which has been shown to contain a naphthoquinone-45 (demethyl K_2-type), although cytochromes supposedly are completely absent in this bacterium. Furthermore, naphthoquinones that satisfy the nutritional requirements of *Bacteroides,* and therefore presumably of the vitamin K type, also have been detected in *Peptostreptococcus magnus* and *Veillonella alcalescens,* both of which are strictly anaerobic cocci.

Function of the glyoxylate cycle during growth on acetate. Although discussions of pathways employed in bacterial metabolism generally center around utilization of carbohydrates, in nature a variety of other substrates are utilized, such as amino acids and other organic acids. A bacterium respiring acetate as its sole energy and carbon source must use it both as a source of biosynthetic precursors and as an energy substrate. As in other organisms, bacterial respiration of acetate proceeds via the Krebs tricarboxylic acid (TCA) cycle. However, as usually formulated the TCA cycle *per se* accomplishes simply the combustion of acetate to two molecules of carbon dioxide and no provision is made for the withdrawal of intermediates for use in biosynthetic reactions. If a molecule of α-ketoglutarate is converted to glutamic acid instead of being processed through the remainder of the cycle, then the oxaloacetate needed for the next turn of the cycle will not have been regenerated. It has been found that growth on acetate involves a somewhat altered version of the usual TCA cycle, which is referred to as the *glyoxylate cycle* (Fig. 5.3). In this scheme one molecule of acetate (as acetyl coenzyme A) is condensed with oxaloacetate to give citrate, which is in turn isomerized to yield isocitrate in the usual fashion (Fig. 5.2). However, the isocitrate is next cleaved by the enzyme isocitratase to produce glyoxylate plus succinate:

$$
\begin{array}{llll}
\text{CHOH—COOH} & & & \\
\mid & & \text{COOH} & \text{CH}_2\text{—COOH} \\
\text{CH}_2\text{—COOH} & \xrightarrow{\text{isocitratase}} & \mid \quad + & \mid \\
\mid & & \text{CHO} & \text{CH}_2\text{—COOH} \\
\text{CH}_2\text{—COOH} & & & \\
\end{array}
\qquad (5.7)
$$

isocitrate glyoxylate succinate

Following this cleavage another enzyme, malate synthetase, can catalyze the condensation of the resulting glyoxylate with a second acetate molecule to form malate:

$$
\begin{array}{lll}
\text{CHO} & \text{COOH} & \text{COOH} \\
\mid \quad + & \mid \xrightarrow{\text{malate synthetase}} & \mid \\
\text{COOH} & \text{CH}_3 & \text{CH}_2 \\
& & \mid \\
& & \text{CHOH} \\
& & \mid \\
& & \text{COOH} \\
\end{array}
\qquad (5.8)
$$

glyoxylate acetate malate

The malate and succinate formed then can be used to regenerate oxaloacetate via the usual citric acid cycle reactions and thereby com-

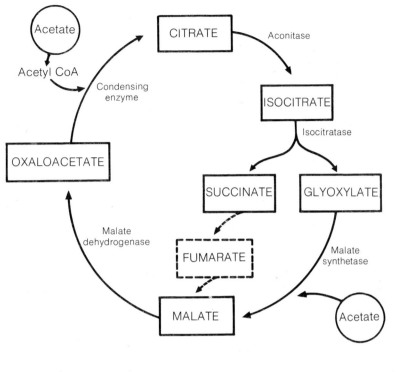

Net: 2 Acetate ⟶ Malate

Fig. 5.3. Conversion of acetate to malate in the glyoxylate cycle. During growth of bacteria on acetate the glyoxylate cycle serves to regenerate the oxaloacetate needed for utilization of acetate.

plete the cycle. In the glyoxylate cycle a bacterium eliminates the decarboxylation steps of the citric acid cycle, and achieves the conversion of two acetates to a 4-carbon acid instead of their combustion to CO_2.

5.4 | Characteristic Bacterial Fermentations of Carbohydrates

In the preceding section it was emphasized that the function of oxygen in respiration is to accept electrons and protons released during oxidation of energy substrates. It is possible to carry out substrate oxidation in the absence of oxygen, provided an alternate electron acceptor is available. Many bacteria ferment energy substrates using an organic compound, often an intermediate in the dissimilation of the substrate itself, as the electron and proton acceptor. The definition of fermentation rests on the nature of the electron acceptor, and not the substrate utilized nor the products formed. A large number of different types of compounds, including carbohydrates, organic acids, alcohols, amino acids, and purines, are fermented by bacteria. The products of these fermentations are equally diverse, but they may be usefully classified into the following several groups, based on methods of separation and analysis: steam-volatile acids (formic, acetic, and propionic acids); nonvolatile acids (lactic and succinic acids); volatile neutral compounds (ethanol, acetone, and isopropanol); nonvolatile neutral compounds (glycerol and 2,3-butanediol); and gases (hydrogen, carbon dioxide, and methane). Despite this complexity of substrates and products, some order is introduced by the fact that the array of fermentation end products formed from a fermentable substance is characteristic for a particular organism. When standardized conditions are employed, a useful way is provided of recognizing and classifying bacteria. An organism which forms almost exclusively lactic acid from the fermentation of glucose *(Streptococcus)* is distinguishable by simple tests from a morphologically similar bacterium which forms a mixture of lactic acid, ethanol, and carbon dioxide *(Leuconostoc)*. Both of these are different from one which produces a mixture of butyric acid, acetic acid, H_2, and CO_2 *(Clostridium)*. A number of end product patterns for glucose fermentation found among the bacteria are outlined in Table 5.6.

Great effort has been expended in careful analysis of the relative proportions of the end products of bacterial fermentations. These often are cast in the form of a "fermentation balance" in which the investigator determines the percentage of the substrate fermented that may be accounted for by the end products found. However, it is now recognized that the relative amounts of various products are not fixed and immutable, but rather are greatly dependent on prevailing environmental conditions. The pH of the medium is especially important, and in earlier studies it usually was not controlled but was allowed to fall as acidic products accumulated. Such conditions gave distinctly different results from those obtained when the pH was maintained constant throughout by periodic additions of sterile alkali. In the "homolactic" fermentation of *Streptococcus* without pH control only traces of products other than lactic acid are found, whereas when an alkaline pH is maintained, significant amounts of formic and acetic acids and ethanol are produced along with the lactic acid.

Table 5.6.   Representative Bacterial Fermentations

Fermentation type	Typical organism	Characteristic products of glucose fermentation
Homolactic	*Streptococcus lactis*	Lactic acid
Heterolactic	*Leuconostoc mesenteroides*	Lactic acid, ethanol, CO_2
Propionic	*Propionibacterium arabinosum*	Propionic, acetic, and succinic acids, CO_2
Formic	*Salmonella typhosa*	Lactic, formic and acetic acids, ethanol, succinic acid
Butanediol	*Bacillus polymyxa*	Butanediol, ethanol, acetic acid, acetoin, CO_2, H_2
Butyric	*Clostridium butyricum*	Butyric and acetic acids, H_2, CO_2
Acetone-butanol	*Clostridium acetobutylicum*	Butanol, acetone, acetic acid, ethanol, acetoin, butyric acid, CO_2, H_2
Isopropanol	*Clostridium butylicum*	Butanol, butyric and acetic acids, isopropanol, CO_2, H_2
Acetic	*Clostridium thermoaceticum*	Acetic, butyric and lactic acids
Ethanol	*Zymomonas mobilis*	Ethanol, CO_2, lactic acid

One of the determining factors with respect to the nature of the products formed, aside from the presence of the requisite enzyme systems, is the need for some of the intermediates to serve as electron acceptors, and thereby to become reduced. In a fermentation one may obtain not only products more oxidized than the starting compound, but also some which are more reduced. An oxidation-reduction balance is valuable for this reason, in addition to a materials balance, in determining whether a significant product has been overlooked in the analysis. If one assigns a value of $+1$ for each hydrogen atom and a -2 for each oxygen atom, then the overall oxidation-reduction value for glucose, $C_6H_{12}O_6$, would be zero; for ethanol, C_2H_6O, the value would be $+4$, indicating that it is at a more reduced state than glucose. On the other hand, carbon dioxide, CO_2, would have a value of -4. In a fermentation such as that of yeasts or *Zymomonas mobilis*, in which essentially the only products are ethanol and CO_2, equimolar amounts would have to be formed in order to achieve a net balance:

$$C_6H_{12}O_6 \longrightarrow 2\ CH_3CH_2OH \quad + \quad 2\ CO_2 \tag{5.9}$$

$1 \times 12 =\ \ 12$ (H)	$1 \times 6 \times 2 =\ 12$ (H)	
$-2 \times 6 = -12$ (O)	$-2 \times 1 \times 2 = -4$ (O)	$-2 \times 2 \times 2 = -8$ (O)
0	8	-8

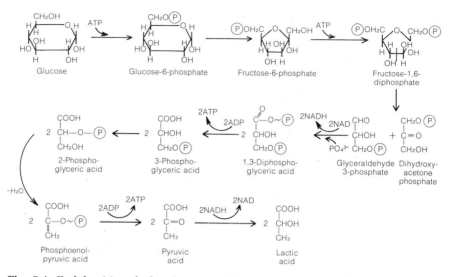

Fig. 5.4. Embden-Meyerhof pathway used by *Streptococcus lactis* for the homolactic fermentation. *S. lactis* obtains energy as ATP by fermenting glucose to lactic acid. Note that glucose is phosphorylated and cleaved into three carbon units before oxidation.

Pathways of carbohydrate dissimilation. There are three major routes utilized by bacteria for the dissimilation of carbohydrates, regardless of whether a respiratory or fermentative process is employed. These are: (1) the Embden-Meyerhof pathway, also known as the hexose diphosphate pathway; (2) the pentose phosphate pathway, also known as the hexose monophosphate pathway; and (3) the Entner-Doudoroff pathway, which is a variation of the hexose monophosphate pathway. Some bacteria utilize only one of these pathways exclusively, whereas others metabolize a certain proportion of carbohydrate via one route and the remainder by a second. The three pathways as used for the homolactic, heterolactic, and *Zymomonas mobilis* fermentations, respectively, are outlined in Figures 5.4, 5.5 and 5.6.

The natural souring of milk is due to the fermentation of the milk sugar, lactose, to lactic acid. One of the organisms responsible for this process, *Streptococcus lactis,* which was perhaps the first bacterium to be isolated in pure culture (by Lister in 1878), carries out a typical homolactic fermentation. Metabolism of glucose by this organism occurs exclusively by the Embden-Meyerhof pathway. The first steps of this pathway involve two phosphorylations in which ATP is expended to provide a more reactive molecule, and an isomerization to yield the hexose diphosphate intermediate, fructose-1,6-diphosphate. This compound then is cleaved by the action of the enzyme aldolase. This enzyme is not present in organisms such as *Leuconostoc mesenteroides* or *Zymomonas mobilis,* which, therefore, do not use this pathway. The oxidation of the resulting two triose molecules provides the only energy-yielding step in the operation of the Embden-Meyerhof pathway proper. This reaction, catalyzed by glyceraldehyde-3-phosphate dehydrogenase, is a classical oxidation of an aldehyde to a carboxyl group, except that it is complicated by the formation of a mixed acid anhydride bond between the carboxyl of the product and an orthophosphate to give 1,3-diphosphoglyceric acid. Such a mixed anhydride bond is much more reactive than a simple phosphate ester bond, and hence can be considered a "high energy" phosphate, usually designated as ~P. Such ~P groups can be transferred readily to an appropriate acceptor, in this case ADP. The generation of a high energy phosphate bond during oxidation of a substrate, in which the oxidized product bears the ~P, and its subsequent transfer to ADP to yield ATP, is referred to as "substrate level phosphorylation." This is in contradistinction to the cytochrome-linked ATP generation associated with respiration and the operation of the electron transport chain, which is termed "oxidative phosphorylation." By means of a rearrangement and a dehydration reaction another high energy phosphate bond can be obtained in the form of phospho-enol pyruvic acid (PEP), and again this can be transferred to ADP to give ATP. On balance it is found that there is a net yield of two molecules of ATP for each molecule of glucose fermented. This may be accounted for as follows: in the preliminary stages the glucose molecule is activated by phosphate additions which expend two molecules of ATP. Subsequently, two molecules of glyceraldehyde-3-phosphate are formed from each glucose molecule, and the oxidation of each of these yields two ATP molecules, or a total of four ATP. Subtracting the two initially used, the net yield becomes two ATP/glucose. The final reactions in this fermentation consist of reduction of the two pyruvic acid molecules by the two NADH molecules formed during the triose oxidation. The overall reaction therefore approximates the following equation:

$$\text{glucose} + 2\text{ADP} + 2\text{PO}_4^{3-} \longrightarrow 2 \text{ lactic acid} + 2 \text{ ATP} \qquad (5.10)$$

The heterolactic fermentation is important in the preparation of certain fermented foods, such as pickles and sauerkraut, in which lactic acid developed by certain species of *Lactobacillus* and *Leuconostoc* exerts a preservative action. In this fermentation, as carried out by *Leuconostoc mesenteroides,* glucose is metabolized by the pentose phosphate pathway (Fig. 5.5). Contrary to the situation in the hexose diphosphate pathway, after the initial addition of phosphate to give

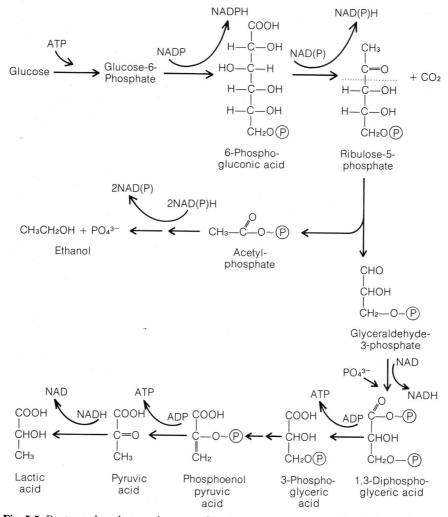

Fig. 5.5. Pentose phosphate pathway used by *Leuconostoc mesenteroides* for the heterolactic fermentation. In this scheme *L. mesenteroides* phosphorylates and oxidizes glucose to 6-phosphogluconic acid before cleavage.

glucose-6-phosphate, this compound is immediately oxidized to 6-phosphogluconic acid. Ordinarily, the enzyme catalyzing this reaction, glucose-6-phosphate dehydrogenase, is specific for NADP rather than NAD as the coenzyme. Following this step there is a further oxidation and decarboxylation to give CO_2 and the pentose, ribulose-5-phosphate. After isomerization (to xylulose-5-phosphate), the pentose is cleaved by the enzyme phosphoketolase, which produces acetylphosphate and glyceraldehyde-3-phosphate. Although acetyl phosphate possesses a high energy phosphate, it is not used to form ATP, and instead the energy is dissipated in a reduction to ethanol. The lone triose molecule is oxidized by the same sequence of reactions used in the Embden-Meyerhof pathway to produce a molecule of pyruvic acid and two ATP. The reductions of pyruvate to lactate and of acetylphosphate to ethanol provide for the regeneration of the oxidized form of the nicotinamide coenzymes. It may be deduced from the foregoing that the net yield of ATP in this fermentation is less than in the hexose diphosphate pathway, namely only one ATP/glucose. The overall heterolactic fermentation would conform to the equation:

$$\text{glucose} + \text{ADP} + \text{PO}_4^{3-} \longrightarrow \text{lactate} + \text{ethanol} + CO_2 + \text{ATP}$$
$$(5.11)$$

Calculations of ATP yields are not mere formalisms; available energy limits the cell crop. Growth yields (Y_{ATP}) in fermentations average 10 g dry cells/gram-mole ATP. For *Leuconostoc* the cell crop per mole of glucose should be half that for *Streptococcus*. It is actually 0.7, which may indicate other energy-yielding steps. The ATP formed in aerobic respiration is uncertain and Y_{ATP} cannot be determined for aerobic bacteria.

Although alcoholic beverages characteristically are produced by yeast fermentations, the production of alcohol in the pulque fermentation is accomplished, in part at least, by a bacterium, *Zymomonas mobilis* (formerly called *Pseudomonas lindneri*), which was first isolated from this source. This bacterium almost exclusively utilizes the Entner-Doudoroff pathway (Fig. 5.6) for the anaerobic fermentation of glucose, and it differs from other bacteria which form ethanol in that it possesses a pyruvic acid decarboxylating system resembling that of yeasts. The use of the Entner-Doudoroff pathway by a fermentative organism is definitely unusual, inasmuch as this pathway is more characteristically used for nonfermentative glucose metabolism in the aerobic species

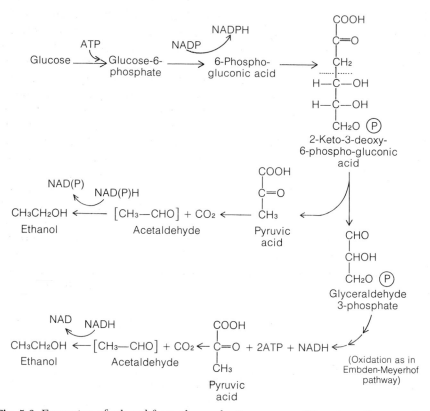

Fig. 5.6. Formation of ethanol from glucose by *Zymomonas mobilis* using a fermentative version of the Entner–Doudoroff pathway. *Z. mobilis* is one of the few bacteria that produce ethanol as the major fermentation end product.

of *Pseudomonas*. In this fermentation glucose is converted by a series of reactions to 2-keto-3-deoxy-6-phosphogluconic acid. This compound is cleaved to yield pyruvic acid plus glyceraldehyde-3-phosphate. The triose can be oxidized to pyruvate as in the Embden-Meyerhof pathway, so that two molecules of pyruvate result from each glucose. These are next decarboxylated to form CO_2 and acetaldehyde, and the reduction of this latter compound permits the reoxidation of the nicotinamide coenzymes. In addition, a small amount of pyruvate appears to be directly reduced to lactate. The overall process corresponds to the equation:

$$\text{glucose} + PO_4^{3-} + ADP \longrightarrow 1.7\,\text{ethanol} + 1.7\,CO_2 + 0.3\,\text{lactate}$$
$$+ ATP \qquad (5.12)$$

As in the heterolactic fermentation, the *Zymomonas mobilis* ethanolic fermentation gives an ATP yield only one-half that of the hexose diphosphate pathway. Yeasts use the hexose diphosphate pathway for the fermentation; thus, although both Z. *mobilis* and yeasts ferment glucose to ethanol plus CO_2, they do so by distinctly different routes and obtain different amounts of useful energy from the process.

5.5 | Fermentation of Noncarbohydrate Materials

Bacteria are by no means limited to carbohydrates as fermentative substrates, but are able to utilize a number of other substances including ethanol, organic acids, and purines. Studies on some of these fermentations have revealed new biochemical pathways previously unsuspected, and have provided explanations for important and familiar occurrences.

The aroma compounds of butter are microbial fermentation products resulting from citric acid metabolism. In the manufacture of butter from sweet cream a "starter culture" consisting of a mixture of *Streptococcus lactis* and *Leuconostoc citrovorum* is added to the cream in order to produce these substances along with the necessary acidity. The homofermentative S. *lactis* produces large amounts of lactic acid, but does not form the desired flavor and aroma compounds. The *Leuconostoc* does not produce as much acid, but, at a low pH, is able to convert substrates to diacetyl (also termed biacetyl), which is the major flavor component. Although cream normally contains some citric acid, additional amounts are sometimes added to increase diacetyl formation.

Clostridium kluyveri is an interesting bacterium, metabolically untypical of clostridia, in that it is unable to ferment sugars or amino acids, but rather is dependent on the fermentation of ethanol. Investigation of this fermentation, however, was instrumental in clarifying the mechanisms involved in the typical butyric and butanol fermentations of other clostridia, especially the role of acetyl CoA and acetoacetyl CoA as intermediates.* Fermentation of ethanol by *Cl. kluyveri* is com-

---

*Coenzyme A (CoA), a carrier of "activated" fatty acids, is composed of adenosine-3'-phosphate linked by a pyrophosphate bridge to the γ-hydroxyl of pantothenic acid, which in turn is joined to 2-mercaptoethylamine by a peptide bond. Fatty acids are "activated" in an ATP-requiring reaction and this leads to the formation of a "high energy" bond between the acyl groups (e.g., acetyl, propionyl, butyryl) and the thiol of CoA. Such acyl CoA derivatives in turn can donate acyl groups to suitable acceptor compounds.

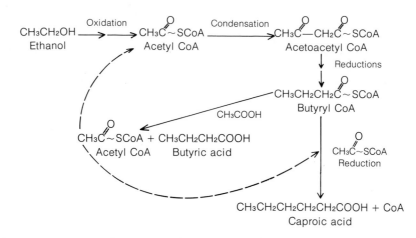

Fig. 5.7. Pathway for fermentation of ethanol by *Clostridium kluyveri*. Butyric and caproic acids are produced in this fermentation by coenzyme A–mediated reactions similar to those used for fatty acid biosynthesis in general.

pletely dependent on the presence of acetate and ceases upon exhaustion of the acetate supply. Mixtures of ethanol plus acetic acid are converted to a mixture of butyric and caproic acids, the proportions of which are related to the ratio of ethanol to acetate. The oxidative steps of this fermentation consist of the oxidation of ethanol to the level of acetic acid, which is actually in the form of acetyl CoA. Condensation of 2 moles of acetyl CoA yields a 4-carbon intermediate, acetoacetyl CoA, that then may be reduced stepwise to butyryl CoA. Finally, butyric acid is produced by a transfer of the CoA from butyryl CoA to a free acetic acid molecule, which then can enter the earlier portion of the pathway to form acetoacetate. When ethanol is in excess, the major product is caproic acid, which is produced by condensation of butyryl CoA with another molecule of acetyl CoA, followed by a reduction. Formation of the more reduced product, caproate, results from the greater amount of reducing power when ethanol is in excess with respect to acetate. Additional relief for this "hydrogen pressure" is afforded by the production of H_2 gas, as well as the reduced carbon compounds. The pathways in this fermentation are outlined in Figure 5.7.

In other *Clostridium* species butyric acid and its reduction product, butanol, arise also from the same acetoacetyl CoA pathway, with the difference that the acetyl CoA originates from carbohydrate instead of ethanol, and further, that the formation of acetone, and its reduction product isopropanol, can be accounted for by a decarboxylation of acetoacetate.

Certain L-amino acids, but not all, may be fermented by anaerobic bacteria as the sole energy source. Among those which are known to be readily fermented are alanine, glycine, aspartic acid, glutamic acid, cysteine, and histidine. The following amino acids appear not to serve alone as fermentable substrates, although some may be reduced or otherwise partially metabolized: proline, hydroxyproline, leucine, isoleucine, valine, and arginine.

A relatively straightforward fermentation of L-alanine is carried out by *Clostridium propionicum,* which produces ammonia, propionic acid, acetic acid, and CO_2. It is believed that the intermediate formed as a result of removal of the amino group from alanine may be acrylate (or acrylyl CoA), $CH_2 = CH-COOH$, which can be directly reduced to propionate. Quantitative measurements show that for each three molecules of alanine attacked, two are converted to propionate and the remaining one to acetate plus CO_2:

$$3CH_3-\underset{\underset{NH_2}{|}}{CH}-COOH + 2H_2O \xrightarrow{\textit{Cl. propionicum}} 2CH_3CH_2COOH +$$

propionic acid

L-alanine

$$CH_3COOH + CO_2 + 3NH_3 \qquad (5.13)$$

acetic
acid

Although these are the same three products formed from glucose by *Propionibacterium* (see Table 5.6) and from lactate by *Veillonella alcalescens,* the pathway for these two propionic fermentations is quite different from that used by *Clostridium propionicum.* The key reaction in the *Propionibacterium* fermentation is a so-called transcarboxylation in which a carboxyl group is transferred from methylmalonyl CoA to pyruvic acid, yielding propionyl CoA and oxaloacetic acid:

$$
\begin{array}{cccccc}
COOH & & CH_3 & CH_3 & & COOH \\
| & & | & | & & | \\
CH_3-CH & + & C=O & \longrightarrow \quad CH_2 & + & CH_2 \\
| & & | & | & & | \\
C\!\sim\!SCoA & & COOH & C\!\sim\!SCoA & & C=O \\
\diagdown O & & & \diagdown O & & | \\
& & & & & COOH
\end{array} \qquad (5.14)
$$

methylmalonyl   pyruvic   propionyl   oxaloacetic
CoA   acid   CoA   acid

The propionyl CoA is then converted to propionic acid, whereas the oxaloacetate is transformed into methylmalonyl CoA via succinate and succinyl CoA as intermediates.

Certain *Clostridium* species, including the botulism organism, *Cl. botulinum,* are able to utilize pairs of amino acids. In these reactions one amino acid is oxidized while the other serves simply as an electron acceptor. This type of fermentation is called the *Stickland reaction* after its discoverer, and may be exemplified by the fermentation of L-alanine by *Clostridium sporogenes* using glycine as the oxidant:

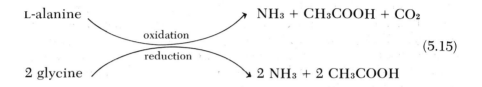

$$\text{L-alanine} \xrightarrow{\text{oxidation}} NH_3 + CH_3COOH + CO_2$$

$$\text{2 glycine} \xrightarrow{\text{reduction}} 2\ NH_3 + 2\ CH_3COOH$$

(5.15)

Amino acids other than alanine that may be fermented in this manner by *Cl. sporogenes* include leucine, isoleucine, and valine. Amino acids replacing glycine as oxidant in the Stickland reaction for *Cl. sporogenes* include proline, hydroxyproline, ornithine, arginine, and tryptophan.

Certain nitrogen-containing heterocyclic compounds, including purines and pyrimidines, also may be fermented, although by a limited group of bacteria. The fermentation of purines by *Clostridium acidiurici* typifies these kinds of reactions. This organism ferments xanthine, hypoxanthine, guanine, and uric acid, all of which are convertible to xanthine, which appears to serve as the common intermediate. Xanthine is attacked with the production of approximately 4 moles of ammonia, 3 of CO_2 and 1 of acetate per mole of xanthine. The fermentative pathway employed by *Cl. acidiurici* is quite distinct from the familiar oxidative pathway in mammals, as neither urea nor allantoin accumulates, nor are these compounds attacked by this organism. The 6-membered ring of xanthine is cleaved to form 4-ureido-5-imidazole carboxylic acid, which is converted stepwise to CO_2, ammonia, and formimino-glycine (FIG). The latter is the source of small amounts of formate found in this fermentation as well as of the acetate. Probably the glycine and formyl portions of FIG are cleaved and then later condensed again to give serine, which is in

turn deaminated to pyruvate. Decarboxylation of pyruvate then would
give acetate plus CO_2:

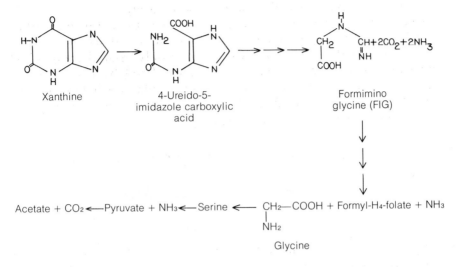

Xanthine   4-Ureido-5-imidazole carboxylic acid   Formimino glycine (FIG)

Acetate + CO_2 ←— Pyruvate + NH_3 ←— Serine ←— CH_2—COOH + Formyl-H_4-folate + NH_3
                                                           |
                                                          NH_2

Glycine

5.6 | Pathways for the Dissimilation of Aromatic Compounds by Bacteria

Although a great number of different aromatic compounds are
metabolized by bacteria, only a limited number of major pathways
appear to be involved. The most extensive investigations concerning
the biochemistry of these pathways have been conducted with aerobic
bacteria of the genus *Pseudomonas*. This has led to the impression that
aromatic compounds are attacked only by mechanisms involving
molecular oxygen. However, there are indications that there may be
anaerobic pathways which permit ring fission without participation
of O_2 *per se*.

The metabolism of various aromatic compounds by aerobic bacteria
is characterized by their conversion to certain common diphenolic
intermediates. In such dihydroxylations of the aromatic ring the -OH
groups may be placed *ortho* to one another as in catechol and protocatechuic acid, or else *para* to each other as in gentisic and homogen-
tisic acids. The hydroxylated intermediates then are cleaved and the
resulting aliphatic products are converted to familiar terminal respira-

tory metabolites such as pyruvate, acetaldehyde, formate, succinate, fumarate, or acetate. These compounds may be oxidized to CO_2 via the Krebs tricarboxylic acid cycle to obtain energy.

Two of the major routes for aromatic oxidations involve either catechol or else protocatechuate as the key intermediate. It is found, for example, that in *Pseudomonas* the two closely related compounds benzoate and *p*-hydroxybenzoate are metabolized by two different pathways. The pathway for benzoate proceeds via catechol as an intermediate, whereas the diphenolic intermediate in the pathway for *p*-hydroxybenzoate is protocatechuate:

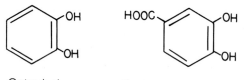

Catechol   Protocatechuic acid

A large number of aromatic compounds are metabolized via either one or the other of these two key intermediates. Among those compounds which are metabolized by the catechol pathway are benzoate, salicylate, naphthalene, phenanthrene, benzene, phenol, *o*-cresol, and mandelate. The protocatechute pathway is employed for *p*-hydroxybenzoate, *p*-aminobenzoate, phthalate, phenylalanine, tryosine, *m*-cresol, *p*-cresol, and hydroxymandelate.

A further complication is introduced by the fact that there are two distinct mechanisms for the oxidative fission of both of these 1,2-diphenols, and for each compound the cleavage of the ring may occur either between, or else adjacent to, the two carbons bearing the hydroxyls.

The first mechanism is referred to as *ortho-cleavage*, an example of which is provided by the attack on catechol catalyzed by the enzyme catechol-1,2-dioxygenase. In this reaction a molecule of O_2 is incorporated across the double bond between the hydroxylated carbons to yield a colorless compound, *cis, cis*-muconic acid:

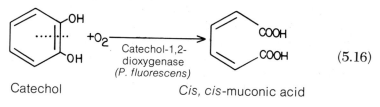

Catechol   *Cis, cis*-muconic acid

(5.16)

In other pseudomonads a second type of reaction, termed *meta-cleavage,* is catalyzed by another enzyme, catechol 2,3-dioxygenase. The product of this type of ring fission of catechol is the yellow substance 2-hydroxymuconic semialdehyde:

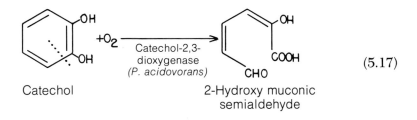

Catechol   2-Hydroxy muconic
                             semialdehyde

(5.17)

The two different cleavage mechanisms result in differing reaction sequences leading from the initial products of ring scission. Further metabolism of *cis, cis*-muconic acid proceeds via β-ketoadipic acid, which in turn is converted to succinate and acetyl CoA. In the case of *meta*-cleavage, the pathway consists of the conversion of 2-hydroxymuconic semialdehyde to 2-keto-4-hydroxy valerate plus formate. The 5-carbon intermediate then is cleaved to produce acetaldehyde plus pyruvate:

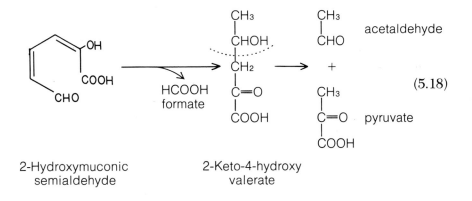

2-Hydroxymuconic   2-Keto-4-hydroxy
semialdehyde   valerate

(5.18)

Essentially the same reaction pathways are available for protocatechuate, and its two cleavage products, which will differ from those for catechol only in the possession of the additional carboxyl group. Thus *ortho*-cleavage of protecatechuate yields the β-carboxy-*cis, cis*-muconic acid:

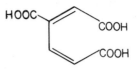

The *meta*-cleavage of protocatechuate produces 4-carboxy-2-hydroxy-muconic semialdehyde.

5.7 | The Role of Inorganic Compounds in Bacterial Energy Metabolism

Energy metabolism of bacteria involves inorganic compounds in several different ways, of which we note the following: (1) as oxidizable substrates employed as the primary source of energy for chemolithotrophs; (2) as the source of reducing power in photolithotrophic bacteria; (3) as the final electron acceptors, or oxidants, in place of molecular oxygen; and (4) as enzyme cofactors or other direct participants in enzymic reactions involved in generation and utilization of energy.

Among the inorganic substances whose oxidation may be utilized by chemolithotrophic bacteria as the sole energy source are ammonia, nitrite, hydrogen, ferrous iron, elemental sulfur, thiosulfate, sulfide, and thiocyanate. In addition, it has been reported that carbon monoxide serves as an energy substrate for one of the hydrogen-oxidizing bacteria, which is for this reason maintained in a separate genus, *Carboxydomonas* (Section 3.13). Despite the interesting nature of this organism it has been accorded but little attention. Some of the energy-yielding reactions utilized by these organisms are summarized in Table 5.7. Certain aspects of chemolithotrophy and photolithotrophy are discussed in more detail in Sections 5.9 and 5.10, respectively.

5.8 | Inorganic Compounds as Terminal Electron Acceptors

In the absence of molecular oxygen, or in those organisms lacking a complete electron transport chain, some other substance can and must be employed as the final acceptor for electrons (and protons) produced in energy-yielding oxidations. Most fermentative organisms rely for this purpose on organic compounds produced during the course

Table 5.7. Energy-Yielding Reactions Employed by Chemolithotrophic Bacteria

Inorganic energy substrate	Formal equation for reaction	Bacterium	
NH_3	$NH_4^+ + \frac{3}{2}O_2 \longrightarrow NO_2^- + H_2O + 2H^+$ $\Delta G^0 = -66$ kcal	*Nitrosomonas*	(5.19)
NO_2^-	$NO_2^- + \frac{1}{2}O_2 \longrightarrow NO_3^-$ $\Delta G^0 = -17.5$ kcal	*Nitrobacter*	(5.20)
H_2	$H_2 + \frac{1}{2}O_2 \longrightarrow H_2O$ $\Delta G^0 = -57$ kcal	*Hydrogenomonas*	(5.21)
S^0	$S^0 + \frac{3}{2}O_2 + H_2O \longrightarrow H_2SO_4$ $\Delta G^0 = -118$ kcal	*Thiobacillus*	(5.22)
Fe^{2+}	$Fe^{2+} \longrightarrow Fe^{3+} + e^-$ $\Delta G^0 = -11$ kcal	*Ferrobacillus*	(5.23)
CO	$CO + \frac{1}{2}O_2 \longrightarrow CO_2$ $\Delta G^0 = -66$ kcal	*Carboxydomonas*	(5.24)

Table 5.8. Inorganic Compounds and Ions as Final Electron Acceptors for Bacteria

Terminal electron acceptor	End product(s)	Bacterium
NO_3^-	NH_4^+, NO_2^-	*Escherichia coli*
NO_3^-	N_2	*Pseudomonas denitrificans*
NO_3^-	$N_2 + N_2O$	*Pseudomonas aeruginosa*
NO_3^- or NO	N_2O	*Corynebacterium nephridii*
SO_4^{2-}	H_2S	*Desulfovibrio desulfuricans*
CO_2	CH_4	*Methanosarcina barkeri*
CO	CH_4	*Methanosarcina barkeri*
$HTeO_3^-$	Te^0	*Staphylococcus aureus*
$HSeO_3^-$	Se^0	*Salmonella typhosa*
$S_4O_6^{2-}$	$S_2O_3^{2-}$	*Salmonella typhosa*
H_2VO_4	$VO(OH)_2$	*Veillonella alcalescens* (*Micrococcus lactilyticus*)
$UO_2(OH)_2$	$U(OH)_4$	*Veillonella alcalescens*
Se^0	Se^{2-}	*Veillonella alcalescens*
S^0	S^{2-}	*Veillonella alcalescens*
Au^{3+}	Au^0	*Veillonella alcalescens*

of dissimilation of the primary substrate. However, many bacteria are able to utilize certain inorganic compounds as electron acceptors in what can be considered a kind of "anaerobic respiration." Although this is a contradiction in terms, it emphasizes the fact that in these reactions there is often an interaction of the inorganic oxidant with a partial or truncated electron transport chain. Some compounds and ions which have been shown to serve as electron acceptors in bacteria are listed in Table 5.8. Among these substances are carbon dioxide and nitric oxide, as well as the following ions: nitrate, nitrite, sulfate, sulfite, tellurite, selenite, and tetrathionate. In several cases physiologically specialized groups of bacteria have developed which make major or exclusive use of one of these inorganic electron acceptors. There are, for example, certain organisms which specialize in the use of nitrate or sulfate ions, or carbon dioxide, that is, the denitrifiers, sulfate-reducers, and methane-producers.

Nitrate reduction. Many organisms are able to assimilate relatively small quantities of nitrate, which is reduced to the level of ammonia for incorporation into amino acids. However, the ability to grow anaerobically using nitrate in place of oxygen is more restricted. Such dissimilatory nitrate reduction takes two forms. In the case of certain nitrate reducing bacteria, such as *Escherichia coli, Aerobacter aerogenes,* certain *Bacillus* and *Clostridium* species, and the nitrogen-fixing bacterium, *Azotobacter vinelandii,* the products formed from nitrate are soluble and no gas is produced. In these organisms nitrate is converted to nitrite and, under some conditions, to ammonia.

In certain other bacteria termed "denitrifiers," the products of nitrate reduction are evolved in gaseous form and lost from the culture, resulting in an overall decrease in the nitrogen content. It is believed that such denitrifications also take place in nature under anaerobic conditions, for example in water-logged soils, and thereby contribute to a loss of fertility through nitrogen depletion. The denitrifying bacteria consist of various metabolic types of organisms, including chemolithotrophs utilizing inorganic substrates, such as thiosulfate (*Thiobacillus denitrificans*) and H_2 (*Micrococcus denitrificans*) as well as chemoorganotrophs, such as *Pseudomonas aeruginosa* and *Bacillus licheniformis.* Such organisms generally reduce nitrate to elemental nitrogen, N_2, or a mixture of N_2 and nitrous oxide, N_2O. Although denitrification processes generally involve N_2 evolution, at least one organism forms exclusively nitrous oxide. *Corynebacterium nephridii,* a gram-positive rod originally found in the medicinal leech (*Hirudo medicinalis*), reduces either nitrate, nitrite, or nitric oxide (NO) stoichiometrically to nitrous oxide without N_2 being produced.

It is presumed that N_2 formation from nitrate proceeds via a multistep reduction sequence such as the following:

$$NO_3^- \longrightarrow NO_2^- \longrightarrow NO \longrightarrow N_2O \longrightarrow N_2$$

It is thought that the initial step, catalyzed by the enzyme nitrate reductase, is common to all organisms reducing nitrate and that in all cases the initial product is nitrite. The subsequent fate of nitrite varies with the properties of the enzymes present in the denitrifiers and other nitrate-reducing bacteria. Bacterial nitrate reductase preparations have been reported to be simple flavoproteins or metallo-flavoproteins which couple electron transfer from the cytochrome chain to nitrate. In some cases nitrate reduction may be coupled to the oxidation of NAD or NADP-linked substrates. Nitrate reduction also can be achieved using cytochrome-linked substrates, such as formate, or by molecular hydrogen when the organism possesses the enzyme hydrogenase. In $E.$ $coli$ the nitrate reductase complex is closely associated with formate dehydrogenase and cytochrome b_1, and formate serves as a preferred electron donor for nitrate reduction. Presumably the pathway of electron flow involved in this reaction is:

$$\text{formate} \longrightarrow \text{formate dehydrogenase} \longrightarrow \text{cyt } b_1$$
$$NO_3 \longleftarrow \underset{(Mo^{2+}, Fe^{2+}?)}{\text{nitrate reductase}} $$

An additional feature noted in this situation is that anaerobic growth of $E.$ $coli$ with nitrate yields significant quantities of a cytochrome c_{552}. This cytochrome, which has a low oxidation-reduction potential of approximately -0.2 v, is not formed under aerobic conditions and may, in fact, be involved in some way in nitrate reduction.

There are indications that cytochromes, especially cytochrome oxidase, may function as the "nitrite reductase" for further reduction of nitrite. Highly purified cytochrome oxidase preparations from the denitrifying bacterium $P.$ $aeruginosa$ reduce nitrite to nitric oxide when reduced cytochrome c_{551} is employed as the electron donor. If nitrate is supplied it is reduced to nitrite by nitrate reductase which interacts with the respiratory chain at cytochrome 560:

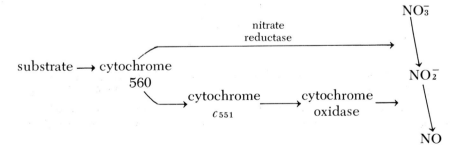

These formulations relative to involvement of the cytochromes offer a satisfactory explanation for activities of most nitrate-reducers, which are facultative anaerobes known to possess cytochromes and other ETS components when grown anaerobically with nitrate. However, it can be deduced that cytochromes may not be obligatory participants in all cases. Nitrite is produced from nitrate by *Clostridium perfringens*, which supposedly lacks cytochromes. The same organism, when supplied with H_2, reduces nitrate to ammonia. In this organism electron carriers other than cytochromes must be responsible.

Assimilatory sulfate reduction. A large number of bacteria are able to utilize inorganic sulfate as the sole sulfur source for growth. In such organisms small amounts of sulfate are reduced to the level of sulfide and used for synthesis of the amino acids cysteine and methionine, as well as other sulfur-containing metabolites and biosynthetic intermediates. Such assimilatory sulfate reduction in bacteria employs as a key intermediate the same form of "active sulfate" used for sulfation of steroids, mucopolysaccharides and other substrates in mammalian organisms. In the initial reaction of the assimilatory reduction process, shown in Fig. 5.8, the enzyme ATP-sulfurylase catalyzes an attack on ATP by inorganic sulfate resulting in the elimination of pyrophosphate and formation of the adenosine-5'-phosphosulfate (APS), which has the following structure:

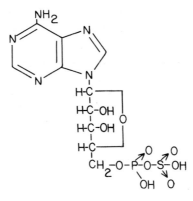

This is followed by a phosphorylation of APS at the 3' position to yield the 3'-phosphoadenosine-5'-phosphosulfate (PAPS). This compound is the "active sulfate" used in most sulfate transfer reactions in bacteria as well as in eukaryotes. In a series of oxido-reductions,

Scheme 1. Assimilatory sulfate reduction in *Salmonella*

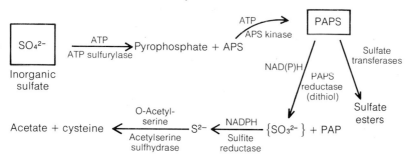

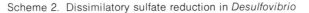

Scheme 2. Dissimilatory sulfate reduction in *Desulfovibrio*

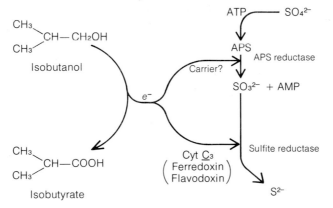

Fig. 5.8. Assimilatory and dissimilatory sulfate reduction in bacteria. Two separate pathways for sulfate reduction, each involving the nucleotide, adenosine phosphosulfate (APS), are found in bacteria. Sulfate assimilation into cysteine and other thiols requires conversion of APS to phosphoadenosine phosphosulfate (PAPS) before reduction (Scheme 1). Use of sulfate as terminal oxidant by *Desulfovibrio* proceeds with the reduction of APS itself (Scheme 2).

the sulfate group of PAPS is reduced to the level of sulfite by a multicomponent enzyme system termed PAPS-reductase. The PAPS-reductase system involves a flavin and a dithiol in the transfer of electrons from NADH (or NADPH) to PAPS. This reduction product, probably an enzyme-bound intermediate rather than free inorganic sulfite *per se*, is then further reduced to the level of sulfide. Sulfide can be used directly in formation of cysteine via the cleavage of

O-acetylserine in which the sulfhydrase enzyme catalyzes a reaction analogous to a hydrolysis except that H_2S replaces H_2O.

Dissimilatory sulfate reduction. In contrast to the situation with nitrate, the ability to use sulfate as the major electron acceptor for cellular oxidations after the fashion of the anaerobic nitrate respiration of the denitrifiers is restricted to only a few species of the so-called sulfate-reducing bacteria (Section 3.17). These comprise two groups of strict anaerobes, the gram-negative *Desulfovibrio,* and the gram-positive spore-formers of the genus *Desulfotomaculum.* One of the latter, an organism formerly called *Clostridium nigrificans,* is thermophilic and occasionally is encountered as a high temperature sulfide-spoilage agent in canned foods. The specificity and distinct biochemical differences between nitrate and sulfate reduction processes are emphasized by the fact that nitrate will not substitute for sulfate as the terminal electron acceptor for *Desulfovibrio,* which therefore fails to form nitrite from nitrate.

Five *Desulfovibrio* species are widely distributed in marine and fresh water muds, which in consequence frequently are laden with the hydrogen sulfide produced by these organisms during dissimilatory sulfate reduction. These organisms are able to utilize as energy substrates various organic compounds, such as pyruvate, lactate, malate, choline, ethanol, and isobutanol. In addition, *Desulfovibrio desulfuricans,* the best known species, is able to utilize molecular hydrogen, H_2, for the reduction of sulfate:

$$SO_4{}^{2-} + 4H_2 \xrightarrow{\text{\textit{D. desulfuricans}}} S^{2-} + 4H_2O \tag{5.25}$$

D. desulfuricans is not solely dependent on the sulfate reduction process, but can grow fermentatively on certain substrates, such as pyruvate or choline, even in the absence of sulfate. It is interesting that *Desulfovibrio* often incorporates little or none of the carbon from the energy substrate. When *Desulfovibrio* is grown on isobutanol, most of the cellular carbon (ca. 90%) is derived from small amounts of yeast extract, which must be included in the growth medium, and the remainder from carbon dioxide fixation. Under these conditions isobutanol, serving as the sole energy source, is converted quantitatively to isobutyric acid, which can be recovered from the spent culture fluid. Certain other substrates may be used to supply both carbon and energy.

The initial step of the dissimilatory sulfate reduction in *Desulfovibrio*

(Fig. 5.8) is the same as in the assimilatory process, namely the formation of APS via the ATP-sulfurylase reaction. However, in the sulfate-reducing bacteria the APS is not converted to PAPS and extracts of these organisms show little or no PAPS-reductase activity. Instead, these organisms reduce APS itself, using an enzyme, APS-reductase, which is found only in sulfate-reducers and in some chemolithotrophic sulfur oxidizers (Thiobacillus). The products of this reaction are AMP and sulfite, the latter being further reduced to hydrogen sulfide.

Several electron carriers which may participate in sulfate reduction have been demonstrated in Desulfovibrio species. One of these is a unique hemoprotein, cytochrome c_3, which possesses a very low standard electrode potential $(E_0' = -0.205$ v). Until the discovery of cytochrome c_3 it was thought that cytochromes were absent in all strict anaerobes, with the exception of the anaerobic photosynthetic bacteria which were obviously a special case. The cytochrome c_3 of D. desulfuricans, which exhibits characteristic absorption peaks at 525 and 553 nm, possesses three heme groups and has a molecular weight of 12,200. Another porphyrin-containing protein, desulfoviridin, has been found in some but not all strains of Desulfovibrio. In view of this nonuniform distribution, the relationship of desulfoviridin to sulfate reduction is problematical. In addition, other low potential electron carriers, such as ferredoxin and flavodoxin, have been found in sulfate reducing bacteria. Bacterial ferredoxins are a group of brownish-colored small proteins with molecular weights of approximately 6,000, which contain 3–7 atoms of loosely bound nonheme iron per molecule, as well as an equimolar amount of inorganic sulfide. Ferredoxins are reversibly oxidized and reduced, and participate in a variety of oxido-reduction reactions in various anaerobic bacteria. In addition, ferredoxins are known to take part in photosynthesis in both eukaryotes and prokaryotes. The standard electrode potentials of the ferredoxins $(E_0' = -0.420$ v) lie considerably below that of the nicotinamide coenzymes $(E_0' = -0.32$ v). Reduced ferredoxin therefore acts as an electron and hydrogen donor for reduction of nicotinamide coenzymes and various other metabolites with oxidation-reduction potentials above -0.400 v; also, reduced ferredoxin can, in conjunction with the enzyme hydrogenase, donate electrons to [H$^+$] and thereby generate molecular hydrogen. For a full understanding of the sulfate reduction process it is essential that the role of these substances be clarified. Cytochrome c_3 and desulfoviridin are not present in Desulfotomaculum, although it is reported that this organism does in fact contain a porphyrin, or protoheme, which could be involved in this system. Although it is obvious that carriers of this sort are required to couple oxidation of substrate to reduction of APS and sulfite, elucidation of the exact sequence of electron flow, and the immediate electron donor for APS–reductase reaction awaits further study.

Methanogenesis. The biological formation of methane is a process analogous to nitrate and sulfate reductions, inasmuch as methane·is produced by reduction of carbon dioxide, which serves as the ultimate oxidant for a group of strictly anaerobic bacteria. These organisms are found in feces, sewage sludge, rumen contents, soil, and pond mud. Neither sulfate nor nitrate will replace CO_2 as terminal electron acceptor for the methane-forming bacteria. The extreme sensitivity to oxygen of this heterogenous collection of methanogenic rods, micrococci, and sarcinae has hampered bacteriological studies, with the result that their taxonomy remains poorly developed. Perhaps it is illustrative of the difficulties involved that only after some decades of study has one of the better known "species," *Methanobacillus omelianskii,* been recognized to be a synergistic mixture of two distinct organisms. Possibly the methodology devised for the study of strict anaerobes of the rumen, a number of which are in fact methane producers, will permit greater progress with these organisms.

The methanogenic bacteria are quite circumscribed with respect to substrates utilized for growth and methane formation. The usual and abundant substances such as cellulose, starch, sugars, and proteins are not directly attacked by these organisms, and only give rise to methane in mixed populations which include other organisms with the ability to convert them to the substrates used by the methane bacteria. These include simple alkanoic acids, lower primary alcohols, hydrogen and carbon monoxide, and carbon dioxide. One group of methane bacteria, which includes *Methanobacterium formicicum, M. ruminantium,* and *M. mobilis*, uses primarily H_2 or formate as oxidizable substrate. Another group utilizes almost exclusively organic acids, and includes *Methanobacterium propionicum* (propionate), *M. söhngenii* (acetate, butyrate) and *M. suboxydans* (butyrate, valerate, caproate). "*Methanobacillus omelianskii*" uses H_2, ethanol and a few other alcohols, but not organic acids. *Methanosarcina barkeri* is restricted to H_2, carbon monoxide, methanol, and acetate as energy sources.

The most straightforward process for biogenesis of methane is the reduction of carbon dioxide with molecular hydrogen, which can be carried out by several organisms including *Methanosarcina barkeri,* *Methanobacterium ruminantium* and *Methanobacterium mobilis.* Presumably this consists of two simple coupled reactions in which hydrogenase and ferredoxin participate in the activation of hydrogen and transfer of electrons to carbon dioxide:

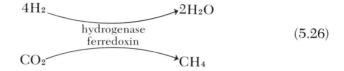

$$\tag{5.26}$$

The formation of methane from CO_2 during the oxidation of ethanol to acetate by "*M. omelianskii*" was originally considered to consist of an analogous pair of coupled reactions:

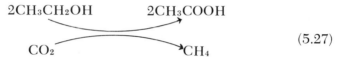

$$\tag{5.27}$$

It now seems possible that this consists of the conversion of ethanol to acetate plus molecular hydrogen by one organism, with the subsequent utilization of the hydrogen by a second bacterium for reduction of carbon dioxide. In any case, tracer studies have shown that all the methane is derived from CO_2, which must be supplied in order to permit ethanol utilization. An unusual situation is presented in the case of methane formation from carbon monoxide by *M. barkeri.* This appears to consist of an initial oxidative hydrolytic reaction in which carbon monoxide is converted to a mixture of carbon dioxide and hydrogen, followed by a reduction of the CO_2 produced:

$$4CO + 4H_2O \longrightarrow 4CO_2 + 4H_2 \tag{5.28}$$

$$\underline{CO_2 + 4H_2 \longrightarrow CH_4 + 2H_2O} \tag{5.29}$$

$$\text{sum: } 4CO + 2H_2O \longrightarrow 3CO_2 + CH_4 \tag{5.30}$$

Methane is also produced in the fermentation of some simple organic acids, including formate, acetate, propionate, and benzoate. The fermentation of *n*-butyrate by *Methanobacterium suboxydans*, like that of "*M. omelianski*" with ethanol, requires a supply of CO_2 which is reduced to methane as the butyrate is converted to acetate:

$$2CH_3CH_2CH_2COOH + 2H_2O + CO_2 \longrightarrow 4CH_3COOH + CH_4 \tag{5.31}$$

In some cases CO_2 need not be supplied, and it could be argued that methane results from the reduction of the CO_2 which is produced in the course of the fermentation. Formate can be utilized, abeit poorly, by *M. ruminantium* and *M. mobilis* in a reaction conforming to the following equation:

$$4HCOOH \longrightarrow CH_4 + 3CO_2 + 2H_2O \qquad (5.32)$$

The utilization of propionate by *M. propionicum* yields a mixture of acetate, CO_2, and methane, the latter probably arising from reduction of CO_2 generated during the attack on propionate:

$$4CH_3CH_2COOH + 2H_2O \longrightarrow 4CH_3COOH + CO_2 + 3CH_4 \quad (5.33)$$

Although the more familiar mechanisms for utilization of aromatic compounds involve ring cleavage via oxygenases utilizing molecular oxygen, benzoate also serves as a substrate for methane production. Enrichment cultures from sewage produce methane from benzoate under stringently anaerobic conditions which preclude operation of O_2 enzymes. Although this process has not been studied with pure cultures, the stoichiometry of product formation estimated by use of ring-labeled ^{14}C-benzoate suggests the following equation:

$$4 \ ^{14}C_6H_6-COO^- + 18H_2O \longrightarrow 15 \ ^{14}CH_4 + 9 \ ^{14}CO_2 + 4CO_2 \qquad (5.34)$$

The reduction of CO_2 *per se* is not the sole mechanism involved in methanogenesis. Methanol is converted to methane by *M. barkeri,* apparently without the intermediate formation of CO_2 and in the absence of exogenous supply of CO_2. In the case of acetate utilization by *Methanococcus mazei* and *M. barkeri,* it was found that the methane produced arises exclusively from the methyl carbon and the carboxyl carbon yields only CO_2:

$$^{14}CH_3COOH \longrightarrow ^{14}CH_4 + CO_2 \qquad (5.35)$$

Although the exact details of the pathway remain vague it seems probable that the actual intermediates in methanogenesis are folic acid and vitamin B_{12} derivatives. Methane bacteria are known to be rich in B_{12}, which accounts for the high levels of the vitamin found in sewage sludge. The final carrier is CoM (unknown structure):

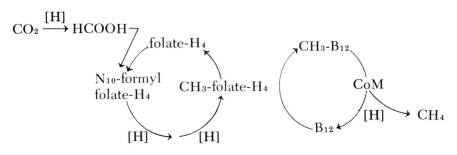

The conversion of the methyl group of acetate to methane presumably would proceed via essentially the same pathway, except that the methyl group would yield a derivative such as methyltetrahydrofolate more directly than would CO_2.

Reduction of other inorganic compounds. In addition to the important nitrate, sulfate, and CO_2 reductions, the bacterial reduction of many other inorganic compounds is known to occur. Although these reactions may not involve specialized groups of organisms as does sulfate reduction and methane production, they may well be of importance in natural cycles of the elements. Furthermore, the ability to carry out specific reductions is a basis for practical identification procedures in diagnostic bacteriology. Recognition of *Staphylococcus aureus* is facilitated by its formation of black colonies on culture media containing tellurite. The black coloration is due to the deposition within the cells of elemental tellurium produced by reduction of tellurite:

$$HTeO_3^- + 4H^+ + 4e^- \longrightarrow Te^0 + 2H_2O + OH^- \tag{5.36}$$

An indication of the wide variety of such reactions is given by the ability of extracts of the bacterium *Veillonella alcalescens,* also called *Micrococcus lactilyticus,* to couple the oxidation of elemental hydrogen with the reduction of over twenty-five inorganic compounds. This organism, which contains hydrogenase, ferredoxin, and vitamin K, but not cytochromes, is a strict anaerobe. Arsenate is reduced to arsenite and vanadate yields vanadyl ion. Uranyl hydroxide is reduced to the tetravalent state, and molybdate to a mixture of reduced molybdenum compounds. Cupric and ferric hydroxides, as well as manganese and lead dioxides, all are reduced to the divalent state. Tellurite, selenite, bismuth hydroxide, bismuthate, and osmium dioxide are readily reduced to the free element, which accumulates in the reaction mixture. In a similar fashion colloidal gold and silver may be produced in small quantities from auric gold and silver phosphate, respectively (Table 5.8).

5.9 | Some Aspects of the Biochemistry of Chemolithotrophy

Since their discovery in 1890 by Winogradsky, chemolithotrophic bacteria, such as nitrifiers, have been regarded with considerable interest. It seemed as though their utilization of inorganic energy substrates such as ammonia or thiosulfate was distinctly "out of the ordinary," and that possibly these forms represented a very primitive physiology. There was a possibility that as such they would employ metabolic processes radically different from "conventional" heterotrophs. This is not the case; our view now is that chemolithotrophs confirm, not confound, the concept of biochemical unity. The earlier literature is replete with proofs of the existence in these bacteria of ATP, NAD, glucose phosphates, phosphoglyceric acid, Krebs' cycle intermediates, amino acids, an array of B vitamins, and, in fact, the whole panoply of intermediary metabolism. Indeed, it seems more reasonable to view such aerobic CO_2-utilizing chemolithotrophs as evolutionarily advanced rather than primitive forms, inasmuch as neither CO_2 nor O_2 is likely to have been present in the primeval earth's environment prior to development of chemoorganotrophic and photosynthetic organisms.

There are four areas of current interest with regard to the biochemistry of chemolithotrophy: (1) pathways employed for the oxidation of primary energy substrates, (2) mechanisms for energy coupling and ATP generation, (3) the CO_2-fixation process and the source of reducing power, and (4) biochemical basis for obligatory chemolithotrophy.

A word of caution perhaps should be extended regarding the subject at hand. These organisms have attracted the attention of many eminent investigators whose biochemical efforts have far overstrained the modest bacteriological underpinnings of the subject. The not insignificant technical problems involved in the study of these organisms are aggravated by a thoroughly muddled taxonomy, the usage of poorly characterized or mixed cultures, and a noticeable penchant for making generalizations based on data from a single strain, coupled with a general reluctance to undertake comparative studies. Possibly this will be resolved as new information continues to be brought forward.

As will be apparent in subsequent discussions, study of the biochemistry of chemolithotrophy has largely been via the classical "solution chemistry" approach. Thus, efforts have been made to resolve multicomponent enzyme systems into their individual members, and to identify stable intermediates in reaction pathways. Although such information is helpful, it may be insufficient to explain the phenomena. Relatively little is known regarding the actual state or *in situ* arrangement of these metabolic systems. In this respect it is to be noted that extensive lamellar arrays have been observed in cells of nitrifying bacteria and thiobacilli, the significance of which is as yet unappreciated (Fig. 2.23A). One is reminded of similar structural configurations in photosynthetic bacteria and is prompted to ask if there is not a peculiar relationship between the needs of chemo- and photolithotrophy and the presence of such internal membranous arrangements. A further unusual and unexplained feature of possible relevance has been reported for marine nitrifying bacteria, which appear to possess but few fatty acids. In the marine ammonia-oxidizers two fatty acids, palmitic and palmitoleic, account for 96-100% of the total fatty acid content. A wider spectrum of fatty acids is seen in the nitrite-oxidizers, but still in these organisms two to four fatty acids comprise more than 80% of the total. Inasmuch as the bulk of bacterial lipids, and, hence, fatty acids, are present in the membranes, this fatty acid composition may reflect some peculiarities of these membranes relating to their function in chemolithotrophy. In any case, it seems likely that the energy transductions involved in chemolithotrophy do not occur in simple free solution and that more efforts to relate structure and function are required.

Nitrification and the nitrifying bacteria. The conversion of ammonia to nitrate, which is termed "nitrification," is accomplished largely, but not exclusively, by a group of obligately chemolithotrophic bacteria, the nitrifiers (Section 4.17). Although some of the exact details of the process remain obscure, it is clear that nitrification takes place in two primary stages: (1) the oxidation of ammonia to nitrite by bacteria such as *Nitrosomonas* and *Nitrosocystis,* and (2) the subsequent oxidation of nitrite to nitrite by *Nitrobacter* and others.

The oxidation of ammonia to nitrite ($N^{3-} \longrightarrow N^{3+}$) is a thermo-dynamically favorable reaction, and in fact involves a greater free energy change than the anaerobic fermentation of glucose to ethanol and CO_2:

$$NH_4^+ + 3/2O_2 \longrightarrow NO_2^- + 2H^+ + H_2O \qquad (5.37)$$

It has been suggested that ammonia oxidation requires a preliminary activation of ammonia, possibly involving a high energy phosphate or adenylate derivative. This is based on energetic considerations as well as certain experimental observations. It seems likely that the pathway proceeds through several intermediates, at least one of which is at an oxidation state equivalent to hydroxylamine, NH_2OH (that is, with a valence of -1). The oxidation of ammonium ion to hydroxylamine is an endergonic reaction ($\Delta G = +4.7$ kcal) and would require an input of energy. This postulated need for ammonia activation is substantiated by the finding that ammonia oxidation by cell-free extracts of *Nitrosocystis oceanus* requires added ATP and Mg^{2+}. These same extracts also catalyze oxidation of hydroxylamine (to nitrite), but in this case no ATP is required. In addition to a requirement for energy, there appears to be a certain amount of structural organization needed for ammonia oxidation by *Nitrosocystis* membrane fragments. Ammonia-oxidizing activity is associated exclusively with a large particle fraction, whereas hydroxylamine may be oxidized both by the large particles and a small particle fraction. Cell-free extracts of *Nitrosomonas* failed to oxidize ammonia in earlier studies, although hydroxylamine was oxidized by particulate fractions containing cytochromes of the *b, c,* and *a*-types. Inhibition of hydroxylamine oxidation by quinacrine would tend to implicate participation of a flavoprotein as well as cytochromes. It seems likely that the "soluble" hydroxylamine-cytochrome *c* reductase preparations which have been reported are actually derived from membrane fragments during sonic disruption. It is possible, of course, that multiple forms of the enzyme could exist. In any case electron flow may be assumed to proceed from hydroxylamine via a flavin to the cytochrome chain and ultimately to oxygen by means of a cytochrome oxidase. Although there seems good reason to presume that hydroxylamine is an intermediate, identity of the other intermediates remains in doubt. It has been argued that the process consists of a series of one-electron changes, and may involve free radical-like intermediates which remain enzyme-bound.

These are shown in parentheses in the following hypothetical pathway:

$$NH_4^+ + energy \longrightarrow (NH_2) \longrightarrow NH_2OH \longrightarrow (NHOH) \searrow$$
$$NO_2^- \longleftarrow NO \longleftarrow (NOH)$$

In *Nitrobacter* the energy-yielding process appears to consist of a single step oxidation of nitrite to nitrate:

$$NO_2^- + H_2O \longrightarrow NO_3^- + 2H^+ + 2e^- \qquad \Delta G = -17 \text{ kcal} \qquad (5.38)$$

The electrons released in this oxidation probably are transferred directly to a cytochrome of the *c*-type without mediation of NAD. A respiratory particle fraction containing the nitrite-oxidizing enzyme as well as cytochromes and cytochrome oxidase activity has been prepared from *Nitrobacter*. It has been pointed out that the failure of these cell-free preparations to reduce NAD is a natural consequence of the high standard electrode potential of the NO_2^-/NO_3^- couple at pH 7 ($E_0' = +0.54$ v) compared to that for NAD/NADH ($E_0' = -0.27$ v). Thus, NAD would not be expected to participate as a cofactor in nitrite oxidation. But, in fact, the lack of involvement of NAD in primary oxidations of chemolithotrophs is of wider generality, and is not simply a peculiarity of the nitrite oxidation.

Oxidation of reduced sulfur compounds by thiobacilli. A variety of reduced sulfur compounds are oxidized by species of the genus *Thiobacillus*. They utilize as energy-yielding substrate: sulfide, thiosulfate, thiocyanate, elemental sulfur, and tetrathionate (Section 3.16). Some species, for example, the *T. thioparus* group, are sensitive to the acidity resulting from these oxidations and may be inactivated before the substrates are utilized to completion, whereas the *T. thiooxidans* group is extremely resistant to acid. Some thiobacilli, including *T. thioparus* and *T. thiooxidans*, are obligate chemolithotrophs and will not grow on conventional peptone media, whereas others, such as *T. novellus* and *T. intermedius,* are facultative chemolithotrophs and will grow either with organic energy substrates or with sulfur compounds.

Sulfite probably occupies a position as the key intermediate in all these oxidations (see Fig. 5.9), although there is disagreement over the pathways actually employed. Contributing to this uncertainty is the intricacy of sulfur chemistry and the possibility of different pathways in different species.

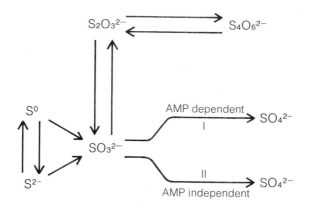

Fig. 5.9. Oxidation of sulfur compounds by thiobacilli. Sulfide (S^{2-}), elemental sulfur (S^0) and thiosulfate ($S_2O_3^{2-}$) are oxidized via sulfite (SO_3^{2-}) as a common intermediate. Possibly another reaction pathway (II) exists for sulfite oxidation in addition to the AMP-dependent APS-reductase pathway (I).

Most thiobacilli can utilize thiosulfate, $S_2O_3^{2-}$, as a growth substrate, and they convert this largely to sulfate, although frequently there is also a substantial, but variable, accumulation of elemental sulfur and tetrathionate in the culture fluid. The amounts of these products apparently depend on cultural conditions, such as pH, aeration, and substrate concentration, and also on the bacterial strain employed. Thus, *T. neapolitanus* may be distinguished from *T. thioparus* by the former's production of tetrathionate and a somewhat greater acidity during growth on thiosulfate.

On the basis of this formation of tetrathionate during growth, and the ability of some strains to oxidize added tetrathionate, it has been proposed that tetrathionate participates as a primary intermediate in the main pathway for thiosulfate metabolism:

$$2S_2O_3^{2-} \longrightarrow S_4O_6^{2-} \longrightarrow SO_3^{2-} + S_3O_6^{2-} \longrightarrow 4SO_3^{2-} \longrightarrow 4SO_4^{2-}$$

Presently there is little direct enzymological evidence for this pathway, and studies on *T. thioparus* indicate that neither elemental sulfur nor tetrathionate is actually intermediate in that species. Instead, the initial reaction in *T. thioparus* is thought to be a reductive dissimilation of thiosulfate to yield one of the sulfur atoms at the level of sulfide and the other at a level equivalent to sulfite. The two sulfur atoms of thiosulfate differ in valency; the "outer" sulfur is divalent, whereas the "inner" sulfur which bears the oxygens is hexavalent. The outer divalent sulfur atom is converted to sulfide, and thence to elemental sulfur, while the inner hexavalent sulfur atom yields sulfite:

$$2\overset{*}{S}\text{-}SO_3{}^{2-} + 4H^+ + 4e^- \xrightarrow{\underset{\text{reductase}}{\text{thiosulfate}}} 2SO_3{}^{2-} + 2H_2\overset{*}{S} \qquad (5.39)$$

$$2H_2\overset{*}{S} + O_2 \xrightarrow{\text{sulfide oxidase}} 2\overset{*}{S^0} + 2H_2O \qquad (5.40)$$

This would account for the observation that when cultures were supplied with thiosulfate labeled with ^{35}S in the outer S sulfur atom, radioactivity was preferentially converted into elemental sulfur.

It seems probable that in all these transformations a crucial role is played by an enzyme-bound thiol or dithiol, which in extracts is replaceable by glutathione. Related to this is the interconversion of sulfide and elemental sulfur, which can be accomplished via a reaction with glutathione (GSSG, oxidized glutathione; GSH, reduced glutathione):

$$H_2S + GSSG \longrightarrow S^0 + 2GSH \qquad (5.41)$$

An oxidation of this sort would generate the reducing power needed for the thiosulfate reductase reaction.

The best known pathway for the further oxidation of sulfite in thiobacilli is via an AMP-dependent system involving the enzyme APS-reductase:

$$2SO_3{}^{2-} + 2AMP \xrightarrow{\text{APS-reductase}} 2APS + 4e^- \qquad (5.42)$$

In this reaction sulfite is oxidized to the sulfate stage and at the same time a high-energy phosphosulfate anhydride bond is formed, linking the sulfate and AMP to yield adenosine phosphosulfate (APS). This same enzyme operating in the reverse direction is involved in dissimilatory sulfate reduction in the sulfate-reducing bacterium *Desulfovibrio*. The formation of a high energy compound such as APS makes possible generation of an ATP via the operation of two additional enzymes also present in thiobacilli:

$$2APS + 2PO_4{}^{3-} \xrightarrow{\text{ADP-sulfurylase}} 2ADP + 2SO_4{}^{2-} \qquad (5.43)$$

$$2ADP \xrightarrow{\text{adenylate kinase}} AMP + ATP \qquad (5.44)$$

This reaction sequence, with the participation of a nucleotide intermediate, accounts for the observed stoichiometry of incorporation of [18]O from inorganic $P^{18}O_4{}^{3-}$ into sulfate. When bacteria are supplied with ^{18}O-labeled orthophosphate along with thiosulfate, some 22-23%

of the ^{18}O of the phosphate is found in the sulfate which is produced. This is very close to the 25% which would be predicted by the operation of the APS-reductase pathway.

The evidence for operation of the APS pathway, at least in $T.$ $thioparus,$ is quite convincing. However, it is by no means certain that this is the sole mechanism involved in the oxidation of sulfite. The single substrate-level phosphorylation which takes place in the APS pathway seems insufficient to account for the growth yields observed. More directly, extracts have been prepared which catalyze an AMP-independent oxidation of sulfite which is linked to the reduction of cytochrome c. The thiobacilli contain a complete and functional electron transport chain including cytochromes, quinones, and flavins, and when whole organisms are supplied with thiosulfate there is a rapid reduction of cytochromes.

The usual in $vitro$ assay for sulfite oxidation via the APS-reductase reaction requires the addition of ferricyanide to serve as the electron acceptor. The natural acceptor is unknown, but none of the following will substitute for ferricyanide: methylene blue, 2,6-dichlorophenolindophenol, benzylviologen, phenazine methosulfate, FMN, FAD, NAD, NADP, ox heart cytochrome c, cytochrome c_3, or ferredoxin. Probably the unknown natural acceptor donates the electrons to the bacterial cytochrome chain and ultimately to oxygen. An interaction with the respiratory chain at the cytochrome c level should permit the formation of two ATP molecules per electron pair, although firm data are lacking regarding the actual P/O ratios.

Some thiobacilli, such as $T.$ $thiooxidans,$ $T.$ $concretivorus,$ and $T.$ $thioparus$ carry on an aerobic oxidation of elemental sulfur to yield sulfate, a reaction which involves a free-energy change of approximately 118 kcal (see Table 5.8).

It has been reported that cell-free extracts of $T.$ $thiooxidans$ catalyze a glutathione-dependent oxygenation of elemental sulfur by O_2 to yield sulfite. Oxidation of sulfur by these extracts in the presence of $^{18}O_2$ resulted in the formation of thiosulfate with an enriched ^{18}O content. Later evidence indicated that the immediate product was sulfite, and that thiosulfate actually arises via a secondary nonenzymatic reaction of sulfite and elemental sulfur:

$$S^0 + O_2 + H_2O \xrightarrow[\text{oxygenase + GSH}]{} H_2SO_3 \tag{5.45}$$

$$H_2SO_3 + S^0 \xrightarrow[\text{non-enzymatic}]{} H_2S_2O_3 \tag{5.46}$$

These results were taken to mean that the initial attack on sulfur is catalyzed by an oxygenase, which from inhibitor studies with chelating agents is postulated to contain a non-heme iron cofactor. Although the operation of such an oxygenase is feasible in the aerobic species, it would seem incompatible with the reported oxidation of sulfur by *T. denitrificans* growing with nitrate as the final electron acceptor. The oxidation of elemental sulfur by membrane fragments in the absence of GSH also has been reported for another strain of *T. thiooxidans*. Presumably in this case some other enzyme-bound thiol was operative instead of GSH. Oxygen uptake by whole organisms and membrane fragments is completely inhibited by cyanide and azide, which would indicate involvement of the cytochrome system. This likelihood is strengthened by data indicating the essentiality of the ubiquinone-40 component of the electron transport chain for sulfur oxidation.

Growth on sulfide has been reported for some thiobacilli, although this situation is complicated by the possibility that elemental sulfur might be produced nonbiologically in such sulfide solutions. It has been shown that cell-free extracts will oxidize sulfide to thiosulfate, but this activity could be nonenzymatic in view of its considerable heat stability. It is thought that both sulfide and sulfur are metabolized via interaction with an enzyme-bound thiol, followed by oxidation to sulfite:

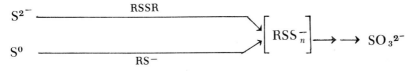

However, it must be stressed that the enzymes involved have not been sufficiently purified and characterized, so that their precise nature and composition is not known with certainty.

One problem not resolved by such formulations based on soluble intracellular enzyme systems, or even on those located in the membrane itself, is that elemental sulfur is not a water-soluble substrate. The manner in which a large particle of elemental sulfur can be attacked by membrane-bound or intracellular enzymes, despite the not inconsiderable barrier presented by the cell wall, has not been satisfactorily explained. No evidence has been adduced for exoenzymes, nor excreted thiols, which could be involved in the mobilization and transport of elemental sulfur. It has been contended that the cells must be permitted direct contact with the sulfur particle, with such contact being aided by phospholipids which increase the hydrophilic properties of the "flowers of sulfur" generally employed as substrate. However, the exact mechanism for sulfur penetration into the cell remains elusive.

Certain acid-tolerant thiobacilli are able to use the oxidation of ferrous iron as an alternative energy source. With respect to usable energy sources it is obvious that three conditions would be possible, and bacteria growing under all three have been reported to occur, that is, some strains can utilize both sulfur compounds and iron, some only sulfur compounds, and some only iron. At least certain cultures can oxidize both iron and sulfur simultaneously, which would be of importance in the bacterial attack on sulfide ore stuffs such as pyrite, marcasite, and chalcopyrite. These organisms are relatively insensitive to the concentration of iron, high concentrations of which must be supplied to achieve even meager growth. This is in accord with the low free-energy yield of the reaction ($\Delta G = -11$ kcal), which involves probably the smallest free energy change for the primary energy-yielding reaction used by any organism (see Table 5.7). The enzyme system for this oxidation, which has been reported variously as constitutive or inducible, couples the oxidation of ferrous iron to the cytochrome chain, probably at the level of cytochrome c.

Metabolism of molecular hydrogen. From a chemical standpoint an obvious source of energy from an inorganic reaction would be the oxidation of molecular hydrogen (Section 3.12). The combustion of hydrogen by O_2 is strongly exergonic and proceeds spontaneously at room temperature (ca. 20° C). In view of the well-known importance of dehydrogenations and hydrogen transport in heterotrophic energy metabolism, one would predict that enzymic mechanisms for dealing with H_2 also should have evolved. This has been realized in the genus *Hydrogenomonas*, which consists of several species of facultative chemolithotrophs capable of obtaining energy from the oxidation of H_2 and of assimilating CO_2 as the sole carbon compound. Alternatively, these organisms grow as conventional heterotrophs on organic substrates. The ability to metabolize molecular hydrogen is not limited to these chemolithotrophs, but rather is fairly widespread among different groups of bacteria, including autotrophs and heterotrophs, aerobes and anaerobes.

Reactions involving H_2 are of two types, consisting of either its consumption or its evolution. In both cases the activity depends on the enzyme hydrogenase, which catalyzes a reversible reaction of the type:

$$H_2 \rightleftarrows \underset{\text{hydrogenase}}{} 2H^+ + 2e^-$$ (5.47)

where one or more of the protons probably exists as an enzyme-hydride of some sort (E-H or H-E-H). Hydrogenase activity has been observed in a variety of bacteria including *Hydrogenomonas;* the strictly anaerobic sulfate-reducer, *Desulfovibrio;* the purple and green photosynthetic bacteria (Section 5.10); certain strict anaerobes such as *Clostridium, Peptostreptococcus,* and *Veillonella alcalescens;* certain facultative anaerobes as the coliforms; and the strongly aerobic nitrogen-fixing bacterium, *Azotobacter.*

Operating in the right-hand direction, the hydrogenase reaction could furnish reducing power for cellular reductions via coupling with appropriate hydrogen carriers. In the opposite direction, the hydrogenase reaction provides a mechanism for anaerobes to relieve the "hydrogen pressure" arising from the dehydrogenation of organic substrates in the absence of O_2 or other suitable hydrogen acceptor. Because of the low oxidation-reduction potential ($E_0' = -0.420$ v) of the hydrogenase reaction, any hydrogen carriers interacting with hydrogenase also should possess a low redox potential. This requirement is satisfied in the case of ferredoxin and related substances such as flavodoxin found in *Clostridium,* and also by the cytochrome c_3 of *Desulfovibrio* ($E_0' = -0.205$ v). Artificial carriers such as benzylviologen ($E_0' = -0.359$ v) and methylviologen ($E_0' = -0.446$ v) are useful for *in vitro* assays of hydrogenase. Since the potential of methylviologen is more negative than that of hydrogenase, incubation of reduced methylviologen with this enzyme results in the evolution of H_2, which can be measured conveniently by manometry.

Evolution of hydrogen occurs during certain fermentation reactions, and also during photosynthesis by *Rhodospirillum rubrum* under particular conditions. In fermentations by saccharolytic *Clostridium* species the hydrogen evolved arises via the "phosphoroclastic split" of the pyruvate which is produced during glycolysis:

$$\text{Pyruvate} + PO_4^{3-} \rightleftarrows \text{acetylphosphate} + CO_2 + H_2 \quad (5.48)$$

Clostridial hydrogenase activity is closely associated with several other oxidation-reduction cofactors including a flavin (FAD, FMN), ferredoxin, and possibly another, unidentified substance. Undoubtedly the electrons and protons from the oxidative cleavage of pyruvate are passed via such intermediate carriers to hydrogenase where they are combined to form molecular hydrogen. In certain of these bacteria the hydrogenase reaction permits the utilization of H_2 for the reduction of some particular metabolite. For example, *Clostridium kluyveri* is able to reduce acetaldehyde to ethanol using hydrogen:

$$H_2 + \text{acetaldehyde} \xrightarrow{\text{Cl. kluyveri}} \text{ethanol} \qquad (5.49)$$

In all these cases utilization of hydrogen does not involve participation of molecular oxygen. However, there are also bacteria, sometimes designated "Knallgasbakterien," which are able to catalyze the oxyhydrogen reaction:

$$H_2 + \tfrac{1}{2}O_2 \longrightarrow H_2O \qquad \Delta G^0 = -57.4 \text{ kcal} \qquad (5.50)$$

Although this capability has been reported among certain gram-positive sporeforming and nonsporeforming bacilli, and possibly in some acid-fast mycobacteria as well, the best known of such hydrogen oxidizers are the gram-negative pseudomonads of the genus *Hydrogenomonas*. Some taxonomists maintain that separate generic rank on the basis of H_2 utilization is unwarranted and that these organisms should be regarded simply as species of the genus *Pseudomonas*. It is interesting that one of these bacteria, investigated extensively as *H. eutropha,* now has been found to exhibit peritrichous, not polar, flagellation, and accordingly has been reclassified as *Alcaligenes eutrophus.*

These bacteria are able to grow as heterotrophs on organic substrates such as lactate, fructose, or amino acids. In addition, they can grow as chemolithotrophs using the energy derived from the oxidation of hydrogen to drive the chemosynthesis of cellular material from carbon dioxide. Thirty years ago it was claimed that *Hydrogenomas* is able simultaneously to utilize both an organic substrate and hydrogen as energy sources (mixotrophic conditions), and this now has been verified in several laboratories. Although these organisms are strict aerobes, several species require somewhat reduced oxygen tensions for growth. The first studied species (*H. flava, H. vitrea,* and *H. rhulandii*) were reported to require 10% or less O_2, and, when cultivated on organic substrates under air, to rather quickly lose permanently the capability for chemolithotrophic growth on hydrogen. Subsequently, other species (*H. facilus, A. eutrophus*) were found which were not sensitive to 20–25% O_2 and which were genetically more stable. These retain the potential for chemolithotrophy even when grown heterotrophically under air. The extensive growth shown by these organisms, especially *A. eutrophus,* under chemolithotrophic conditions makes them particularly favorable objects for biochemical studies.

In essence the oxidation of hydrogen by *Hydrogenomonas* consists of the activation of H_2 by the enzyme hydrogenase, followed by a series of hydrogen transfer and electron transfer reactions leading ultimately to a cytochrome oxidase where protons and electrons are combined with O_2 to give water:

$$H_2 \longrightarrow X \longrightarrow \text{flavoprotein} \longrightarrow \text{cytochrome system} \longrightarrow O_2$$
$$\downarrow$$
$$NAD$$

In this scheme X represents one or more intermediate carriers. One major biochemical problem is the identification of these carriers, especially the one initially interacting with hydrogenase. Although there has been no demonstration of ferredoxin in *Hydrogenomonas,* the initial H-acceptor might be expected to have somewhat similar properties. In this regard, the oxygen sensitivity of some *Hydrogenomonas* is of interest, as ferredoxin is known to be inactivated by oxygen. It seems probable that X does not include NAD or NADP.

The study of hydrogenase has been hampered by the fact that in many cases hydrogenase activity cannot be dissociated from the membrane. The occurrence in some extracts of both particulate and "soluble" hydrogenases, often with differing acceptor specificities, has been taken to mean that there are multiple hydrogenase enzymes in these organisms. However, it is possible that the soluble enzyme is an artifact resulting from its release from the membrane during cell breakage. Differences in acceptor specificity then might reflect the presence or absence of natural acceptors in the enzyme complex released during this breakage.

It is found that hydrogenase preparations from *Hydrogenomonas* generally will catalyze a reduction of cytochrome *c*, as well as of artificial electron acceptors such as methylene blue or ferricyanide. Despite the favorable standard electrode potential of the hydrogenase reaction, NAD and NADP usually are not directly reduced by such preparations. A reduction of NAD can be accomplished, but only secondarily via a flavin-dependent step. A hydrogenase activity linked directly to NAD has been reported in *H. rhulandii,* and in *Pseudomonas saccharophila,* which despite its taxonomic assignment grows chemolithotrophically with hydrogen. The true significance of the NAD-linked enzyme, which has been termed a "hydrogen dehydrogenase," remains uncertain, as it does not appear to occur in other species of *Hydrogenomonas*.

Energy coupling and CO₂ fixation in chemolithotrophs. In the foregoing discussion of substrate oxidations by chemolithotrophs it was implicit, if not always explicit, that the free energy released was trapped in some way and used subsequently for biosynthesis and other energy-requiring reactions. The energy-trapping system in chemolithotrophy undoubtedly involves ATP as the major, if not the exclusive, high-energy product. It is well known from studies with chemoorganotrophs that ATP can be generated by either substrate-level phosphorylation and ~P group transfer, or via cytochrome-linked oxidative phosphorylation mechanisms. With respect to chemolithotrophs, only in the case of the APS-reductase reaction for sulfite oxidation by *T. thioparus* has a substrate-level phosphorylation been satisfactorily documented. Presumably the ATP needs of these organisms are met by coupling substrate oxidation to the electron transfer system which permits oxidative phosphorylation. Evidence for the occurrence of oxidative phosphorylation during chemolithotrophic oxidations, although largely indirect, has been accumulating and is now very suggestive, if not conclusive.

Addition of the energy substrate to whole cells of various genera (for example, *Thiobacillus, Hydrogenomonas*) results in a rapid reduction of endogenous cytochromes as revealed by low-temperature-difference spectra. It was shown that in intact cells of *T. thioparus*, 2,4-dinitrophenol (DNP) abolished both CO₂ fixation and ATP formation. Cell-free extracts of *Nitrobacter agilis, Thiobacillus neapolitanus* and a *Hydrogenomonas* species show a small but definite esterification of orthophosphate during electron transport, achieving P/O ratio of approximately 0.5. In the case of the latter organism phosphorylation was uncoupled by standard uncoupling agents such as DNP, dicumarol, and carbonyl cyanide *m*-chlorophenyl hydrazone (CCCP), although, in fact, the system was not particularly sensitive to DNP.

The lack of participation of NAD as the proximate electron and hydrogen acceptor in these oxidations has already been mentioned. If the substrate oxidation does not yield NADH, one is constrained to ask what is the source of reducing power used for CO₂ fixation? It is known from studies with mitochondria that it is possible to obtain reduced NAD by means of a reversed electron transport which is

ATP-dependent. Thus, although the substrate oxidation in chemolithotrophs is linked directly to the electron transport system, the ATP resulting from the accompanying oxidative phosphorylations could be used in part to drive a reversed electron flow to obtain the NADH needed for CO_2 fixation reactions. Such an ATP-dependent reduction of NAD now has been demonstrated in a number of chemolithotrophic bacteria including *Nitrobacter, Nitrosomonas, Ferrobacillus,* and *Thiobacillus.*

Carbon dioxide fixation in the chemolithotrophic bacteria involves essentially the same reductive pentose phosphate pathway, or Calvin cycle, employed by photosynthetic organisms (Section 5.10 discusses the Calvin cycle). Evidence for operation of this pathway, consisting of the direct demonstration of ribulose diphosphate carboxylase (carboxydismutase) activity in extracts, or chromatographic analysis showing short-term $^{14}CO_2$ fixation primarily into 3-phosphoglyceric acid (3-PGA), has been presented for *Thiobacillus thioparus, T. thiooxidans, T. novellus, T. ferrooxidans, Hydrogenomonas* species, and *Nitrobacter agilis.* One point of difference between the CO_2 fixation process in chemolithotrophic bacteria and the classical Calvin cycle of green plants is the utilization of NADH instead of NADPH for the reductions involved.

An interesting feature of this system, in the thiobacilli at least, is the inhibition by AMP of $^{14}CO_2$ fixation via ribulose diphosphate carboxylase. This may serve as a control mechanism to prevent exhaustion of ATP by continued CO_2 fixation. Should the generation of ATP become impaired, conversion of ATP to AMP in chemosynthesis would tend to close off the ATP drain to CO_2 fixation.

In addition, CO_2 may be fixed by means of the phosphoenolpyruvate carboxylase reaction in at least some of these organisms. The presence of PEP-carboxylase has been reported for *T. thiooxidans, T. novellus,* and *Hydrogenomonas facilis.* This enzyme catalyzes the addition of CO_2 to PEP to yield oxaloacetate:

$$^{14}CO_2 \searrow \quad \begin{array}{c} COOH \\ | \\ C-O\sim P \\ || \\ CHOH \end{array} \xrightarrow[\text{carboxylase}]{\text{PEP}} \begin{array}{c} COOH \\ | \\ C=O \\ | \\ CH_2 \\ | \\ ^{14}COOH \end{array} + PO_4^{3-} \quad (5.51)$$

phosphoenol
pyruvate (PEP)   oxaloacetate

The significance of this pathway for CO_2 fixation during chemolithotrophic growth is still unclear, but quantitatively the Calvin cycle is the more important. In whole organisms short-term exposure to $^{14}CO_2$ for two seconds results in very heavy labeling (greater than 80%) in 3-PGA, and only about 8% of the CO_2 fixed appears in aspartic acid which is derived from the oxaloacetate formed in the PEP carboxylase reaction. Another indication of the more intimate connection between chemolithotrophy and the Calvin cycle is the finding that in the facultative chemolithotrophic *Thiobacillus novellus* growth on organic substrates strongly represses synthesis of ribulose diphosphate carboxylase, whereas the level of PEP-carboxylase remains the same. Furthermore, the PEP-carboxylase of this organism does not exhibit the inhibition by AMP which is seen with the ribulose diphosphate carboxylase.

The biochemical basis of obligate chemolithotrophy. The existence of facultative chemolithotrophs such as *Thiobacillus novellus* and the species of *Hydrogenomonas,* and especially the ability of the latter to utilize simultaneously both hydrogen and an organic compound for energy (mixotrophy), emphasizes that there is no fundamental incompatability or mutual exclusiveness between these two processes. There are, however, certain organisms, such as *T. thioparus* and *T. thiooxidans,* which cannot grow on conventional organic media under the usual conditions, and hence have been considered strict or obligate chemolithotrophs. Are these in fact obligately chemolithotrophic by definition, which is to say absolutely incapable of utilizing organic compounds as an energy source? Or do they fail to use organic substrates only by virtue of a particular experimental regime? If there is, in fact, obligate chemolithotrophy, what accounts for this property?

It has been shown repeatedly that these bacteria are not completely impermeable to exogenous organic compounds, small quantities of which are taken into the organism and assimilated into cell substance. This does not mean, however, that organic substrates are used for energy. There have been reports of the growth of *T. thiooxidans* and similar "obligate chemolithotrophs" on glucose in the absence of the inorganic energy substrate. It seems reasonable to reserve judgement on these claims until certain technical objections are satisfied. There are numerous examples to indicate that exclusive reliance on liquid media without adequate cloning on solid medium may mask the existence of mixed populations. The classical methanogenic "organism"

Methanobacterium omelianskii (Section 5.8) was studied for some decades before it was shown to be a mixture. In the case of the thiobacilli and ferrobacilli, it is now known that there exist in the same habitats other acidophilic chemoorganotrophic bacteria, some of which also are incapable of growth on nutrient broth. In addition, the complete removal of elemental sulfur from the inoculum is a difficult problem, and substrate carryover is likely to occur. All claims for the growth of "obligate chemolithotrophs" on organic substrates should demonstrate that this is a general property of all the organisms in the culture, and not the result of the outgrowth of a few mutants or contaminants. Furthermore, it must be shown that it is possible to make serial transfers of the culture under such conditions, and that there is a net increase in total cell mass sufficient to rule out nutrient carryover or simple residual growth due to endogenous stored substances (for example, poly-β-hydroxybutyrate).

Explanations for obligate chemolithotrophy have taken two positions. One explanation holds that such bacteria lack one or another enzyme (or enzyme system) necessary for utilization of organic energy substrates. It has been reported that in certain bacteria such as *T. thioparus* and *T. thiooxidans,* NADH-oxidase activity was absent or practically so. Without ability to couple the oxidation of organic substrates to the electron transport system, growth obviously would be impaired, although possibly not prevented. However, other studies have indicated that NADH-oxidase activity, in fact, could be demonstrated in other strains of these same two species, and in various other chemolithotrophs as well. One is forced to conclude that lack of NADH-oxidase cannot be offered as a general explanation for obligate chemolithotrophy. A somewhat similar argument has been presented with respect to the tricarboxylic acid cycle: α-ketoglutarate dehydrogenase was undemonstrable in extracts of *T. thioparus* and several other thiobacilli. Whether this metabolic defect is universal among obligate chemolithotrophs remains to be seen.

A second type of explanation involves an inhibitory effect of the organic substrate or its metabolic products. The older concept of a generalized toxicity of organic matter for chemolithotrophs has not been substantiated. The inhibitory effects of organic substances are quite specific with respect to which compounds and concentrations produce effects on given organisms, and, as previously mentioned, many organic compounds may be incorporated into cell substance. High levels of peptones may on occasion be inhibitory to certain organisms such as the nitrifiers, although the explanation of this action is unknown. Supplying single amino acids may produce in the chemolithotrophs the same metabolic imbalances as observed in chemoorganotrophs. Thus, when valine alone is supplied to *E. coli* the initial enzymes in the branched pathway leading to valine and isoleucine are inhibited, which results in an isoleucine deficiency and prevents growth. The inhibitory effect of valine is overcome by supplying isoleucine in the medium.

Intermediates of the tricarboxylic acid cycle often are found to inhibit particular chemolithotrophs. The inhibition of *T. thiooxidans* by oxaloacetate (or pyruvate) has been claimed to be the basis of its inability to grow on glucose under the usual conditions. Glucose is reportedly metabolized by this organism with accumulation of oxaloacetate in sufficient concentrations to prevent continued growth. Removal of the oxaloacetate via a dialysis system is reported to permit growth of *T. thiooxidans* on glucose.

In sum, there is as yet no satisfactory general explanation for obligate chemolithotrophy. It could be that such a unitary hypothesis is not possible. Inasmuch as these organisms are of uncertain phylogenetic relationship, there is no necessity to suppose that a common metabolic lesion exists in all strains.

5.10 | Energy Transduction in Bacterial Systems

Our discussion of energy metabolism has centered primarily on some of the pathways and mechanisms involved in the generation of chemical bond energy, chiefly in the form of ATP, by means of the oxidation of other chemical compounds. The processes of oxidative phosphorylation and substrate-level phosphorylation represent simply the conversion of chemical energy to a more immediately useful form of chemical energy. Although this is doubtless an important consideration, it may be useful to suggest that the bacterium can be viewed as a kind of energy transducer. Like nonliving transducers, it can receive energy in a particular form from one system, and supply energy in a different form to a second system. Several examples of such energy-transducing systems may be cited, of which none is exclusively limited to bacteria: biothermogenesis (chemical to thermal energy), motility (chemical to kinetic energy), bioluminesence (chemical to radiant energy), and photosynthesis (radiant to chemical energy).

The release of energy as heat during chemical reactions is, of course, well known. Calorimetric measurements of heats of combustion and of other reactions form the basis of thermochemical analysis and chemical thermodynamics. Similar analyses of the reactions catalyzed by bacteria and other organisms can be made, and the development of sensitive microcalorimeters has stimulated a revival of interest in this approach to bioenergetics.

Ordinarily the heat evolved in these biochemical reactions is dissipated to the surrounding environment and goes unnoticed. But under some conditions where heat exchange is retarded, the temperature can be increased dramatically as a result of bacterial action (Section 4.9). Growth and metabolism of bacteria and fungi in hay stacks, straw piles, and manure heaps may raise the temperature to 60°–70° C. This is the explanation for the older method of preparing hot beds for starting seedlings. In order to prevent freezing in the early spring, the frames in which the seeds had been planted were heaped with fresh manure. The heat generated by microbial reactions in the manure kept the bed warm.

The "spontaneous combustion" of haystacks is not a rare occurrence, and actually may be related to bacterial thermogenesis. Once the temperature has been elevated by metabolic reactions of bacteria, purely nonbiological chemical processes may be accelerated to such an extent that "spontaneous combustion" results. It has been suggested that flammable bacterial fermentation products such as hydrogen, methane, and ethanol contribute to the final ignition in such situations.

Several types of mechanical work are performed by living organisms and cells. Among these biomechanical processes one might mention muscle contraction, amoeboid movement, motion of flagella and cilia, cytoplasmic streaming, chromosome migrations, and cellular constriction during division. In principle, there is no difficulty with the concept of harnessing chemical energy to do mechanical work. In biological systems the difficulty lies in understanding the detailed molecular process by which this is accomplished. Most explanations invoke energy-linked alterations of protein structure, such as aggregation of subunits or conformational changes, which result in a change in length. In the case of bacteria the action of flagella is the clearest example of the transduction of chemical energy into mechanical work (Section 2.5). However, the explication of bacterial motility has scarcely been accomplished in gross mechanical terms, much less at the molecular level.

5.11 | Bacterial Luminescence

The conversion of chemical to radiant energy is a striking example of the process of energy transduction. The widespread occurrence of bioluminescent systems in organisms of diverse phylogenetic origin attests to its utility (Section 3.18). Most of the biochemical studies of bacterial luminescence have been conducted with *Photobacterium fischeri* and *Photobacterium phosphoreum*. Although *P. fischeri* is a facultative anaerobe, the bioluminescent reaction itself is dependent on the presence of O_2. The light evolving system is quite active in these organisms, and under optimum conditions the emission rate has been calculated to be approximately 10^4 protons per second per cell. If it is assumed that the emission of one photon requires one molecule of O_2, then the energy flux involved in bioluminescence would account for about 1% of the oxygen consumed by the cell.

The general term used to denote the substance which emits the light is *luciferin*, and the enzyme catalyzing the reaction is referred to as *luciferase*. The reaction involves the induction of an excited state in the luciferin molecule. When the excited molecule (L*) returns to the ground state, energy is released as a photon:

$$L \xrightarrow{\text{luciferase}} L^* \qquad (5.52)$$

$$L^* \longrightarrow L + h\nu \qquad (5.53)$$

Studies of the light-yielding reaction in cell-free extracts indicate that bacterial bioluminescence differs considerably from the better known firefly system. Bacterial luciferin has not been identified conclusively, although it is suggested that it is reduced flavin mononucleotide ($FMNH_2$) or a reaction product of $FMNH_2$. Although no other suitable candidate has been found for this function, the wave length of the emitted light (490 nm) is rather far removed from the known fluorescence maximum of $FMNH_2$. The bacterial luciferase has been partially purified and characterized. It is found to be a protein whose molecular weight is approximately 60,000. It appears to be comprised of two nonidentical subunits, an alpha and a beta polypeptide chain. Although the bacterial luciferase contains some eight thiol groups, there are no disulfide linkages, metal cofactors, or organic coenzymes. This enzyme is present in rather large amounts, constituting as much as 2–5% of the soluble proteins of the cell.

The dim light given off by cell-free extracts is stimulated by addition of $FMNH_2$, or in crude systems NADH + FMN. Obviously electron flow from oxidizable substrates can be directed through NAD to supply reduced FMN for bioluminescence:

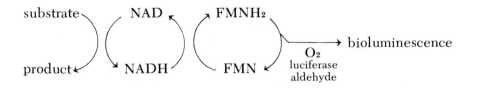

The light yield in such reactions is greatly enhanced by long chain aldehydes, such as decanal, the precise function of which needs to be clarified. One idea was that the oxidation of such aldehydes might supply energy needed to permit formation of an excited state, but no firm evidence has been adduced for the oxidation of aldehyde in purified preparations, nor for the occurrence of such compounds in the organism. Another suggestion is that the aldehyde acts not as a substrate, but rather it acts to introduce some conformational change in the luciferase.

It is apparent that in marine animals and insects luminescence may serve a postive biological function. In bacteria the possibility of any such biological role seems remote, and no definite biochemical function has suggested itself. Since bioluminescence involves expenditure of energy, which is diverted from the usual electron transport system and cytochrome-linked phosphorylation, it would seem to be a wasteful process. One suggestion has cast bacterial luminescence in the role of a vestigial oxygen scavenging system which was originally evolved to protect O_2-sensitive anaerobes. A more interesting possibility is that it is related in some way to the formation of an excited singlet oxygen. Although there is no definitive evidence regarding the participation of such excited oxygen species in biochemical systems, some reactions such as the hydroxylation of aromatic rings have been interpreted in this fashion. One suggested scheme would have the formation of singlet oxygen (O_2^*) as an alternative to the emission of light by luciferin. In this view the actual biochemical function of luminescence would be to generate the excited oxygen needed for certain types of reactions. Once formed, the singlet oxygen could then enter into chemical reactions with other substrates, in which case the extra energy simply would be evolved at heat.

$$L^* \begin{array}{l} \nearrow L + h\nu \\ \searrow L + O_2^* \end{array} \quad O_2 \tag{5.54}$$

$$O_2^* + X \longrightarrow X\text{--}O_2 + \text{Heat} \tag{5.55}$$

Another scheme postulates that luminescence actually requires a simultaneous transition of two excited molecules, one of which would be luciferin and the other singlet oxygen:

$$L^* + O_2^* \longrightarrow L + O_2 + h\nu \tag{5.56}$$

In this case substrates reacting with excited oxygen as in reaction 5.55 would be competitive with luminescence since they would remove one of the necessary reactants. The validity of such imaginative interpretations remains to be seen, but none the less they suggest the potentialities which may lie in further study of bacterial luminescence.

5.12 | Bacterial Photosynthesis

The great range and diversity of bacterial life is nowhere better exemplified than by the photosynthetic types (Section 3.19). Bacterial photosynthesis differs in a number of respects from the process as it occurs in eukaryotes and in blue-green algae. In all cases so far known bacterial photosynthesis is an anaerobic process which utilizes some substance other than water as the ultimate reductant for carbon dioxide assimilation and consequently does not result in the evolution of molecular oxygen. The rationalization of such seeming paradoxes has led to improved insight regarding the fundamental nature of the photosynthetic mechanism. Thus, we have come to appreciate the green plant photosynthesis as simply a special case of a more general mechanism for photophosphorylation and CO_2 reduction in which O_2 production is dispensable and the use of electron donors other than H_2O is equally plausible.

Photophosphorylation. The absorption spectrum of bacteriochlorophyll as it exists *in situ* within the intact cell is different from that of the purified substance. Such a spectral displacement is taken to indicate some sort of alteration due to the particular arrangement of the bacteriochlorophyll in the lamellar or vesicular membranous structures in which the photochemical reactions take place. Interactions between chlorophyll and other components, such as lipids and proteins, as well as between the closely packed chlorophyll molecules themselves, seem likely.

The current dogma holds that some forty or more molecules of bacteriochlorophyll constitute an individual reaction domain (Fig. 5.10). Light quanta absorbed by any of these molecules are funneled to one particular chlorophyll molecule which is the reaction center for initiation of electron transfer. This "reaction center chlorophyll" is recognizable spectroscopically as it undergoes a light-induced, oxido-reduction reaction. In the purple nonsulfur bacterium, *Rhodospirillum rubrum,* a light-dependent bleaching occurs at approximately 890 nm, and therefore this photoactive component, which is assumed to be reaction center chlorophyll, is designated P890. In *Chlorobium* it may be that the small amount of bacteriochlorophyll *a* which is present functions as the reaction center chlorophyll, whereas the bacteriochlorophylls *c* and *d* act primarily as the quanta gathering system.

The energy of the absorbed photons, when received by the reaction center chlorophyll, is used to trigger the ejection of an electron from the chlorophyll molecule. This electron is transferred to some adjoining unknown molecule, designated X in Figure 5.10. The loss of this electron produces on the P890 molecule a net positive charge, which in turn is quickly neutralized by an electron donated by an associated cytochrome of the *c* type. One of the early events in the photochemical process, which can be detected in experiments with intermittent light, is this transient oxidation of the cytochrome component. The electron from P890 is transferred first to X and then sequentially along a special photosynthetic electron transport system (ETS) which includes quinones and cytochromes, until it is finally used for reduction of the oxidized cytochrome *c*. In this way there ensues from the absorption of light energy by bacteriochlorophyll a closed cycle of electron flow from cytochrome *c* to chlorophyll to X and back again to cytochrome *c* via quinones and other cytochromes. Because useful chemical bond energy in the form of ATP is generated during the passage of the

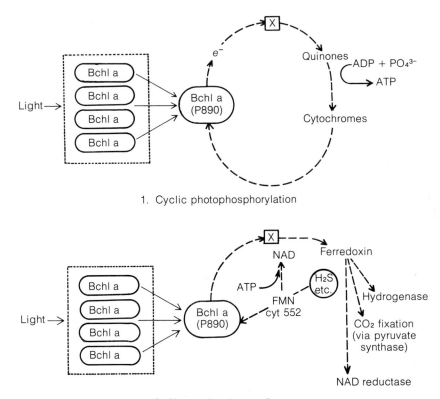

Fig. 5.10. Reaction sequences in bacterial photosynthesis. Light quanta absorbed by bacterial chlorophylls (Bchl a) funnel into a "reaction center" (P890) and trigger both cyclic and noncyclic electron flows. In cyclic photophosphorylation the passage of a high energy electron (e^-) through the photosynthetic electron transport system of unknown carriers (X), quinones, and cytochromes results in synthesis of ATP. In the noncyclic pathway an exogenous reductant such as H_2S supplies electrons needed to replace those ejected from P890 and used for reduction of NAD or CO_2.

electron along the ETS, this process is termed *cyclic photophosphorylation.* Although the identity of compound X, the immediate acceptor of the electron from P890, is unknown, one of the earliest participants in the pathway has been identified as ferredoxin. It is possible that compound X could be ferredoxin itself, but there is little evidence on this point.

Photosynthetic electron transport systems. The electron transport systems of photosynthetic bacteria differ from the conventional type found in aerobic chemoorganotrophic bacteria. Inasmuch as bacterial photosynthesis takes place under anaerobic conditions there is no need for a cytochrome interacting with molecular oxygen. Accordingly, cytochrome *a* and other types of cytochrome oxidase are not part of the photosynthetic ETS. Similarly there has been no demonstration of a cytochrome *b* as part of the photosynthetic ETS, although possibly the presence of a small amount of cytochrome *b* could be masked by other components. Only in the facultative photosynthesizers, such as *Rhodospirillum rubrum,* is there formed an ETS for aerobic respiration of the usual type, including cytochrome *b* and cytochrome oxidase, and this only during growth with organic substrates in the dark.

Several unusual substances are found as components of the ETS of photosynthetic bacteria. A number of different cytochromes of the *c* type have been discovered, along with certain distinctive quinones. The purple nonsulfur bacterium *R. rubrum* contains a cytochrome c_2 which, like the cytochrome *f* of green plant chloroplasts, has a high oxidation-reduction potential of greater than 0.28 v. This organism also produces a distinctive cytochrome *cc'* containing two different heme moieties. This component was formerly designated simply "Rhodospirillum heme protein." In addition to ubiquinone (UQ-45), another closely related benzoquinone, rhodoquinone, is present in this organism. Rhodoquinone resembles ubiquinone, except that the methoxyl group in the 2 position is replaced by an amino group:

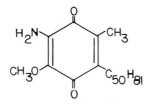

Purple sulfur bacteria, such as *Chromatium,* contain both the cytochrome c_2 and cytochrome *cc'* types, and in addition, another interesting cytochrome component, cytochrome c_{552}, which contains two heme groups and one flavin (FMN) per molecule of molecular weight 72,000. *Chromatium* employs both a naphthoquinone (vitamin K_2-35) and a benzoquinone (UQ-35) as part of the photosynthetic ETS.

The green sulfur bacterium *Chlorobium* possesses three cytochromes of the *c* type, but none of these has a high oxidation-reduction potential like that of the cytochrome *f* of green plants or the cytochrome c_2 of the purple bacteria. An unique naphthoquinone has been found in *Chlorobium*. The structure of this compound, chlorobium quinone, differs from the usual vitamin K types which are commonly found in other bacteria and is as follows:

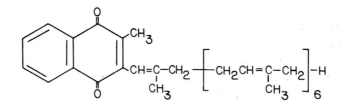

Utilization of CO_2 in photosynthesis. Until bacterial photosynthesis was subjected to detailed study it had been thought that all photosynthetic organisms used a single mechanism for carbon dioxide reduction, namely the ribulose diphosphate carboxylase reaction and the so-called reductive pentose or Calvin cycle. It is now recognized that in addition to the reductive pentose pathway, a second pathway for CO_2 assimilation assumes a significant or even major position in bacterial photosynthesis.

In cell-free extracts of both *Chlorobium* and *Chromatium* it is found that carbon dioxide is used to form pyruvate by means of a pyruvate synthase reaction in which acetyl CoA accepts the CO_2 and is reduced by ferredoxin (see Fig. 5.11). The nicotinamide coenzymes are not involved in this reduction. Reduction of ferredoxin is accomplished in a light-dependent reaction with H_2S acting as the ultimate reductant. A similar α-ketoglutarate synthase reaction involving carboxylation of succinate has also been observed. The relatively low activity of Calvin cycle enzymes in these organisms and the labeling patterns obtained for $^{14}CO_2$ fixation by whole cells suggest that these ferredoxin-mediated systems are quantitatively important mechanisms for CO_2 assimilation.

In the reductive pentose cycle, which appears to operate in varying degrees in all the photosynthetic bacteria, CO_2 is used to carboxylate the compound ribulose-1,5-diphosphate. The unstable 6-carbon intermediate so formed is immediately converted to two molecules of 3-phosphoglyceric acid:*

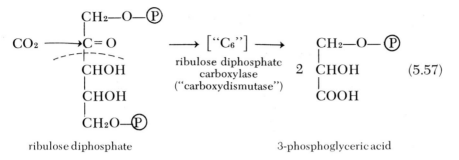

ribulose diphosphate   3-phosphoglyceric acid

This compound can be reduced to yield glyceraldehyde phosphate via a two step process involving the expenditure of an ATP and use of NADPH as the reductant:

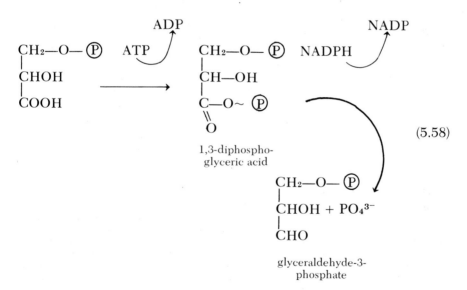

A complex series of interconversions then is required to regenerate the ribulose diphosphate needed for continuing CO_2 fixation (Fig. 5.11). A molecule of glyceraldehyde-3-phosphate can react with a 7-carbon sugar (sedoheptulose-7-phosphate) to yield two molecules of

*In Eqs. (5.57) and (5.58) (P) signifies a phosphate group (PO_4^{3-}).

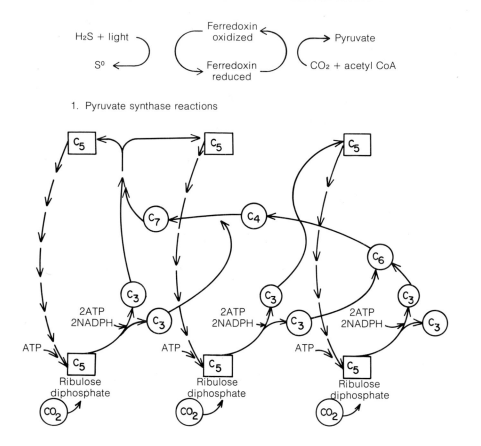

1. Pyruvate synthase reactions

2. Reductive pentose cycle

Fig. 5.11. Mechanisms for carbon dioxide assimilation in bacterial photosynthesis. Reduction of carbon dioxide is accomplished by two different systems in photosynthetic bacteria. In photosynthetic sulfur bacteria pyruvate is produced by addition of CO_2 to acetate (as acetyl CoA) in the pyruvate synthase reaction (1), the reducing power for which is provided by a light-dependent oxidation of H_2S. A reductive pentose or Calvin cycle like that of green plants also operates in bacteria for conversion of CO_2 into carbohydrate. As shown here 3 turns of the cycle (2) result in a net synthesis of a 3 carbon fragment from CO_2.

pentose phosphate. These then are phosphorylated to give ribulose diphosphate. The C_7 compound is produced via a condensation of a C_3 fragment (as dihydroxyacetone phosphate, which is derived from glyceraldehyde) with a C_4 unit (erythrose phosphate). The C_4 unit in turn arises in the transketolase reaction in which fructose-6-phosphate plus a C_3 unit combine to yield a pentose plus tetrose (i.e., C_3 + $C_6 = C_5 + C_4$). Again, the pentose can be converted to ribulose diphosphate by expenditure of an ATP. The hexose needed for this reaction is formed by a condensation of two triose molecules. Three successive turns of the cycle involving three carboxylations will be required to permit the net synthesis of one molecule of glyceraldehyde from 3 CO_2; also needed for this synthesis are 9 ATP and 6 NADPH. After six turns of the cycle it will be possible to produce a molecule of hexose by condensation of the two triose molecules, thus accounting for the observed formation of carbohydrate during photosynthesis. Although the ribulose diphosphate carboxylase reaction is characteristically limited to photosynthetic organisms and chemolithotrophs, a number of the other reactions of this cycle will be recognized as the same as those used in the familiar hexose monophosphate pathway for glucose dissimilation.

The generation by photophosphorylation of the ATP needed for such a cycle has been described previously, but no mechanism for the formation of the reduced nicotinamide coenzyme has been indicated thus far. One possibility would be to couple the oxidation of reduced ferredoxin to the reduction of NAD or NADP. A soluble ferredoxin-NAD reductase has been detected in cell-free extracts. Another method would be via an ATP-dependent reversal of electron flow through the ETS. Such a reaction, which is essentially the reverse of the events in respiration, has been observed both in mammalian mitochondria and in bacterial extracts, and is presumed to take place in the living cell as well.

References

Arnon, D. I., Ferredoxin and Photosynthesis, *Science* **149,** 1460 (1965).

Barker, H. A., *Bacterial Fermentations,* Wiley, New York, 1957.

Bartsch, R. G., Bacterial Cytochromes, *Ann. Rev. Microbiol.* **22,** 181 (1968).

Calvin, M., and J. A. Bassham, *The Photosynthesis of Carbon Compounds,* W. A. Benjamin, New York, 1962.

Cohen, G. N., *The Regulation of Cell Metabolism,* Holt, Reinhart and Winston, New York, 1968.

Dagley, S., Catabolism of Aromatic Compounds by Microorganisms, in A. H. Rose and J. F. Wilkinson (eds.), *Adv. Microb. Physiol.* **6,** 1 (1971).

Doelle, H. W., *Bacterial Metabolism,* Academic Press, New York, 1969.

Evans, M. C. W., and F. R. Whatley, Photosynthetic Mechanisms in Prokaryotes and Eukaryotes, in H. D. Charles and B. C. J. G. Knight (eds.), *Organization and Control in Prokaryotic and Eukaryotic Cells,* 20th Symp. Soc. Gen. Microbiol. Cambridge University Press, Cambridge, England, 1970.

Gunsalus, I. C., and R. Y. Stanier (eds.), *The Bacteria, A Treatise on Structure and Function,* Vol. II., *Metabolism,* Academic Press, New York, 1961.

Hastings, J. W., Bioluminescence, *Ann. Rev. Biochem.* **37,** 597 (1968).

Kamen, M. D., and T. Horio, Bacterial Cytochromes: I. Structural Aspects, *Ann. Rev. Biochem.* **39,** 673 (1970).

Lehninger, A. L., *Bioenergetics,* 2nd ed., W. A. Benjamin, New York, 1971.

Loewy, A. G., and P. Siekevitz, *Cell Structure and Function,* 2nd ed., Holt, Rinehart and Winston, New York, 1969.

Olson, J. M., The Evolution of Photosynthesis, *Science,* **168,** 438 (1970).

Payne, W. J., Energy Yields and Growth of Heterotrophs, *Ann. Rev. Microbiol.* **24,** 17 (1970).

Peck, H. D., Jr., Energy-Coupling Mechanisms in Chemolithotrophic Bacteria, *Ann. Rev. Microbiol.* **22,** 489 (1968).

Pfennig, N., Photosynthetic Bacteria, *Ann. Rev. Microbiol.* **21,** 285 (1967).

Rittenberg, S. C., The Roles of Exogenous Organic Matter in the Physiology of Chemolithotrophic Bacteria, in A. H. Rose and J. F. Wilkinson (eds.), *Adv. Microb. Physiol.* **3,** 159 (1969).

Roy, A. B., and P. A. Trudinger, *The Biochemistry of Inorganic Compounds of Sulphur,* Cambridge University Press, Cambridge, England, 1970.

Sokatch, J. R., *Bacterial Physiology and Metabolism,* Academic Press, New York, 1969.

Stadtman, T. C., Methane Fermentation, *Ann. Rev. Microbiol.* **21,** 121 (1967).

Trudinger, P. A. Assimilatory and Dissimilatory Metabolism of Inorganic Sulphur Compounds by Microorganisms, *Adv. Microb. Physiol.* **3,** 111 (1969).

Wolfe, R. S., Microbial Formation of Methane, in A. H. Rose and J. F. Wilkinson (eds.). *Adv. Microb. Physiol.* **6,** 107 (1971).

six | Bacteria as Environmental Determinants

"There is nothing permanent except change."
Heraclitus (530–470 B.C.?)

It is a biological truism that life processes are superimposed upon the background of the physical environment. Certain environmental parameters including light, temperature, and oxygen, as discussed in Chapter 4, play key directive roles in the physiology of individual bacteria, and in the development of biotic communities. However, living organisms are not passive and their activities influence the character of the physical environment itself.

Surface features of the earth are not fixed and static; Heraclitus based his philosophy on this point. These changes usually are gradual and often imperceptible within the lifetime of the individual, but occasionally they are sudden and spectacular. Over the eons mountains are pushed up, only to be worn away by erosion, and silt is carried back to the sea. Whole continents are lifted up, or sink, or drift apart. Glaciers slowly advance over vast areas and then retreat. Volcanoes arise out of the ocean in a day, to last a month or perhaps a millenium before being destroyed in a cataclysmic eruption. These purely physical tectonic processes are primary factors in the shaping of the environment of the biosphere, but in addition, there are further alterations designated as "biogeochemical changes," produced as a result of the metabolism of living organisms.

Bacteria make important contributions to these biogeochemical processes, among which are: (1) participation in the cyclic transformations of elements, such as in the sulfur, nitrogen, and carbon cycles, especially with reference to degradation and decomposition of organic compounds; (2) formation of mineral deposits such as limestone, calcite, gypsum, elemental sulfur, metallic orestuffs, and fossil fuels; (3) solubilization and removal of mineral deposits; and (4) the "weathering" of rocks and formation of soil.

6.1 | The Age of the Earth and the Beginning of Life

Current estimates place the age of the earth at approximately 5×10^9 years. The pre-biotic period of the earth's existence probably occupied the first $1-2 \times 10^9$ years, during which time the necessary chemical and physical prerequisites of life evolved. Among these were the formation of a relatively stable crust, condensation of sufficient water in the liquid state, and the generation of appropriate organic compounds for the construction of cells. Life as we know it is based on aqueous systems, which would have been impossible until the prevailing temperatures were at or below 100° C. Under conditions presumed to have occurred on the primeval earth certain of the compounds utilized in living systems, including amino acids, purines and pyrimidines, can be produced nonbiologically from simple precursors by purely chemical reactions. Thus, following a period of prebiotic chemical evolution, it is thought that life began about $3.0-3.5 \times 10^9$ years ago. It is widely presumed that the first organisms were structurally very simple, possibly similar to the modern *Mycoplasma,* and that present-day bacteria represent descendants of an early, if not the first, life form.

Opinion concerning the mode of origin of the earth is divided, but there is general agreement that conditions now prevailing have not always existed, but are the result of a long evolutionary process. One view that seems reasonable is that the earth was formed by the accretion of smaller bodies of solidified materials resembling meterorites (*planetesimals*). In fact, meteorites could be residual planetesimals not used up in forming the planets. Over a period of possibly 600 million years heating of the growing earth by energy released in radioactive decay raised temperatures sufficiently to begin liquefaction of the components, thereby initiating a physical and chemical differentiation that resulted in their separation into a dense iron-rich core, a thick overlying mantle of silicate rocks, and a 10–35 km thick outer crust of lighter silicate rocks.

Formation from solid planetesimals would suggest that the earth initially lacked an atmosphere, as these smaller objects would not have retained an external gaseous phase because of their slight gravitational attraction. Generation of the atmosphere from materials contained in the solid phase would perhaps serve to explain certain features of its composition, especially with respect to the inert gases. All the inert gases, helium, neon, argon, krypton, and xenon, are present on earth in much smaller amounts than in the universe as a whole. Direct condensation of the earth from primordial solar material should have resulted in a terrestrial occurrence of these elements in amounts reflecting their cosmic abundance. Atmospheric helium, and to a lesser extent neon, would tend to escape from earth's gravitational field because of their low atomic mass, resulting in minute residual levels regardless of initial amount, but the heavier krypton and xenon would not have escaped so readily, and it is concluded that they probably never were present in quantity. Occurrence of krypton and xenon in amounts some 10^6 times less than their cosmic abundance lends support for a planetesimal hypothesis. An anomalously high level of argon (mainly ^{40}Ar) in the present atmosphere is probably attributable to continual enrichment with argon produced by radioactive decay of ^{40}K.

The chemical composition of the earth reveals a number of interesting features, all leading to the major conclusion that the distribution of elements on earth does not reflect their cosmic abundance (Section 2.14). It is known from spectroscopic evidence that in the universe as a whole, and in the sun, the predominant elements are hydrogen and helium. The general trend observed is an almost exponential decrease in cosmic abundance with increasing atomic weight for the first forty elements, with roughly equal amounts of each of the elements above that atomic number. Only ten elements, all with atomic numbers less than 30, occur in appreciable amounts in the sun or the cosmos: H, He, C, N, O, Ne, Mg, Si, S, and Fe. The most prevalent elements in the earth as a whole, on a weight basis, are iron (35%), oxygen (30%), silicon (15%), magnesium (13%), nickel (2.4%), sulfur (1.9%), calcium (1.1%), and aluminum (1.1%). Each of the remaining known elements is present in amounts less than 1%. This distribution, while differing from that of the sun, is remarkably similar to that for stony meterorites *(chondrites)*. Three of the elements of crucial importance to living organisms are present in the earth in relatively small amounts: hydrogen, nitrogen, and carbon. Relative to silicon (atoms per 10^4 atoms of Si) the estimated cosmic abundances of H, N, and C are 4×10^8, 6.6×10^4, and 3.5×10^4, respectively, whereas for their terrestrial occurrence the values are 8.4×10^1, 2×10^{-1}, 7×10^1. Thus, utilization of these elements for the construction of biological materials was not caused simply by their ready availability, but must have been dictated by other chemical requirements, such as the character of covalent bonding between them.

The primary atmosphere formed from gaseous materials expelled from the molten earth before the separation of mantle and core would have reflected the reducing conditions provided by the presence of large amounts of iron. During the initial phase the atmosphere possibly had hydrogen and methane as major constituents. The gases released probably were predominantly H_2, H_2O (as steam), and CO, with smaller amounts of N_2 and H_2S. Secondary reactions could have resulted in the formation of ammonia, methane, cyanide, and possibly formic acid. Condensation of water vapor as temperatures dropped would have led to the development of progressively larger bodies of water, and ultimately the oceans. Although less salty, the primitive ocean would have been more alkaline than it is now because of its contact with basaltic igneous rocks. Addition of powdered basalt to distilled water gives a pH of about 9.6. After the iron was largely sequestered by the formation of the core, the gases would have more closely resembled those of modern volcanoes, water being the dominant material, with CO_2, CO, H_2, SO_2, and N_2 as minor components. The continuing release of materials from the earth's interior plus secondary chemical reactions including the dissolving of certain components in the ocean, especially CO_2, resulted in the further development of the second stage atmosphere. It was during this phase, when the atmosphere was composed largely of N_2, with smaller amounts of

CO_2 and water, that life probably began.

Once living organisms evolved, certain other profound changes in the environment became possible, the most significant of which was deferred until the development of photosynthesis. It seems likely that the earliest organisms would have been nonphotosynthetic, possibly fermenting organic substrates which had been produced during the prebiotic period of chemical evolution. There was no free O_2 in the atmosphere at this time, so the first organisms would have been anaerobes. Once the organic material became depleted, organisms would have turned to the use of inorganic energy substrates, and ultimately to radiant energy. Possibly the first photosynthetic processes would have resembled those used by modern purple and green sulfur bacteria. Finally, the photosynthetic mechanisms typical of algae and green plants, in which molecular oxygen is produced, emerged at some point at least 1.7–2.0×10^9 years ago. At first the O_2 produced by these early photosynthetic organisms did not result in any significant increase in the atmospheric levels, because it was rapidly utilized in the chemical oxidation of crustal materials, notably the ferrous iron which had been carried into the sea. Extensive iron ore deposits of marine origin, typified by the Precambrian formations of the Lake Superior region, resulted from this new availability of O_2. By 1.2×10^9 years ago the amount of atmospheric O_2 had increased to a level permitting creation of nonmarine iron oxide deposits (so-called "red beds"), but not until about 6×10^8 years ago (the beginning of the Paleozoic era) was there sufficient O_2 to allow development of metazoan forms dependent on aerobic respiration.

6.2 | Nonbiological Geochemical Processes

For convenience, the earth can be considered to consist of four major zones or geochemical sectors: the atmosphere or gaseous phase; the hydrosphere or aqueous phase; the lithosphere or solid phase, including crust, mantle, and core; and the biosphere, made up of all the living organisms found in the other three spheres. The atmosphere comprises about 5×10^{21} g of a gas mixture whose approximate composition by weight is N_2 (755,100 ppm), O_2 (231,500 ppm), Ar (12,800 ppm), CO_2 (460 ppm), Ne (12.5 ppm), and Kr (2.9 ppm). Most of the hydrosphere is attributable to the ocean, which covers an area of 361 $\times 10^6$ km^2 and has a total mass of some 1.4×10^{24}g. In contrast, the mass of fresh water is estimated at only 0.5×10^{21} g and that of continental ice at 33×10^{21} g. For our purposes the most significant portion of the lithosphere is the upper 35 km of the crust (about 0.4% of the total mass). We have already alluded to the geochemical history of the earth as one of differentiation and separation of the primary components, ultimately resulting in the abovementioned spheres. This developmental process began before life originated, and both its purely nonbiological phases and those involving bacteria and other organisms are continuing. Obviously living organisms do not participate in the high temperature processes such as the formation of primary igneous rocks from molten material. However, organisms play a highly significant part in the subsequent alteration of such rocks to form sedimentary rocks. The general nature of the geochemical transformations occurring on the earth today is outlined in Figure 6.1.

Radioactive decay. Radioactive decay is an important geochemical process from several standpoints: (1) it probably was the original source of heat responsible for both the initial melting and separation of materials and the continuing volcanic and igneous activities; (2) it has been responsible for alterations of the chemical composition of the earth

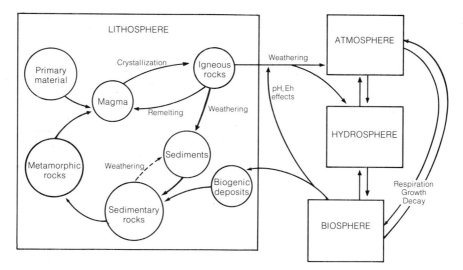

Fig. 6.1. General geochemical cycle. (Adapted and redrawn from B. Mason, *Principles of Geochemistry*, p. 286, John Wiley and Sons, New York, 1966. Courtesy B. Mason and permission of John Wiley and Sons.)

through the contribution of elements which are the products of decay; and (3) it provides a time scale for estimating the age of rocks and the earth. Among the reactions of prime interest are the decay of radioactive potassium to give argon and calcium, and that of uranium or thorium to yield helium and lead:

$$^{40}K \longrightarrow \,^{40}Ar \qquad\qquad \text{half-life} = 1.3 \times 10^9 \text{ years} \qquad (6.1)$$

$$^{235}U \longrightarrow \,^{207}Pb + 7\,^4He \qquad \text{half-life} = 7.1 \times 10^8 \text{ years} \qquad (6.2)$$

$$^{238}U \longrightarrow \,^{206}Pb + 8\,^4He \qquad \text{half-life} = 4.5 \times 10^9 \text{ years} \qquad (6.3)$$

The usefulness of these reactions for dating purposes may be illustrated as follows. It is observed that ^{238}U is about 140 times more plentiful than ^{235}U. If both isotopes were present originally in equal amounts, the shorter half-life of ^{235}U would cause it to become depleted more rapidly. The time required to arrive at a concentration 1/140 that of ^{238}U would be about 5×10^9 years, which is, in fact, the basis for the estimate of the age of the earth given earlier. A slightly more complex analysis, utilizing the ratios of lead isotopes, can be employed to calculate the age of formation of individual rocks. There are four isotopes of lead, ^{208}Pb, ^{207}Pb, ^{206}Pb, and ^{204}Pb, of which only the last is not a product of radioactive decay. Determination of $^{207}Pb/^{204}Pb$ and $^{206}Pb/^{204}$ Pb ratios in meteorites indicates that they were formed about 4.5×10^9 years ago.

Formation of igneous rocks. In the primitive earth a separation of the elements would have taken place on the basis of their chemical properties in relation to those of the major components. Oxygen and silicon would combine to give silicates, and, being present in such a large amount, iron would have existed not only as iron silicate, but also as iron sulfide and as the free element. Other elements would tend to be distributed between these three phases through displacement reactions occurring on the basis of thermodynamic equilibria. Elements less electronegative than iron, such as magnesium, would accumulate in the silicate phase, and those more electronegative would tend to be transferred into the metallic phase as they were displaced from ionic compounds by iron. Elements such as Hg and Ag, which have a greater affinity than iron for sulfur, would be enriched in the sulfide phase. The occurrence of specific mineral types and many of the geochemical processes which still are taking place may be understood on the basis of such displacement reactions, and this has led to the characterization of elements in terms of their relative affinity for iron (so-called "siderophile elements," such as Au, Pt, Pd, and Os), sulfur ("chalcophile elements," such as Hg, Ag, Zn, and Cd), or silicates ("lithophile elements," such as Li, Na, K, and Ca).

Igneous rocks are formed by the cooling and solidification of molten, predominently silicate, materials from the interior termed *magma*. Much of the magma solidifies slowly beneath the surface to yield macrocrystalline rocks, such as granite and diorite. These consist mostly of quartz (SiO_2) and an array of silicate minerals, including feldspars ($K_2O \cdot Al_2O_3 \cdot 6SiO_2$ and $Na_2O \cdot Al_2O_3 \cdot 6SiO_2$) and pyroxenes ($CaMg(SiO_3)_2$). Magma, which finds its way to the surface through volcanoes, cools rapidly from a temperature of 500–1200°C, forming microcrystalline tuffs, lavas, and basalts. There are also volatile components in magma, and these are important in the transport and deposition of certain minerals. The composition of the volatile fraction is uncertain but it is probably largely water, which is thought to be present in magma in amounts ranging from 0.5 to 8%. Some inferences regarding the volatile fraction have been drawn from analysis of volcanic gases, but these have proven highly variable from place to place, and vary with time even from the same source. Furthermore, much of the material in volcanic gases is probably of secondary origin, being incorporated into the magma from surrounding rocks during its upward passage. Additions of water and other volatiles to the surface and atmosphere through volcanic activity must be considered a significant factor when projected over the long course of geological history.

Sedimentation. The minerals which make up igneous rock, in general, are not stable under surface conditions; hence exposure begins a sequence of alterations, referred to as *sedimentation,* leading to the formation of new minerals of greater stability. Sedimentation processes include the "weathering" of rocks to produce soil; the erosion of soil by wind and water; the solubilization, transport, and deposition of minerals by surface waters; and the creation of new rocks (sedimentary rocks) and minerals from igneous precursors. Certain mechanical effects play a part, such as the fracturing of rocks by ice formation and thermal expansion, and the grinding and scouring action of wind and flowing water. Chemical reactions, such as hydrolyses, oxidations, and reductions, probably are more important. The general course of events in sedimentation consists of a separation of the components somewhat as follows: (1) the most resistant portion of igneous rock, quartz, which will accumulate as a quartz sand or be converted into a sandstone; (2) the chemical breakdown of aluminosilicate components leads to the solubilization of certain elements plus the formation of so-called "clay-minerals," which are insoluble and settle out as fine muds, considerably enriched in aluminum; (3) iron is removed in solution, but it is ultimately oxidized and precipitated as ferric hydroxide; (4) calcium is precipitated from solution by the action of CO_2 to give calcium carbonate (calcite); (5) certain cations, such as sodium (Na^+) and to lesser degree potassium (K^+) and magnesium (Mg^{2+}) remain in solution to be accumulated in the ocean, from which they are deposited only after extensive evaporation. Still further modifications may take place, such as the adsorption of potassium by clay minerals, or the formation of dolomite from calcite by the replacement of calcium by magnesium:

$$2CaCO_3 + Mg^{2+} \longrightarrow CaMg\,(CO_3)_2 + Ca^{2+} \qquad (6.4)$$

The dependence of many of these reactions on conditions of pH and oxidation-reduction potential provides a means whereby actions of living organisms exert an influence on sedimentation processes.

6.3 | Participation of Bacteria in Cyclic Geochemical Transformations of the Elements

Living organisms mediate geochemical changes in two ways, directly and indirectly. Direct effects include involvement of the element or compound in the metabolism of the organism, as in the oxidative transformations by chemolithotrophic bacteria of the inorganic compounds used as energy substrates, or in the bacterial reduction of compounds such as O_2, nitrate, and sulfate, which are used as terminal electron acceptors. Indirect effects, such as lowering of pH or oxidation-reduction potential because of metabolic activity, also are of great importance in promoting geochemical processes.

These biogenic processes are of great magnitude, and, as previously indicated, have been responsible for permanent changes in the earth's environment. The various compounds of a given element are subject to a large number of individual transformations, the overall result of which is a biological cycle for that element. Three important cycles, those of carbon, nitrogen, and sulfur, will be described as examples.

In addition to the simple tracing of the pathway by which an element is cycled through the biosphere, including possible detours via the atmosphere or hydrosphere, two other points are worthy of attention, that is, the quantity of materials involved, and the rate at which they are passed through the cycle. Unfortunately, these two features are largely matters of conjecture, and estimates of the rates of turnover, or the actual amounts of substance in each sector or pool in the cycles, have little precision. Furthermore, it is essential to appreciate that these so-called "cycles of the elements" are actually not closed and self-contained entities, but instead there is a continued input of primary or juvenile material derived from the earth's interior via igneous or magmatic activities.

6.4 | The Carbon Cycle

All life as we know it is based on carbon compounds, despite the comparative scarcity of that element, and consequently the chemical transformations of carbon are of great importance. First, we need to identify the principal geochemical reservoirs of carbon, as well as the amounts and predominant chemical forms occurring in each. Then the processes involved in the transfer of carbon between these sectors can be delineated. Interactions with the biosphere are our primary interest. Practically all naturally occurring forms of carbon, except the free element itself, are susceptible to biological attack. Elemental carbon, whether amorphous (carbon black) or crystalline (graphite, diamonds), apparently is not metabolized by living organisms.

In the atmosphere carbon is present almost exclusively as carbon dioxide, the total amount of which is estimated to be approximately 2.3×10^{18}g. The smallest segment with respect to the amount of this element is the biosphere, which includes only about 2.6×10^{17}g of carbon. The great bulk of the earth's carbon is contained in the lithosphere. In sedimentary rocks this element is present mostly as inorganic carbonates in the form of limestone, chalk, dolomite, calcite, siderite, and shell marls, and these account for about 1.23×10^{22}g of carbon. A considerable amount of organic carbon, amounting to about 3.6×10^{21} g, also occurs in sedimentary rocks, partly as massive deposits of coal and petroleum, but the greater portion in highly dispersed form, referred to as "kerogen." Igneous and metamorphic rocks also contain carbon, and the amount in these materials has been estimated to be about 1.3×10^{22} g.

Photosynthesis. The geochemical cycle of carbon is shown in Figure 6.2. The element enters the biosphere primarily by way of the fixation (reduction) of carbon dioxide by the photosynthetic reactions of terrestrial plants and aquatic algae, and especially the unicellular marine diatoms and dinoflagellates. Estimates of the magnitude of this process vary widely, but it is probable that on the order of $5\text{--}15 \times 10^{16}$ g of carbon enters the biosphere annually by photosynthesis, which amounts to from 8 to 23% of the total carbon content in the atmosphere.

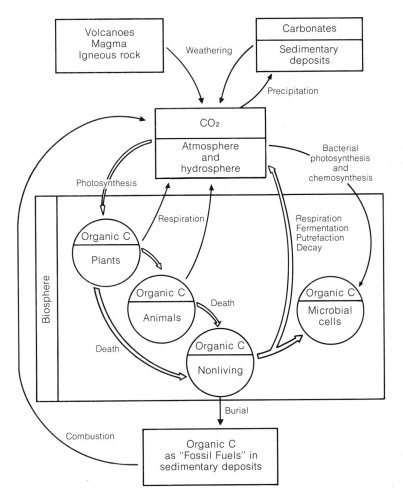

Fig. 6.2. The biological carbon cycle.

Efficient operation on this vast scale is possible because of the large oceanic reservoir of bicarbonates and dissolved CO_2. On a global scale it seems clear that the photosynthetic activities of bacteria quantitatively must be of much less importance than those of algae and higher plants. Because of their requirements for anaerobic conditions and for hydrogen sulfide, photosynthetic bacteria are restricted to limited habitats, and consequently the amount of CO_2 fixed by them must be small. Likewise, CO_2 fixation of chemolithotrophic bacteria probably is inconsequential because of the restricted occurrence of conditions permitting their development.

Mineralization. The carbon reduced in photosynthesis appears in the form of organic compounds, especially plant biopolymers such as cellulose, starch, pectins, and proteins, which are then passed along through the food chain from plants to animals. Ultimately the organic carbon is reoxidized to CO_2 by aerobic respiratory processes, and thereby returned to the atmosphere. This occurs by respiration during the life of the plant or animal, and also by the bacterial decomposition of its waste products and, after death, its dead remains. Biodegradation of organic carbon by chemoorganotrophic bacteria takes place on a grand scale and is of utmost importance in insuring the continued flow of carbon through the cycle. In some cases bacterial action results in a greater availability of the organic material to the biological community as a whole. Insoluble polymers such as cellulose are hydrolysed to monomers which are more easily attacked by all the bacteria present. Removal of a protective layer or shielding matrix by one bacterium increases the vulnerability of the remaining material to exploitation by other organisms.

The most important consequence for the carbon cycle of bacterial degradative activities is the transformation of organic carbon to CO_2, which is lost to the atmosphere. During anaerobic fermentations only a fraction of the material is converted to CO_2 by means of decarboxylations. However, bacterial respiration of organic substrates accomplishes their complete combustion to CO_2. A portion of the carbon (often about one-third) will be used to synthesize new bacterial cellular materials, but in turn these will become substrates for respiratory attack by other bacteria after death of the original cells. The transformation of elements from an organic into an inorganic form is given the general term *mineralization*. The mineralization of organic residues must be considered as one of the key biological and biogeochemical functions of bacteria in nature. In soil the mineralization process can be detected by CO_2 production. The organic content of an average soil is about 2%, and there is about 0.8% in the underlying subsoil. This material is attacked by bacteria and fungi with the release of 10–70 mg CO_2/kg of soil per day. Controlling effects on the rate of CO_2 evolution are exerted by the moisture content and the pH of the soil.

It is apparent from the foregoing that the cycles of carbon and oxygen must be interrelated. During green plant photosynthesis oxygen is evolved from water as the CO_2 is assimilated and introduced into the biosphere, and then subsequently atmospheric oxygen is consumed during the mineralization of organic C to CO_2:

$$H_2O \underset{\text{mineralization}}{\overset{\text{photosynthesis}}{\rightleftharpoons}} O_2 \qquad (6.5)$$

This emphasizes that the role of oxygen in respiratory processes is that of the final acceptor of the protons and electrons removed from the compound being oxidized. There are, to be sure, a few oxidation reactions in which molecular oxygen is added directly via covalent bonding to the compound being oxidized (so-called oxygenase reactions). However, for the most part the O_2 used in mineralization appears as water, not in the CO_2 produced.

Mineralization takes place in aqueous as well as terrestrial environments. The availability of oxygen is a limiting factor in both, but the limited solubility of oxygen in water poses special problems for aqueous systems. Oxygen solubility is inversely related to temperature, so that there is less dissolved oxygen (DO) at higher than at lower temperatures. The DO levels in distilled water saturated with air at 760 mm pressure for 5°C, 15°C, and 25° C are 12.8, 10.2, and 8.4 ppm, respectively. As bacteria attack the readily oxidizable organic matter in water their respiratory activities tend to deplete the DO. In summer months oxygen depletion will be most rapid because at the higher temperatures metabolic reactions will tend to be at a faster rate, and O_2 solubility is least. Reoxygenation of water takes place at the air-water interface, so that mechanical agitation, which constantly renews the material at the interface, is of prime importance. The mixing action and entrainment of air in rapidly moving streams facilitate replenishment of the oxygen consumed in mineralization. In some lakes of moderate depths there are seasonal occurrences of thermal mixing (the spring and autumn "over-turns") which serve to reoxygenate the deeper portions. The onset of lower autumn temperatures leads to cooling of O_2-laden surface waters with a corresponding increase in density. As these sink to the bottom they are replaced by the warmer, less dense, O_2-poor bottom layers, which in turn become oxygenated. If the amount of organic matter is high and there are limited means for reoxygenation, as in slow-moving streams or in deeper portions of lakes, conditions will become anaerobic and respiratory activities will be succeeded by fermentative and putrefactive reactions. This results in foul smelling and sometimes toxic products, and the accumulation of excess organic matter, since complete mineralization is prevented. Fish and other respiring forms cannot survive under these conditions, and generally the entire original biotic community is destroyed and replaced by a new one, characterized by an extremely limited array of species and the absence of macrofauna.

The amount of oxygen required to permit complete bacterial oxidation of readily attackable organic matter can be measured in the laboratory. This is frequently used as an index of the amount of biologically available carbon in water. This quantity is expressed as the "Biochemical Oxygen Demand" (BOD), which is generally given as the amount of oxygen (in ppm or mg/liter) consumed during five days' incubation of the sample in the dark at 20° C in a sealed vessel. The initial and final dissolved oxygen levels are measured by titrimetric or polarographic methods and the difference represents the BOD. The five-day BOD of relatively clean streams is usually less than 1 ppm, but raw sewage has an average BOD of 200 ppm and some industrial wastes as those from canning and dairy processes may reach BOD values of 2,000–5,000. Addition of such concentrated organic wastes to streams or lakes can have only disastrous results. Hence, biological treatment processes are interposed, which permit mineralization of these materials within the confines of the treatment facility.

Under certain conditions, such as the deep burial of dead plant and animal residues, the mineralization process is bypassed, and a portion of the organic carbon is shunted out of the mainstream of the biological cycle to give rise to sedimentary deposits of modified carbonaceous materials, such as petroleum and various types of coal, collectively known as "caustobioliths" (inflammable minerals of biological origin), or, less technically, "fossil fuels." The relative stability of such deposits against bacterial attack is due mostly to their physical separation from more active aerobic zones, because their exposure to the surface results in degradation and re-entry into the cycle. Formation of other sedimentary deposits of inorganic carbonates such as limestone and dolomite also withdraws large amounts of material from the cycle. The total amount of sedimentary carbonates far exceeds the amount of carbon in the atmosphere, hydrosphere, and biosphere combined. When such limestone deposits are subjected to weathering, they are solubilized and ultimately transported into the ocean in the form of bicarbonates, which accounts for the nearly saturating levels of calcium bicarbonate in sea water. Carbon dioxide is, in effect, the anhydride of carbonic acid, and when CO_2 dissolves in water it forms carbonic acid, which in turn dissociates to give bicarbonate and carbonate ions and protons:

$$CO_2 + H_2O \rightleftharpoons H_2CO_3 \rightleftharpoons H^+ + HCO_3^- \rightleftharpoons 2H^+ + CO_3^{2-} \qquad (6.6)$$

These exist in a series of equilibria poised by the CO_2 content of the overlying gas phase. Rainwater charged with carbonic acid from the atmosphere and respiratory CO_2 in soil attack limestone by forming the more soluble calcium bicarbonate, whose removal in some cases has carved out gigantic caverns:

$$CaCO_3 + CO_2 + H_2O \rightleftharpoons Ca(HCO_3)_2 \qquad (6.7)$$

In the sea this same equilibrium reaction occurs and maintains the calcium bicarbonate in solution. This provides a mechanism whereby living organisms interact with these inorganic processes, because utilization of CO_2 by photosynthetic organisms and alterations of pH tend to displace the equilibrium to the left, and because of its limited solubility, calcium carbonate is precipitated. The formation of limestone in marine sediments is thus partly, if not largely, the result of biological actions (see Section 6.11). In view of the large amount of carbon sequestered in sedimentary deposits the question arises as to how a balance could have been maintained in the cycle. There have been no major alterations of the atmosphere since Paleozoic times (6×10^8 years ago), as far as can be determined, so there must have been a continuing addition of juvenile carbon to the cycle. The bulk of this probably arises from gases released during crystallization of magma and from volcanic emissions. A smaller amount (possibly 1% as much) is added by the weathering of igneous rocks. Bacteria undoubtedly play some part in the weathering process through the production of organic acids, especially those with chelating properties such as citric acid.

There has been some concern voiced regarding effects of the addition to the atmosphere of the CO_2 produced by the large-scale combustion of fossil fuels by modern industrial processes. The total annual world coal production of about 1.63×10^9 tons, if combusted, would yield about 4.7×10^{15} g of CO_2 and the burning of the total annual world petroleum output of 3.4×10^9 barrels would add another 1.53×10^{15} g of CO_2. This gives a total annual addition equal to about 0.3% of the atmospheric content of CO_2. The fate of this increment actually is unknown. Some have assumed that it would simply expand the atmospheric CO_2 pool, and claims have been made that shifts in atmospheric CO_2 levels were already detectable. Increases in atmospheric CO_2 could result in rising temperatures, because CO_2 tends to trap infrared radiation. Temperature stability demands that energy received by the earth as solar radiation be balanced by that re-emitted as infrared radiation. Greater CO_2 would prevent this loss and therefore raise the prevailing temperature. But this ignores the role of the ocean in regulating the atmospheric CO_2 level. Probably the greater portion of the CO_2 added would dissolve in the ocean, possibly resulting in a slight shift in the CO_2-carbonate-bicarbonate equilibrium. The effects of such displacements are unknown. More exact information on photosynthetic productivity and the extent of CO_2 "sinks" in the sea would be useful in assessing these effects. The danger from thermal alterations in this specific instance seems to have been overstated, but certainly recent experience suggests caution regarding any deliberate perturbations of the environment.

6.5 | The Nitrogen Cycle

The major steps in the cyclic transformation of nitrogen are shown in Figure 6.3. Organic nitrogen compounds are of crucial importance to living organisms and are found in essential cellular components such as proteins, nucleic acids, and mucopolysaccharides. We may consider proteins as an example of these because proteins comprise a substantial proportion of all organisms, amounting to 60-75% of the dry matter in the case of bacteria. As the "primary producers," green plants and algae originate large amounts of proteins as well as other organic N compounds. These photolithotrophic organisms utilize inorganic nitrogen compounds, such as nitrate or ammonia,

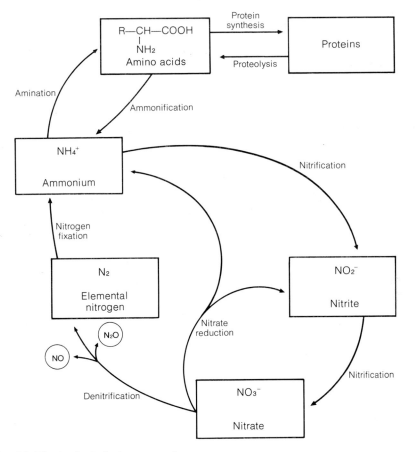

Fig. 6.3. The biological nitrogen cycle.

so that, in terms of geochemistry, photosynthesis may be viewed as a process for conversion of both C and N from inorganic to organic forms.

A major route for assimilation of inorganic nitrogen compounds is by *amination* of α-keto acids in a reaction with ammonium ion that yields amino acids. Typical of such aminations is the reaction catalyzed by the enzyme glutamic acid dehydrogenase:

$$\alpha\text{-ketoglutarate} + NH_4^+ + NADH \quad \underset{\substack{\text{glutamic}\\\text{dehydrogenase}}}{\overset{\longrightarrow}{\longleftarrow}}$$

$$\text{glutamic acid} + NAD + H_2O \qquad (6.8)$$

This and analogous reactions of ammonium ion with pyruvate and oxaloacetate to give alanine and aspartate provide some of the entry points for the introduction of inorganic nitrogen into the biosphere. Once in the amino form the nitrogen is transferable to other keto acids by transamination reactions which give rise to a variety of other amino acids. In addition, ammonium ion is converted directly or indirectly, via amino acid intermediates, to other important organic nitrogen compounds including amino sugars (glucosamine, muramic acid), purines and pyrimidines (adenine, guanine, cytosine, uracil), phosphatides (phosphatidylethanolamine) and various coenzymes and prosthetic groups of enzymes (nicotinic acid, biotin, heme).

Utilization of ammonium ion is not restricted to photosynthetic organisms, and many, but not all, nonphotosynthetic bacteria are able to employ ammonium salts as their sole source of nitrogen. Organisms utilizing more oxidized forms, such as nitrate, simply reduce them to the level of ammonium ion. This process is termed *nitrate reduction,* and is observed in two forms. Green plants and many bacteria reduce nitrate on a limited scale in order to assimilate amounts sufficient for multiplication. Such assimilatory nitrate reduction does not result in a net increase in the ammonium content in the external environment. There are some bacteria which conduct the process on a larger scale, resulting in significant extracellular accumulations of ammonium ion. This dissimilatory nitrate reduction is favored by anaerobic conditions, which demand employment by the organism of an electron sink other than oxygen. Thus, certain bacteria use the reduction of nitrate to ammonium as a biochemical alternative to the reduction of oxygen to water. Some aspects of this process are discussed further in Section 5.8.

The importance of bacteria in the degradation of biopolymers has already been emphasized in discussing the carbon cycle, and nitrogen-containing biopolymers are susceptible to the same sort of attack. Many bacteria produce extracellular hydrolytic enzymes which depolymerize proteins, nucleic acids, and other nitrogenous macromolecules. These molecules are not readily taken into the bacterial cell because they are large while intact, but the monomeric subunits made available by these hydrolases are readily utilized. Some of the better known proteolytic enzymes of bacteria are those of typical soil forms such as *Bacillus, Clostridium,* and *Streptomyces* species.

Bacillus subtilis forms a proteolytic enzyme, subtilisin, which has found employment in analysis of protein structure. Subtilisin (mol wt 25,000–30,000) has been obtained in crystalline form and acts on casein most rapidly in alkaline solution (pH 10). This enzyme exhibits broad specificity in the type of peptide linkages cleaved, and can attack both terminal and internal peptide bonds. The treatment of small peptides with subtilisin reveals that a number of different bonds are susceptible to hydrolysis, indicated in the following example by vertical dotted lines:*

ile-val-⫶-ser-asp-gly-asp-⫶-gly-met-asn-⫶-ala-try-val-ala-⫶-try-arg

Subtilisin has an interesting limited action on ovalbumin which consists of the release of three small peptides, leaving a residual, still crystallizable protein termed plakalbumin because of its plate-like crystal structure:

$$\text{ovalbumin} \xrightarrow{\text{subtilisin}} \text{plakalbumin}$$
$$+ \text{ala-gly-val-asp-ala-ala} + \text{ala-ala} + \text{ala-gly-val-asp} \qquad (6.9)$$

The proteolytic enzymes of the streptomycetes show a wide-ranging attack on different proteins; for example, *Streptomyces fradiae* forms a keratinase capable of digesting the keratin of wool and hair. Keratin resists the attack of trypsin and many other proteolytic enzymes. *Streptomyces griseus* produces a proteolytic enzyme, available commercially, which shows the broadest known specificity for attackable peptide linkages. This enzyme produces a very extensive hydrolysis, cleaving up to 87% of the peptide bonds present in ovalbumin, which results in a high yield of free amino acids. Studies with synthetic substrates indicate the *S. griseus* enzyme is capable of catalyzing the hydrolysis of practically all di- and tripeptides, including those with asparagine and glutamine.

Many *Clostridium* species are highly active in the degradation of proteins and produce a number of different types of proteolytic enzymes. One interesting example of these enzymes is the collagenase of *Clostridium histolyticum,* a pathogenic organism which causes extensive digestion of living animal tissues. This enzyme is quite specific for collagen and gelatin, which are hydrolyzed to peptides with molecular weights around 600. It has no action on fibrin, elastin keratin, albumin,

---

*The abbreviations used for amino acids are: ile, isolencine; val, valine; ser, serine; asp, asparate; gly, glycine; met, methionine; asn, asparagine; ala, alanine; try, tryptophan; arg, arginine; pro, proline.

or casein. Analysis of the action of this enzyme on defined peptides suggests that its specificity for collagen arises from a requirement for proline residues situated near the point of attack. The *Cl. histolyticum* collagenase preferentially cleaves the peptide bond linking two amino acids positioned between two proline residues:

$$—pro—aa_1 \vdots aa_2—pro—$$

It is apparent from the foregoing that under both aerobic and anaerobic conditions animal and plant proteins are subjected to bacterial attacks, which render the constituent amino acids available for assimilation by the bacterial cells. Most bacteria which are able to make use of ammonium or nitrate salts also readily assimilate preformed organic nitrogen compounds, especially in the form of amino acids. This permits bypassing the biosynthetic steps required when ammonium salts are the nitrogen source, with attendant gains in growth efficiency. There are also some bacteria (*Streptococcus* and *Lactobacillus,* for example) which are unable to use inorganic nitrogen compounds as their sole nitrogen source and require an exogenous supply of amino acids or other organic nitrogen compounds. Some bacteria need NH_4^+ in addition to amino acids. Often amino acids are incorporated directly into bacterial protein without alteration, but under some conditions they are extensively metabolized by bacteria. In the present context the reactions of most significance are *deaminations* which result in the liberation of ammonium ion. The overall process of ammonium production from proteins is termed *ammonification,* and consists of the hydrolysis of protein by proteolytic enzymes and the deamination of the resulting amino acids. Whether or not there will be a net accumulation of ammonium ion during bacterial attack on natural materials is dependent on the prevailing C:N ratio. Bacterial cells are composed of carbon and nitrogen in approximately a 10:1 ratio, so that the assimilation of nitrogen requires at least a ten-fold greater amount of assimilable carbon. Furthermore, not all of the carbon present is available for assimilation, because a large fraction (possibly two-thirds) must be employed as a source of energy.

Deaminations of several types occur in bacteria. One of these is an oxidative deamination, which produces the corresponding keto acid plus ammonia. Oxidative deaminations include actions of specific amino acid dehydrogenases, such as the glutamic acid dehydrogenase described previously in connection with ammonia utilization (equation 6.8), and also less specific L- and D-amino acid dehydrogenases which act on more than one substrate. A second type of deamination is exemplified by the aspartase and tryptophanase reactions, in which the oxidation state of the product is unchanged from that of the substrate:

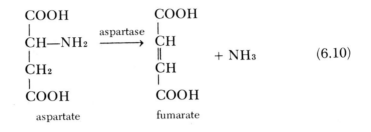

$$\underset{\text{aspartate}}{\begin{matrix} COOH \\ | \\ CH-NH_2 \\ | \\ CH_2 \\ | \\ COOH \end{matrix}} \xrightarrow{\text{aspartase}} \underset{\text{fumarate}}{\begin{matrix} COOH \\ | \\ CH \\ \| \\ CH \\ | \\ COOH \end{matrix}} + NH_3 \qquad (6.10)$$

In addition, anaerobes frequently carry out reductive deaminations which yield ammonia plus a saturated fatty acid.

Under aerobic conditions ammonia seldom accumulates to any great extent because it is subject to bacterial oxidation to nitrate in a process called *nitrification* (see Sections 4.17 and 5.9). This takes place at substantial rates in soils treated with ammonium salts, the quantity of nitrate-N formed by nitrifying bacteria amounting to 40-80 lb per acre per day. Nitrification has been considered beneficial or even essential in agriculture. It has been used as an index of soil fertility, as "richer" soils show greater numbers of nitrifying bacteria. This observed correlation undoubtedly reflects a simple case of enrichment of the population of nitrifiers due to previous applications of fertilizers, and it is by no means clear that the activities of these organisms in themselves are the *cause* of greater fertility. Plants will grow with ammonium salts, although some workers contend that their development is somewhat different. Even if nitrate is a superior nitrogen source for plants, nitrification is not of unmixed benefit in agricultural practice. Ammonium ions tend to remain complexed to soil components, but nitrate is more easily leached away by rain or irrigation water. Bacterial nitrification is promoted by application of ammonical fertilizers, but this often results in significant overall losses of nitrogen. The nitrate leached from such fields becomes a pollutant in the streams receiving the runoff and in the ground water as well.

Whether suppression of nitrification would be desirable is unresolved, but the possibility of inhibitory effects on nitrifying bacteria must be considered in connection with applications of agricultural chemicals, such as persistent pesticides. Soil fumigants such as methyl-bromide are found to depress nitrification, with corresponding increases in ammonia levels. It has been found that dinitro-*o*-sec-butylphenol (DNBP) inhibits the conversion of nitrite to nitrate by *Nitrobacter,* resulting in a temporary accumulation of the more toxic nitrite.

Losses of nitrate nitrogen occur not only because of leaching, but also because of bacterial action in the process of *denitrification.* Under anaerobic conditions a number of bacteria, such as *Pseudomonas stutzeri, Micrococcus denitrificans,* and *Bacillus licheniformis,* are able to use nitrate in place of O_2 as the terminal oxidant for their energy metabolism. This results in a successive reduction of nitrate through several stages, first to nitrite and then to nitric and nitrous oxides and finally to elemental nitrogen. The escape to the atmosphere of these gaseous-reduction products constitutes a net loss of nitrogen from the biosphere. Some biochemical aspects of denitrification are discussed further in Section 5.8.

Nitrogen exists in the atmosphere almost exclusively as N_2, in which form it is unavailable for metabolic reactions of metazoa and metaphyta. The entry of N_2 into the biosphere is made possible only by certain microorganisms, which are able to reduce the N_2 molecule to the level of ammonia. This reduction process, referred to as *nitrogen fixation,* is carried on by a number of different types of aerobic and anaerobic bacteria and also by certain blue-green algae and yeasts (see Sections 3.15 and 4.18).

The ability to fix atmospheric nitrogen confers a certain competitive advantage, as indicated by the frequent association of this capacity with pioneer organisms. Alder (*Alnus*), which is a typical pioneer plant species thriving in poor soils, fixes nitrogen symbiotically in conjunc-tion with an actinomycete growing in root nodules. Nitrogen fixation probably is an important factor in the colonization of rocks by lichens, in which one of the partners is generally a nitrogen-fixing blue-green alga. For a world increasingly dependent on high-yield agriculture to feed its hungry billions, the ability of microorganisms to mobilize the vast reservoir of atmospheric nitrogen would seem to hold great potentialities.

The reduction of nitrogen is carried out by a multienzyme complex referred to as "nitrogenase," which includes low-potential electron carriers, along with nonheme iron, labile sulfide, and molybdenum. This reaction appears to consist essentially of the reduction of an electron carrier, such as ferredoxin, by electrons from cellular energy metabolism, and then transfer of the electrons to some member (X) of the nitrogenase complex along with activation of this component by use of ATP. The activated reductant so formed (X_{red}*) then can donate electrons to the N_2 molecule, yielding ammonia:

$$e^- \longrightarrow \text{ferredoxin} \longrightarrow X_{ox} \xrightarrow[\substack{Mo^{2+} \\ Fe^{2+}}]{} X_{red}^* \longrightarrow \begin{array}{c} N_2 \\ \Big\downarrow \\ 2\,NH_3 \end{array} \qquad (6.11)$$

Ferredoxin is found in the anaerobic nitrogen-fixing bacteria, such as *Clostridium,* but the electron carriers in aerobes, such as *Azotobacter,* are not known with certainty.

Measurement of N_2 reduction is rather cumbersome and usually entails mass spectrometric determination of the conversion of $^{15}N_2$ into ammonia or cellular protein. Certain other electron acceptors can be reduced by the nitrogenase reaction, including acetylene, cyanide, and azide. The reduction of acetylene to ethylene has proven useful as a simplified index of the N_2-fixing capacity of organisms in nature.

6.6 | The Sulfur Cycle

The sulfur cycle (Fig. 6.4) occupies a pivotal position from a biogeochemical standpoint because of the great influence exerted by its interactions with other biological activities and chemical transformations. Sulfur is presented to the biosphere mostly as sulfate, in which sulfur has a +6 valence. Green plants and many bacteria assimilate sulfate as their sole sulfur source. As described in Section 5.8, sulfate is converted to an "active" nucleotide form before further metabolism. The sulfate group of "active sulfate" is readily transferable to suitable acceptor compounds bearing hydroxyls to give sulfate esters, but reduction of active sulfate to the level of sulfide (valence of −2) is necessary for its incorporation into cysteine and other thiol compounds. The sulfur-containing amino acids are em-

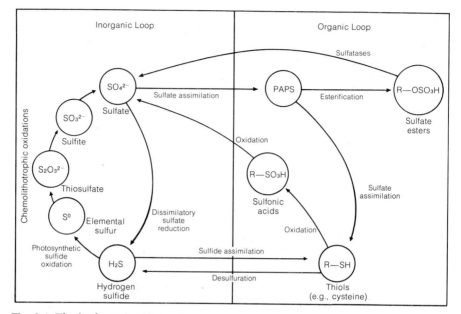

Fig. 6.4. The biological sulfur cycle.

ployed as protein constituents; and a number of other essential meta-bolic participants, such as biotin, coenzyme A, ferredoxin, lipoic acid, and thiamine pyrophosphate, also contain sulfur with a −2 valence. Those organisms unable to utilize sulfate as a sole sulfur source, that is, those defective in sulfate reduction, generally satisfy their sulfur requirement with cysteine, or in some cases, H_2S.

Under aerobic conditions bacteria oxidize the thiol group of cysteine to give the corresponding sulfinic ($R\text{-}SO_2H$) and sulfonic ($R\text{-}SO_3H$) acids, and finally the desulfonation of the latter liberates inorganic sulfite, which in turn is oxidized to sulfate:

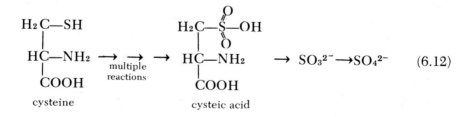

$$\to SO_3{}^{2-} \to SO_4{}^{2-} \qquad (6.12)$$

Anaerobically the attack on cysteine by bacteria, such as *Clostridium, Proteus,* and *Salmonella,* results in the release of hydrogen sulfide in the cysteine desulfhydrase reaction:

$$HSCH_2-CHNH_2-COOH + H_2O$$

cysteine

$$\xrightarrow[\text{desulfhydrase}]{\text{cysteine}} H_2S + NH_3 + CH_3-CO-COOH \qquad (6.13)$$

pyruvic acid

Because proteins contain cysteine, putrefactive reactions of bacteria on protein generally involve generation of H_2S. Similarly, the origin of methylmercaptan (CH_3SH) in putrefaction is traceable to bacterial attack on methionine.

In addition to that generated in protein decomposition, large amounts of H_2S originate bacteriologically via an entirely different type of reaction sequence, the so-called *dissimilatory sulfate reduction* process. As described in Section 5.8, this consists of the use of sulfate as the electron "sink" for cellular oxidations by certain anaerobes, e.g., *Desulfovibrio desulfuricans:*

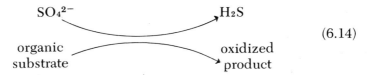

$$(6.14)$$

The nearly universal distribution of putrefactive anaerobes and dissimilatory sulfate-reducing bacteria as well as their substrates insures the presence of hydrogen sulfide in marine and freshwater muds, waterlogged soils, and other anaerobic environments.

Hydrogen sulfide is oxidized under anaerobic conditions by certain photosynthetic bacteria which utilize it as a source of reducing power for CO_2 assimilation and as a sulfur source for biosynthesis of cysteine. The initial product of sulfide oxidation by the photosynthetic anaerobes is elemental sulfur, but they can carry out the complete oxidation of sulfide to sulfate. Further descriptions of these organisms and their ecology are given in Section 3.19, while the biochemistry of bacterial photosynthesis is discussed in Section 5.12.

A number of different bacteria oxidize sulfide to sulfate aerobically. One interesting group of bacteria, the thiobacilli, has specialized in the oxidation of sulfur compounds (see Section 3.16). These organisms couple the energy released in the oxidation of sulfide, elemental sulfur, or thiosulfate, to the chemosynthesis of cellular material from carbon dioxide:

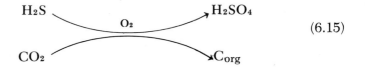

$$H_2S \qquad \qquad \qquad H_2SO_4$$
$$\qquad \qquad O_2 \qquad \qquad \qquad \qquad (6.15)$$
$$CO_2 \qquad \qquad \qquad C_{org}$$

A factor of some significance for geochemical processes is the increase in acidity which results from the generation of sulfate from H_2S or sulfur (see Section 5.9)

6.7 | Phosphorus Transformations

Phosphorus is an essential element for all organisms; it is required for the synthesis of nucleotides, phosphatides, sugar phosphates, and other phosphorylated intermediates. This element is present in nature as phosphate (PO_4^{3-}) which is its major stable form. A typical phosphate mineral is apatite, $Ca_5(PO_4)_3F$ or $Ca_5(PO_4)_3Cl$. It is not known whether there is any bacterial involvement in genesis of geological deposits of apatite, but certain bacteria have been implicated in the occasional formation of the chloride variety of this mineral in animal stomachs. Biological availability of phosphate is mediated to some extent by calcium and iron, both of which form insoluble compounds tending to precipitate from aqueous systems and accumulate in sediments. In the ocean such precipitation results in the formation of phosphate nodules found on the floor. Bacterial acid production again solubilizes the phosphate and renders it available.

Uptake of phosphate is linked to metabolism and little is accumulated unless the organism is actively metabolizing. Two alternatives are possible after entry of phosphate into the cell: introduction into the metabolic cycles by way of conversion to some organic phosphate ester such as glucose-6-phosphate, or ATP, or storage in the form of polyphosphate. This polymerization serves to assure a constant internal supply of phosphate despite external fluctuations. Formation of the polymer is accomplished by transfer of a phosphate group from

ATP to a polyphosphate by a polyphosphate kinase:

$$\text{ATP} + (\text{PO}_4^{3-})_n \rightleftharpoons \text{ADP} + (\text{PO}_4^{3-})_{n+1} \qquad (6.16)$$

Polyphosphate can be mobilized by several mechanisms, including cleavage catalyzed by a "polyphosphatase." Bacteria participate actively in the decomposition of phosphorus compounds in nature, releasing phosphate by means of a variety of phosphomono- and diesterases of differing specificity: the deoxyribonucleases, ribonucleases, nucleotidases, ATPases, and phosphatases.

It has been generally accepted that biological transformations of phosphorus consist only of this interconversion of phosphate between inorganic and organic forms, in which the oxidation state of the element remains unchanged. This may be an oversimplification, the result of the limited state of knowledge of phosphorus transformations in nature, because it is known that phosphate is microbiologically reduced under anaerobic conditions to give phosphite and hypophosphite; furthermore, some common organisms such as *Pseudomonas* and *Serratia* will use phosphite in place of phosphate, and enzymatically catalyze the oxidation of phosphite to phosphate.

6.8 | Bacterial Degradation of Plant Materials

It follows from the dominant position of photosynthesis in the biological cycle of carbon that the decomposition of the products of this synthesis will assume great significance. The ultimate source of carbon for chemoorganotrophic organisms, whether bacteria, fungi, or metazoa, is the carbon that is fixed in photosynthesis. The chemical structures of some major plant polymers, whose degradation is so important for continuation of the carbon cycle, are shown in Figure 6.5.

Cellulose, a water-insoluble "high polymer" of β-1 $\rightarrow$ 4-linked D-glucose units, is a major structural substance of plant cell walls, accounting for at least one-third of the total dry weight of most plant materials. The largest commercial source of cellulose is wood, which has an average cellulose content of about 43%, but the purest sources are textile fibers, such as cotton. Cotton fibers, which are actually unicel-

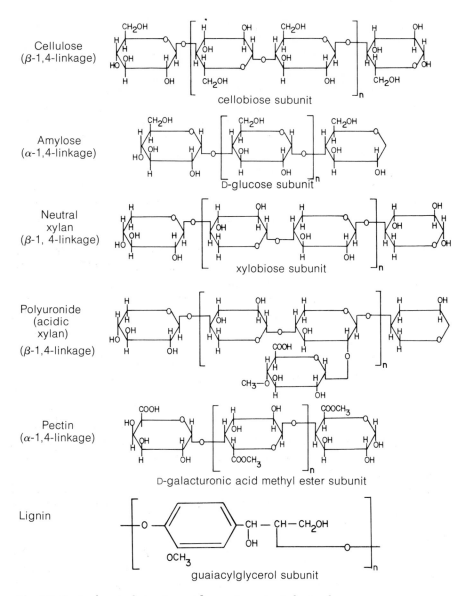

Cellulose (β-1,4-linkage) — cellobiose subunit

Amylose (α-1,4-linkage) — D-glucose subunit

Neutral xylan (β-1, 4-linkage) — xylobiose subunit

Polyuronide (acidic xylan) (β-1,4-linkage)

Pectin (α-1,4-linkage) — D-galacturonic acid methyl ester subunit

Lignin — guaiacylglycerol subunit

Fig. 6.5. Basic chemical structures of some important plant polymers.

lular seed hairs, consist of 91–94% cellulose. Native cellulose molecules contain between 8×10^3 and 1.5×10^4 glucose residues, which is equivalent to a molecular weight of $1–2.4 \times 10^6$. The polymer apparently is a simple linear, unbranched chain, but its secondary structure is ill defined. The cellulose molecules are arranged into microfibrils with highly ordered, crystalline regions detectable by X-ray diffraction analysis. Cellulose is insoluble in dilute acid or alkali, or in organic solvents, but it can be solubilized, with chemical alteration, in concentrated mineral acids or ammonical copper sulfate (Schweitzer's reagent).

In the plant cell wall, cellulose is embedded in a noncellulosic matrix consisting of so-called "hemicelluloses" plus pectic substances and lignin. The overall lignin and hemicellulose of a plant material such as oat straw average about 20% each, compared to a cellulose content of about 35%. In hardwoods, lignin accounts for about 20% of the dry weight, hemicelluloses about 30–35%, and pectins 1–4%. Hemicelluloses form a heterogeneous collection of smaller water-insoluble polysaccharides, with molecular weights between 7×10^3 and 2.5×10^4, which are separable from cellulose on the basis of their solubility in alkali. There are two principal types of hemicelluloses, the acidic polyuronides and the neutral cellulosans. The polyuronides are heteropolymers containing glucuronic or galacturonic acid residues, commonly in a terminal position branching from the main polysaccharide chain, which often is made up of repeating pentose units. Because of the uronic acid components, these substances are readily soluble in cold dilute alkali (4% NaOH). The neutral noncell-ulosic polysaccharides, or cellulosans, include both linear and highly branched polymers, which are often analyzed by means of solubilization in cold concentrated alkali (17.5% NaOH). These include homopolymers of pentose or hexose units, such as the xylans, arabans, mannans, and galactans, and also copolymers, such as galactomannans and arabinogalactans. Pectins are chemically related to the hemicelluloses, and consist of high molecular weight (3×10^5) polymers of galacturonic acid and its methyl ester. Pectins are particularly rich in young tissues and fruits, from which they can be extracted with dilute acid. Lignin is not a single compound, but a class of substances ranging in size and molecular weight from 8×10^2 to 4×10^3. Lignins actually are not polysaccharides, but rather are amorphous high polymers of aromatic subunits, believed to be related to guaiacylglycerol. These phenylpropane subunits probably are not all linked in a single fashion as implied in Figure 6.5, but instead are joined in a variety of ways including double linkages to give multiple condensed ring systems.

Cellulose degradation. The ability to attack cellulose is not universal among bacteria, but it is decidedly not rare. There are 78 currently recognized species spread over 16 widely diverse genera that have been reported to digest cellulose. Among these are gram-positive and gram-negative rods and cocci, aerobes and anaerobes, thermophiles and mesophiles, and unicellular and multicellular forms from animal, marine, and terrestrial habitats. A listing of genera which include cellulolytic species is made in Table 6.1. This is not intended to be a definitive enumeration, but only an illustration of the occurrence of this property in four distinctly different types of bacteria: eubacteria, fruiting and nonfruiting myxobacteria, and streptomycetes. Probably a thorough search would reveal many additional cellulolytic bacteria.

A number of these organisms appear to utilize cellulose normally as their major energy source, and are able to make use of only a limited number of carbohydrates other than cellulose, usually only glucose and/or cellobiose, both of which are derivable from cellulose. It is not surprising that bacteria should have developed an enzymic mechanism to avail themselves of such a ubiquitous energy source. Indeed, the surprising point is that there are so many soil bacteria that appear not to utilize cellulose. The possession of cellulolytic enzymes would seem to have such great adaptive value for bacteria that none would be found without them. The wide distribution of this capability suggests that this solution to the pressure of nutrient deficiency has been arrived at repeatedly and independently. The occurrence of both cellulolytic and noncellulolytic species of the same genus in the same habitat shows that the competitive advantage confered by cellulolytic activity, while probably real, is far from absolute.

Bacteria are the chief cellulolytic organisms in anaerobic environments, such as the rumen and gut of herbivores, in dung heaps and in waterlogged soils. Under aerobic conditions, however, bacteria face severe competition from fungi, many of which are extremely active with respect to cellulose digestion. Most of the cellulose decomposition in soil takes place in the upper 20 cm, which is sufficiently aerobic for both the fungi and bacteria. Bacterial attack is favored by a pH between 6.5 and 8.0 and a soil humidity of 60%. Fungi assume the dominant role in cellulose decomposition in acidic soils, and they have been long recognized as major contributors to the rotting of wood.

Table 6.1. Genera of Bacteria Containing Cellulolytic Species*

Group	Aerobes	Anaerobes
Eubacteria	*Cellulomonas* (10) *Cellvibrio* (4) *Cellfalcicula* (2) *Vibrio* (3) *Pseudomonas* (8)	*Butyrivibrio* (1) *Clostridium* (9) *Bacteroides* (1) *Ruminococcus* (2)
Nonfruiting Myxobacteria	*Cytophaga* (7) *Sporocytophaga* (3)	
Fruiting Myxobacteria	*Sorangium* (3) *Polyangium* (1) *Angiococcus* (1)	
Actinomycetes	*Streptomyces* (20) *Micromonospora* (2)	

*The number of cellulolytic species is given in parentheses following each genus.

Bacterial attack on wood is slow and the organisms spread only with difficulty, apparently by breaching the thin-walled tracheid pits. Digestion of cellulose also occurs during invasion of living plants by plant pathogens. The correlation between cellulolytic activity and phytopathology is well established for fungal pathogens, but for the most part, bacterial phytopathogens do not show much dependency on cellulolytic enzymes. Infection of tomato stems by *Pseudomonas solanacearum* does show a progressive hydrolysis of the cellulose accompanying the advance of the diseased state.

Bacterial cellulases are in many cases surface-bound enzymes not readily released into the surrounding environment, and cellulose digestion generally requires close contact between cells and substrate. Colonies of cellulolytic bacteria often fail to show zones of hydrolysis on cellulose agar, and "spent" growth medium lacks cellulase activity. This would explain why in the soil bacteria are found colonizing the exterior surfaces of plant fragments or other cellulosic substrates. Attack on the insoluble substrate by cellulases bound to the cell surface cannot occur at a distance, and is limited to a slow etching and erosion of those regions where the cells are in direct contact with the cellulose. Generally, bacterial cellulases are inducible enzymes, but how an insoluble high polymer can achieve enzyme induction is not obvious. Presumably some enzyme must be present initially in order to hydrolyze enough cellulose to the soluble fragments which are the actual inducer molecules. Some cellulolytic bacteria do show zones around colonies on cellulose agar and enzyme activity in culture filtrates.

There is some uncertainty with regard to the number of enzymes actually involved. A single hydrolytic enzyme should suffice for a homopolymer such as cellulose, but multiple enzyme systems have been invoked to account for the differing problems of the initial attack on crystalline regions versus simple chain scission in smaller polymers. Frequently, but not invariably, multiple cellulases are demonstrable chromatographically or electrophoretically. *Streptomyces gilvus* cellulase preparations show five individual bands when subjected to electrophoresis. Three of these are identical immunologically, but are unrelated to the other two components. *Cellvibrio gilvus* cellulase consists of four electrophoretically separable components, which are not identical in kinetic properties. Enzymology of bacterial cellulases is still rudimentary, but available evidence suggests that possibly there are two different types of enzymes with regard to their mode of attack. One mechanism shown by the cellulases from *Streptomyces* and *Pseudomonas* resembles that of fungal cellulases and consists of a random hydrolysis of interior β-1 $\rightarrow$ 4-linkages to produce various shorter cellodextrins. This type of enzyme should produce a rapid drop in the degree of polymerization, but only a slow increase in reducing sugar values. The second type is represented by the cellulases of *Cellvibrio*, and possibly *Ruminococcus* and *Clostridium* species, which exhibit a nonrandom attack on the end of the chain. The mechanism of hydrolysis by *Cellvibrio* enzymes has been studied using soluble, short-chain cellodextrins containing 3–6 glucose units. These enzymes were found to be endoglucanases which preferentially attack the second and third glycosidic bond from the nonreducing end of cellopentose and cellohexaose, thereby releasing cellobiose. Presumably the same mechanism applies for the highly polymerized substrate, but such extrapolations are not always warranted.

Attack on pectin. Pectin-decomposing bacteria are common in soil, with numbers ranging from 10^4 to 10^6 per gram. They include several species of aerobic and anaerobic sporeforming bacilli *(Clostridium, Bacillus)*. "Soft-rots" produced by some phytopathogenic gram-negative rods *(Xanthomonas, Erwinia)* result from the attack on the pectin of the middle lamella by these organisms. In the case of flax, hemp, and jute, the useful cellulosic fibers form part of the vascular system of the stem and are embedded in pectin. This is removed by bacterial action in the so-called *retting process.* The retting organisms should be highly active in pectin breakdown, but should have no action on cellulose. Pectin-fermenting clostridia isolated from soil and retting operations, such as *Cl. felsineum, Cl. pectinovorum* and *Cl. multifermentans*, meet this requirement, as do the two species of *Bacillus* which are active in pectin degradation, *B. polymyxa* and *B. macerans*. Such pectinolytic bacteria seem to be less specialized than some of the cellulolytic types, and they are able to utilize a variety of energy sources in addition to pectin.

There are two enzymes involved in the attack on pectin. One of these, pectin esterase, removes the methyl esters from the carboxyl group, but does not act on glycosidic bonds of the resulting polygalacturonic acid. The other enzyme is an exopolygalacturonate lyase, which degrades the polygalacturonic acid by cleaving off digalacturonic acid units from the reducing end of the chain. The lyase does not attack highly esterified pectins, and thus is dependent on prior action of the esterase. In view of the complimentary action of these two enzymes it is interesting that they are formed as a tightly bound unit by *Clostridium multifermentans* during growth on pectin. The esterase and lyase are not resolved during a multistep purification procedure involving gel filtration, ion-exchange chromatography, and centrifugation.

Starch hydrolysis. Starch, the reserve substance of seeds and callus tissue, is composed of two fractions, amylose and amylopectin, both of which are homopolymers of α-D-glucose residues. Amylose consists almost exclusively of linear chains in which the D-glucose units are linked α-1 $\rightarrow$ 4, while amylopectin has a branched structure with α-1 $\rightarrow$ 6-linkages at the branch-points of shorter sequences of α-1 $\rightarrow$ 4-linked glucose units. There are about 20α-1 $\rightarrow$ 4-linkages for each α-1 $\rightarrow$ 6-linkage in amylopectin.

Starch is depolymerized by bacteria, so that this property is routinely determined for taxonomic purposes. However, the enzymes involved have been given only cursory examinations in all but a limited number of cases. A few bacterial amylases, including those of the mesophile, *Bacillus subtilis,* and the thermophile, *Bacillus stearothermophilus,* have been obtained in crystalline form. Bacteria produce several types of starch-degrading enzymes. Most appear to be alpha amylases resembling those of higher plants in their lack of thiol groups at the active center, activation by calcium, and the random mode of attack on the α-1 $\rightarrow$ 4-glycosidic bonds of amylose. These, like alpha amylases in general, are "liquefying" enzymes in the sense that random cleavage results in a rapid decrease in viscosity of starch solutions compared to the slower increase in reducing power (end groups). The final product of prolonged action is maltose plus a small residue of unattackable glucan, the so-called "limit dextrin" resulting from the inability of this enzyme to hydrolyze α-1 → 6-bonds.

An interesting type of attack on starch is exhibited by *Bacillus macerans,* which forms, instead of maltose, high yields of cyclic oligomers, the so-called "Schardinger dextrins," consisting of six or more α-D-glucopyranose units linked α-1 → 4. The *B. macerans* enzyme appears to preferentially attack the nonreducing end of the chain with the release of 6, 7, or 8 glucose units, which are cyclized to give the α, β, and γ Schardinger dextrins:

$$(G\text{-}G)_n\text{-}G \vdots G\text{-}G\text{-}G\text{-}G\text{-}G\text{-}G\text{-}G \xrightarrow{B.\ macerans} (G\text{-}G)_n\text{-}G + \begin{array}{c} \text{-G-} \\ G \quad\quad G \\ \\ G \quad\quad\quad G \\ \\ G\text{---}G \end{array}$$

β-dextrin

$$(6.17)$$

There is no increase in reducing sugar values during this conversion because the Schardinger dextrins have no free carbonyl end group. These substances, which are easily crystallized from propanol-water mixtures, are rather resistant to acid hydrolysis and the action of plant amylases, but are readily cleaved by the fungal and bacterial enzymes. *Bacillus polymyxa* attacks the β-dextrin to yield three molecules of maltose plus glucose.

Bacteria apparently do not form beta amylases such as those of plants, which are characterized by a selective attack on the nonreducing end with the sequential release of maltose units. However, certain bacteria such as *Clostridium acetobutylicum* do produce somewhat analogous glucamylases, which cleave glucose units from the ends of the chains. And, in addition, bacteria are known to form phosphorylases. These accomplish the phosphorolysis rather than hydrolysis of the terminal glycosidic bond to yield glucose-1-phosphate.

Lignin degradation and humification. The great heterogeneity of hemicelluloses unduly complicates any attempt at a systematic discussion of their breakdown by bacteria. Instead we must content ourselves with the statement that bacteria as a whole possess the enzymic prowess to degrade such materials, and that in a number of cases, including the rumen environment, they are the principal agents of hemicellulose biodegradation. Such is not the case with lignin, which proves to be quite resistant to bacterial attack. Decomposition of lignin is accomplished almost exclusively by certain basidiomycetes, the so-called white-rot fungi, which in six to eight months can destroy some 60–75% of the native lignin of plant materials. This relatively slow process is confined to aerobic habitats, since the white-rot fungi, like most other fungi, are aerobes. This lack of bacterial action explains the stability of lignin under anaerobic conditions where fungal attack is precluded. The subsequent fate of the phenolic subunits cleaved from lignin by the white-rot fungi is obscure. Some must be assimilated by the fungi to supply energy and structural material for synthesis of new fungal cells, but at least a portion is converted by unknown oxidative mechanisms to new polymeric forms, collectively called humus. The relative importance of contributions of fungi and bacteria to the humification process is problematical, but it is generally presumed that bacteria are of considerable significance in this transformation. Humus can be subdivided into several fractions, of which the humic acid fraction is the largest. The humic acid fraction consists of polyphenolic material characterized by the presence of aromatic nuclei, and a higher content of carboxyl and carbonyl groups than the supposed lignin precursor material. It is apparently devoid of nitrogen, which serves to distinguish it from dark melanin pigments of similar appearance which are also produced by soil fungi. Possibly the immediate products of lignin hydrolysis are oxidized to phenolic and quinone structures which are then polymerized by fungi and bacteria to yield humic acids. The humic substances, however formed, act as a reserve source of usable nutrients for soil bacteria. Although they can be degraded, their relative persistence in soil indicates that not all of the humic substances are directly accessible at any one time. The greater portion probably is complexed with clay minerals, and thereby rendered unavailable. Desorption by alternate wetting and drying cycles releases fresh supplies for microbial metabolism. In general, we may conclude that because of the voracious and omnivorous nature of the appetites of the soil microflora all available materials will be quickly depleted. Thus the persistence of a substance is tantamount to proof of its inaccessibility.

6.9 | Chitin Degradation

The mucopolysaccharide chitin assumes an importance in the arthropods and filamentous fungi analogous to that of cellulose in higher plants, and billions of tons of chitin, in the form of insect and crustacean exoskeletons and fungal cell walls, become available annually for microbial attack in the soil and marine environments. Chitin, a polymer of β-1 $\rightarrow$ 4-linked N-acetylglucosamine units, is degraded by a considerable number of different bacteria, including streptomycetes, myxobacteria, and eubacteria such as *Serratia, Klebsiella, Pseudomonas, Vibrio, Clostridium,* and *Flavobacterium*. It has been estimated that over 0.1% of all the bacteria in the ocean show chitinolytic activity.

Bacterial degradation of chitin resembles the attack on cellulose: the "chitinase" is an inducible endoglycanase, specifically hydrolyzing the β-1$\rightarrow$4-linked N-acetylglucosamine polymer to the dimer, N,N-diacetylchitobiose. Chitobiose in turn is cleaved by an associated "chitobiase" (N-acetylglucosaminidase) to give N-acetylglucosamine as the final product. Bacterial chitinase preparations show no activity against the deacetylated polymer (the so-called "chitosan") or against cellulose, and generally native chitin is less easily hydrolyzed than "purified" or "swollen" material.

Little is known about the possible significance of chitinolytic activities of bacteria beyond simple participation in the degradation of dead remains. An interesting suggestion has been made that chitinolysis may serve as a control mechanism in the ecology of soil fungi. Addition of chitin to crop land reportedly exerts a beneficial effect by suppressing phytopathogenic fungi. Presumably the added chitin stimulates development of a chitinolytic microflora which is able to attack fungal cell walls.

6.10 | Bacteria in Relation to Fossil Fuels

The existence of large deposits of coal and petroleum is a matter of theoretical and practical importance for geochemistry and bacteriology. It would be interesting to determine how these deposits originated and what bacterial processes might have been involved in their genesis. Bacteria are detected in the earth at great depths ($>10^3$ m) in oil explorations, but it is not known whether these are indigenous or have been introduced by the drilling. If these organisms are part of an indigenous subterranean flora, how did they get there, how long have they been there, and what influences do they exert? Since bacteria have the ability to oxidize hydrocarbons, attempts have been made to develop bacteriological prospecting methods for petroleum (see Section 3.14), and to devise bacterial treatments for the alleviation of pollution problems arising from oil spills and waste oil disposal. Attempts are being made to employ petroleum and coal as substrates for growth of microorganisms for use as food, and for transformation into valuable products by bacterial action. Solutions are needed to practical problems such as the formation of acid drainage waters from coal mines, the bacterial corrosion of drilling apparatus, and the bacterial decomposition of fuels and cutting oils.

Origin of fossil fuels. The prevailing opinion is that both coal and petroleum are of biogenic origin and represent the much altered remains of ancient life forms. In this view each was formed through the conservation of the more stable materials of dead organisms, mostly plants, with petroleum resulting primarily from the lipoidal fractions (hydrocarbons and fatty acids) and coal from woody tissues (cellulose, lignin, cutin). Bacteria undoubtedly participated in the initial biochemical stages of the process, when the various components of the biological raw materials were winnowed out, degraded, and resynthesized, but the actual extent and nature of this involvement are conjectural.

Coal occurs in sedimentary rocks as old as the Precambrian ($1-3 \times 10^9$ years ago), but extensive deposits date only from the Carboniferous period (beginning about 3.45×10^8 years ago) following the widespread development of terrestrial plants. In swamps and bogs these gave rise to accumulations of peat, later metamorphosed into coal. This geological metamorphosis was a slow nonbiological transformation occurring under the influence of elevated pressures and temperatures. The various types of peat and coals (lignite, bituminous, anthracite) can be arranged in a series in which increasing carbon content is accompanied by a decreasing content of hydrogen and oxygen. The carbon contents of peat, lignite, bituminous, and anthracite coal are 55, 73, 83, and 94%, respectively. The oxygen content falls from 36% in peat to 20% in lignite, 8% in bituminous and 3% in anthracite coal. A decrease in oxygen content would be an expected consequence of conversion of the aromatic acids of humic substances into hydrocarbons, with a loss of carboxyl and phenolic hydroxyl groups. One view is that rank in this series respresents the degree of metamorphosis undergone by the original peaty materials, which were essentially similar in all cases, but some geochemists consider rank to be determined by the nature of the original materials. Coal is composed of complex, insoluble polymeric materials. These are predominantly aromatic condensed polycycles and the degree of aromatization increases with rank (carbon content). Simple examples of the type of polycyclic aromatic structures found in coal are naphthalene, anthracene, phenanthrene, and chrysene.

In swamps the course of events following the death of plants is dictated by the prevailing anaerobic conditions, and is therefore different from that in aerobic soils. The principal agents for the degradation of plant remains are bacteria rather than fungi. Anaerobiosis is no bar to bacterial decomposition of cellulose, hemicelluloses, pectin, and other polysaccharides, but it hinders the attack on fatty acids, hydrocarbons, waxes, and aromatic compounds. Because lignin is degraded almost exclusively by aerobic fungi and not by bacteria, lignin also tends to accumulate. The biological activity continues until conditions preventing further action are developed. A high content of humic substances, aromatic acids, tannins, and lignin derivatives may be responsible for the lack of bacterial activity by direct inhibitory effects and by forming complexes with otherwise biodegradable materials. This might explain the existence of cellulose, hemicellulose, amino acids, glucose, and xylose in peat samples. In Scottish moss peats, analysis shows as much as 1–4% protein, and up to 20% cellulose. In some mature peat bogs microbial activity must be essentially nonexistent, as shown by the recovery from Danish bogs of only partially decomposed human bodies dating from Roman times. One of the more remarkable of these is the so-called Tollund man (Fig. 6.6), whose body had been protected from bacterial action for an estimated 2,000 years by burial in the peat of the Tollund Bog.

Oil deposits are found associated with sedimentary formations, the most ancient dating from the late Precambrian and Cambrian. If petroleum was produced in earlier periods, it may have been destroyed, but possibly the process had to await development of suitable biological material, perhaps marine algae. The precise site of origin of a deposit is indeterminable because petroleum is a mobile liquid subject to possibly extensive migrations. One idea is that the parental oleaginous materials moved from the original sediments into porous reservoir rocks through which they were driven by water flows until eventual

Fig. 6.6. Head of the Tollund man, Silkeborg Museum, Jutland, Denmark. This remarkably preserved body, showing little evidence of bacterial attack, was discovered in the Tollund Bog of central Jutland, Denmark, and is estimated to be approximately 2000 years old. (Photograph by permission of the Danish Information Office, New York.)

entrapment. Crude oils vary in composition depending on their source, but they always contain a high proportion of alkane, cycloalkane, and aromatic hydrocarbons, ranging from 50% in high asphaltic types (Mexican and Californian) to over 95% in the geologically older high

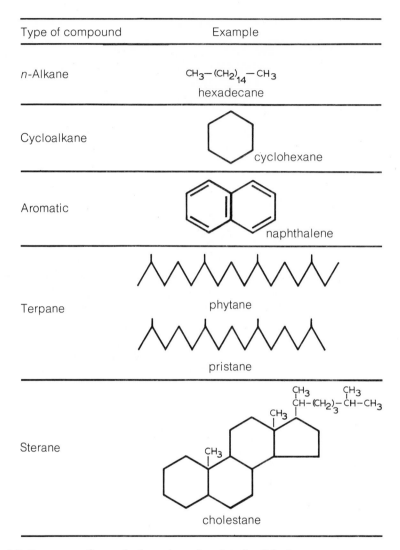

Type of compound	Example
n-Alkane	$CH_3-(CH_2)_{14}-CH_3$ hexadecane
Cycloalkane	cyclohexane
Aromatic	naphthalene
Terpane	phytane pristane
Sterane	cholestane

Fig. 6.7. Structures of some hydrocarbons found in fossil fuels.

paraffinic Pennsylvania types. Some representative hydrocarbon structures are shown in Fig. 6.7. The alkanes are mostly the normal paraffins
and isoparaffins with chain lengths from 1 to 30 carbons without
marked differences in abundance for odd and even numbered chain
lengths. Branching is limited (e.g., methyl groups); thus, only a rela-

tively few of the large number of possible isomers are represented. Nonhydrocarbon compounds containing oxygen, sulfur, and nitrogen also occur in petroleum, along with a small amount of nickel, vanadium, and iron. The biogenic theory was strengthened by the early finding in petroleum of metalloporphyrins of the etio- and deoxyphylloporphyrin types, generally with vanadium or nickel as the metal component. Chlorophyll immediately suggests itself as a ready biological source of the basic porphyrin structure. Several petroleum hydrocarbons, phytane ($C_{20}H_{42}$), pristane ($C_{19}H_{40}$), and farnesane ($C_{15}H_{32}$), could have been derived from the isoprenoid alcohol, phytol, which forms part of the chlorophyll molecule. The corresponding carboxylic acids, phytanic, pristanic, and farnesanic, also have been detected in petroleum, but phytol itself is absent. Similar reasoning would connect the sterane hydrocarbons of petroleum, such as cholestane, with steroidal precursors, such as cholesterol.

In fairness, a *caveat* must be offered regarding this biogenic theory, which is after all simply a theory. Porphyrins have been reported in the Orgueil meteorite, possibly indicating abiogenic synthesis of the porphyrin structure, if contamination of the meteorite by terrestrial material is discounted. Hydrocarbons with a distribution somewhat resembling that in ancient sediments and petroleum have been synthesized nonbiologically in the laboratory in "primitive earth" experiments. These do not prove an abiogenic origin for petroleum, but suggest that a reexamination of the deductions regarding hydrocarbons as biological markers may be warranted.

A relatively small fraction of the carbonaceous material formed in photosynthesis becomes buried or trapped in sediments and is removed from the carbon cycle. In the sea probably not more than 0.5–0.8% of the overall phytoplankton biomass reaches the bottom without decomposition. In bottom sediments the parental materials are in part transformed by bacterial action, resulting in a rather high content of humic substances (15–30% of the total organic fraction). This bacterial attack on natural substrates is not irresistable, however, and a portion remains intact. Marine sediments of the Recent period (less than 10^4 years old) are often rich in organic matter, averaging about 2.5% near shore and 1% in open oceanic regions. Among the identifiable materials found in such sediments are polysaccharides (starch, cellulose, mannans); monosaccharides (glucose, xylose); amino acids (all L-isomers); porphyrins and chlorophyll-like compounds; fatty acids (mostly C_{14} and C_{16}, with low content of odd numbered carbon chains); and hydrocarbons (n-C_{15}, n-C_{17}, n-C_{29} and n-C_{31} alkanes, and pristane).

The overall content of lipoidal substances in these sediments is low, but possibly sufficient to account for petroleum generation, since the total petroleum reserves are estimated to represent only 0.01% of the organic matter present in sedimentary rock. To account for the petroleum hydrocarbons there are two possibilities, either simple preservation of hydrocarbons of the original sedimentary material, or conversion of other substances (e.g., fatty acids) to hydrocarbons. These are not mutually exclusive and both may have occurred. The strong odd-numbered chain preference noted for the n-alkanes of many living organisms and recent sediments is less prominent in ancient sediments, and absent altogether in petroleum. Although pristane occurs in some living organisms and Recent sediments, phytane is absent in both. The phytane found in ancient sediments and petroleum may represent a conversion product, or might possibly have been a hydrocarbon component of ancient but not modern organisms. Fatty acid profiles of Recent and ancient sediments and petroleum are noticeably different: the predominance of C_{14} and C_{16} acids in Recent sediments shifts to the C_{16} and C_{18} acids in ancient sediments along with a higher content of odd numbered fatty acids, while in petroleum odd and even numbered acids are equally represented.

Bacteria contain small amounts of hydrocarbons, which might contribute to the accumulations in sediments. These are still imperfectly known, and some of the more detailed studies have dealt with species unlikely to be important in sediments. Among the hydrocarbons found in bacteria are long-chain unbranched and branched alkanes and alkenes, as well as terpenoid compounds related to squalene. The saturated and mono-unsaturated n-C_{17} paraffins comprise a large proportion of the hydrocarbon fraction of the marine organism *Vibrio marinus*. In *Sarcina lutea* the major hydrocarbons are the C_{27}, C_{28}, and C_{29} paraffins, with 3,25-dimethyl-heptacos-13-ene representing the predominant C_{29} compound.

Utilization of fossil fuels. The ability of a rather large number of bacteria to oxidize hydrocarbons may be of possibly greater significance (Section 3.14). Among the hydrocarbon-oxidizing bacteria are unicellular gram-negative rods (*Pseudomonas, Flavobacterium, Alcaligenes,* and *Achromobacter* species), unicellular gram-positive rods and cocci

(*Brevibacterium, Corynebacterium, Arthrobacter, Bacillus,* and *Micrococcus* species), acid-fast rods (*Mycobacterium*), and multicellular, branching organisms (*Nocardia,* and *Streptomyces* species). The exact manner by which water-insoluble hydrocarbons are attacked by bacteria from the aqueous phase presents an interesting problem, and obviously there must be an intimate association. In the case of *Nocardia* at least, the cells become hydrophobic (coated with hydrocarbon?), and upon centrifugation of hydrocarbon-grown cultures the cells are recoverable at the oil-water interface.

Geologically young areas subjected to folding and uplift develop fractures and ruptures of the rock layers and these allow petroleum to gain access to the surface. Such petroleum "seeps," in the form of asphalt and tar pits, as well as gaseous and liquid discharges, have been known from antiquity and have led to the discovery of rich oil fields, especially in the Middle East. Burning gas seeps, known as "The Eternal Fires," have been aflame in Iraq for centuries. Seepage of petroleum of high paraffin content sometimes results in the formation of a deposit of a yellow to brown waxy solid material termed ozokerite, mined in the past as a source of paraffin. Ozokerite consists of higher alkane fractions (melting above 60° C), somewhat modified by bacterial oxidation and the admixture of bacterial products. Hydrocarbon-oxidizing *Mycobacterium* and *Nocardia* species have been isolated from ozokerite. Gaseous seepage in regions near oil fields presents a rich if somewhat unusual energy source for soil organisms and results in a prolific development of hydrocarbon oxidizers. The heavy growth of these organisms along with their extracellular products imparts a distinctive gummy or waxy quality to the soil in the area of the seepage ("paraffin dirt").

The key to the biochemical ecology of the hydrocarbons lies in the requirement of molecular oxygen for the microbial oxidation process. This means that hydrocarbons will remain stable and nonbiodegradable under anaerobic conditions. The molecular structure of the hydrocarbon is also an important factor in biodegradability. Despite the general similarity of the compounds, bacteria exhibit a distinct substrate specificity with hydrocarbons, even within an homologous series. Many of the cultures isolated on hexadecane, for example, do not attack the gaseous hydrocarbons of this series (methane, ethane, propane, or butane) or certain of the members in the C_5–C_8 range. In general, the straight-chain compounds are more amenable to bacterial oxidation than branched compounds, and branched compounds with only a methyl group for the sidechain are more susceptible than highly branched structures. An exception to this generalization regarding branching are the terpanes, such as pristane, which are readily attacked.

One major route for bacterial hydrocarbon utilization is an attack on the terminal methyl group catalyzed by an oxygenase, which results in the formation of the corresponding primary alcohol. Subsequent oxidation of this alcohol to the carboxylic acid permits the molecule to be handled as a typical fatty acid, that is, to be converted to acetate units. Thus, beyond the initial oxygenation steps, the hydrocarbon loses any distinctive properties and enters the general metabolism of the organism. Some possible biochemical improvisations on this motif, such as formation of the dioic acid by omega-oxidation of the alkanoic acid, or attack on the subterminal methylene to form the methylketone, have been recognized.

A somewhat different problem is presented by cycloalkanes and aromatic hydrocarbons which lack free methyl groups. In the case of naphthalene, bacterial oxidation begins by oxygenation of the ring with O_2 to give a dihydroxylated derivative, 1,2-dihydroxynaphthalene. This is followed by a ring scission and chain shortening resulting in salicylic acid. Salicylic acid is convertible to catechol, thereby allowing entry into the main pathway of aromatic metabolism. Anthracene and phenanthrene also are degraded via salicylic acid as an intermediate.

6.11 | Formation of Mineral Deposits by Bacterial Activities

Sedimentary mineral deposits are classified into two types: those that formed at the same time as the sedimentary rocks in which they are found (primary, or *syngenetic* deposits); and those that were produced subsequently (secondary, or *epigenetic*). Either type may be of biological or nonbiological origin. Bacterial participation in the accumulation of inorganic materials on a geologically significant scale would seem unlikely, but, in fact, bacterial activities clearly are involved in the formation of certain deposits of calcite, gypsum, elemental sulfur, and various metallic ores. An interrelated complex of processes accounting for a number of these arises from bacterial activities in the sulfur cycle, as outlined in Section 6.6.

Several mechanisms, direct and indirect, are involved in such depositional processes. One simple possibility would be the active intracellular accumulation of the element in question, for use as an enzyme cofactor or as a structural component, followed by mineralization upon death of the organisms. Bacteria achieve a many-fold concentration of metal ions and other mineral substances over the levels in their environment, as do other microorganisms, but these are needed in only catalytic quantities, generally too low to be of geological importance. A more favorable situation is obtained by direct involvement of substrate quantities of the inorganic compound or element in the dissimilatory oxidation-reduction reactions of cellular energy metabolism. Appreciable deposits arise by the enzymatic oxidation or reduction of mineral substances in which an insoluble product results. Less direct actions often are equally effective, e.g., the promotion of purely chemical oxidations and reductions by bacterially-caused alterations of the prevailing pH or O-R potential. Certain end products of bacterial metabolism can undergo secondary chemical reactions with inorganic compounds or elements to form insoluble precipitates. In some cases a metal ion is present in the form of a soluble complex with an organic compound, whose utilization by bacteria permits the release and subsequent precipitation of the metal.

Deposition of calcite. As indicated previously (Section 6.4) much of the earth's carbon is in the form of calcium carbonate ($CaCO_3$) and other carbonates. The two major crystalline forms of calcium carbonate are the rhombohedral calcite, and the less stable orthorhombic form, aragonite. Formation of primary calcium carbonate deposits is a major biogeochemical process and takes place mostly in the ocean, largely as a result of the actions of marine animals and algae, but perhaps also of bacteria. The ocean is the great repository for the calcium and magnesium ions liberated from the crust by weathering, and it is the neutralization of such cations by atmospheric CO_2 which accounts for its removal to form the large oceanic reservoir of bicarbonate. Calcium ion is thus present as the more soluble bicarbonate, but this in turn is dependent on the partial pressure of CO_2, and therefore processes that utilize CO_2 tend to shift the equilibrium to give the poorly soluble carbonate. The large-scale withdrawal of CO_2

by algae for photosynthesis has an alkalizing effect, and results in the generation of calcium carbonate, which crystallizes from solution as calcite:

$$Ca(HCO_3)_2 \rightleftharpoons H_2O + \boxed{CO_2} + CaCO_3 \downarrow \qquad (6.18)$$

(calcite)

photosynthesis

In addition to this algal mechanism, various nonphotosynthetic bacteria have been shown to be responsible for calcite depositions in soil, especially in the root zones, and in marine and fresh water habitats. Some marine bacteria actively concentrate calcium ions, and then each individual cell acts as a nidus for crystallization of aragonite, the crystals of which contain one or more entrapped bacteria. In calcium-laden waters issuing from some thermal springs a number of bacteria have been observed to deposit calcium carbonate crystals in extracellular slime formations.

Sulfate-reducing bacteria are implicated in a number of the important biogeochemical processes, including formation of secondary calcite. Gypsum ($CaSO_4 \cdot 2H_2O$) serves as a source of sulfate for these organisms, and is converted to calcium sulfide (CaS). Subsequent interaction of the calcium sulfide with carbon dioxide brings about formation of calcium bicarbonate and hydrogen sulfide. The soluble calcium bicarbonate then is sometimes removed from its point of origin and transported elsewhere by ground waters. Later, calcite can be produced from the calcium bicarbonate in the usual manner, by removal of CO_2:

$$CaSO_4 + 8e^- + 8H^+ \rightleftharpoons CaS + 4H_2O \qquad (6.19)$$

$$CaS + 2CO_2 + 2H_2O \rightleftharpoons Ca(HCO_3)_2 + H_2S \qquad (6.20)$$

$$Ca(HCO_3)_2 \rightleftharpoons CaCO_3 + CO_2 + H_2O \qquad (6.21)$$

The stalactites and stalagmites formed in caves are usually secondary calcite deposits produced according to reaction 6.21.

Deposition of elemental sulfur. Volcanic actions result in the production of small amounts of elemental sulfur, but over 90% of the deposits of elemental sulfur are of sedimentary origin. These include both syngenetic and epigenetic types, and bacteria appear to have been instrumental in the formation of both.

Syngenetic sulfur is laid down in a dispersed manner along with the other components of a sediment and is often undetectable by visual inspection, or else it may be present as thin layers. Production of elemental sulfur in some modern lakes possibly is a contemporary model for the syngenetic process (Fig. 6.8). Among these are certain small saline lakes in Libya, such as Lake Ain-ez-Zauni, the bottoms of which are covered with 20–30 cm of an ochre-colored deposit rich in elemental sulfur (about 50% by weight). Even the water itself has a milky opalescence due to the quantities of sulfur being precipitated. The amount of sulfur generated is sufficient to sustain a local "mining" operation recovering 200 tons per year. Apparently the elemental sulfur arises, at least in part, by the oxidation of hydrogen sulfide by a large population of photosynthetic sulfur bacteria such as *Chromatium* and *Chlorobium*. The waters have a high content of hydrogen sulfide, reaching 100 mg/liter at the bottom and averaging about one-fifth this level at the surface. This sulfide has been attributed to action of *Desulfovibrio* in reducing sulfates, also present in these lakes in large amounts. The organic material necessary for the sulfate reducers would be supplied by the photosynthetic sulfur bacteria, thereby achieving a self-sustaining cycle.

Such processes are not always completely dependent on bacterial sulfate reduction as the source of the needed hydrogen sulfide, as a somewhat similar, if less intense, biogenic production of elemental sulfur occurs in various other lakes fed by sulfide springs ("plutonic" sulfide). Photosynthetic bacteria are not the only organisms producing elemental sulfur from sulfides. *Beggiatoa* accumulates sulfur as intracellular granules when growing on organic material in sulfide-rich waters (Section 2.11), and the oxidation of reduced sulfur compounds by *Thiobacillus thioparus* results in the formation of extracellular sulfur.

In Lake Sernoe in the USSR the mineral spring waters entering the shallow lake contain about 85 mg of H_2S per liter, and this is oxidized to elemental sulfur which accumulates in the sediments. Apparently both aerobic and anaerobic oxidation mechanisms are involved in this process: patches of photosynthetic purple and green sulfur bacteria are visible in some of the regions of sulfur deposition, and bacteriological examination of other colorless sulfurous patches reveals large numbers of *Thiobacillus thioparus*. Laboratory experiments involving incubation of samples of these sulfur deposits with $H_2{}^{35}S$

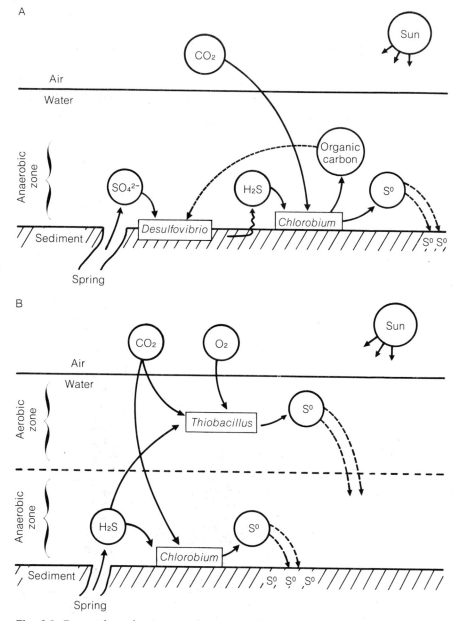

Fig. 6.8. Bacterial mechanisms in deposition of elemental sulfur in lakes. *A,* Fully anaerobic system, with bacterial generation of H₂S from sulfate and photosynthetic oxidation of sulfide to elemental sulfur as in Lake Ain-ez-Zauni (Libya). *B,* Mixed system, with anaerobic photosynthetic and aerobic nonphotosynthetic oxidation of sulfide from mineral springs, as in Lake Sernoe (USSR).

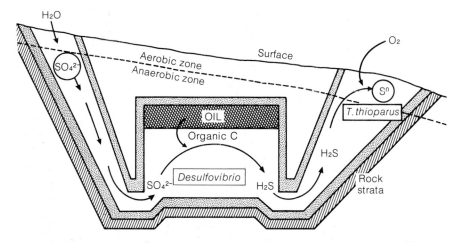

Fig. 6.9. Possible mode of formation of epigenetic sulfur deposits by bacteria. Infiltrating ground waters dissolve sulfates which are reduced to hydrogen sulfide by *Desulfovibrio* in region of oil pocket, and finally sulfide is oxidized to elemental sulfur by *Thiobacillus thioparus* near surface. (From S. I. Kuznetsov, M. V. Ivanov, and N. N. Lyalikova, *Introduction to Geological Microbiology*, McGraw-Hill, New York, 1963. Used with permission of McGraw-Hill.)

confirm the presumption that the bacteria present are actively converting sulfide to elemental sulfur, and point to the aerobic thiobacilli as the major contributors.

Some sulfur deposits are probably epigenetic, the sulfur being laid down after formation of the rocks in which it is found, and it has been suggested that bacterial actions have been responsible to some extent. A possible bacterial mechanism accounting for the formation of epigenetic sulfur is outlined in Fig. 6.9. In this scheme surface water infiltrates into a porous rock layer sandwiched between two impervious layers and dissolves sulfate as it percolates downward through the rock. Hydrostatic pressure carries the water through the aquifer (water bearing rock layer), and as it passes near an oil deposit, the sulfate is reduced to hydrogen sulfide by sulfate-reducing bacteria. When these hydrogen sulfide-charged waters encounter oxidizing conditions nearer the surface, elemental sulfur is formed. An oxidation of sulfide by purely chemical means takes place to some extent, but this is greatly accelerated by thiobacilli, as previously discussed.

Soviet investigators conclude that the sulfur deposits of the Shor-Shu region in the Altai Range (Uzbek SSR) conform to this model. These are near or at the surface adjacent to an oil field, as are most other known epigenetic sulfur occurrences. *T. thioparus* is abundant in the zone of sulfur precipitation, and is estimated to be responsible for at least 40% of the sulfide oxidation taking place, the remainder being the result of chemical oxidation. *Desulfovibrio* has been detected in the deep ground waters in the zone of sulfide formation near the oil deposits at Shor-Shu. It is known that in general the content of dissolved organic matter in ground water increases nearer oil deposits, and this is correlated with greater numbers of sulfate reducers. The obvious conclusion is that the sulfate reducing bacteria are able to utilize this organic matter and multiply in deep subterranean regions.

Bacterial fractionation of isotopes in epigenetic sulfur deposits. Some of the largest secondary sulfur deposits occur in association with massive salt domes in the regions of the Texas and Louisiana oil fields along the Gulf of Mexico. A thorough study of sulfur isotope distribution in these salt domes has shown that this elemental sulfur is of biogenic origin, and has led to the general recognition of the geochemical importance of bacterial isotope fractionation processes.

A diagrammatic cross-sectional representation of a salt dome is provided in Fig. 6.10 for orientation as to the relative positions of the several layers. The core of the salt dome rises from great depths as the immense weight of overlying rock deforms a layer of rock salt, perhaps the evaporated remains of a Jurassic sea from 160 million years ago. The core is over 90% sodium chloride, with the remainder consisting of anhydrous calcium sulfate (anhydrite, $CaSO_4$). Directly above the core is a layer of anhydrite, probably formed by the dissolving away of the sodium chloride by ground water as the dome pushed upward. This is succeeded by an intermediate zone of hydrated calcium sulfate (gypsum, $CaSO_4 \cdot 2H_2O$), which in turn is surmounted by a layer of calcite, clearly of secondary nature. Elemental sulfur is confined to the calcite and gypsum zones, and is not found in the anhydrite layer or core. Petroleum may be contiguous with the outer calcite layer of the cap rock, but it is absent from the core and anhydrite region.

There are four stable isotopes of sulfur, the two most abundant being ^{32}S and ^{34}S. Mass-spectrometric analysis shows a distinct variation in the ratio of these two isotopes in sulfur compounds from different sources. Sulfides of meteorites (e.g., troilite, FeS), presumably formed without reference to biological systems, show a $^{32}S/^{34}S$ ratio

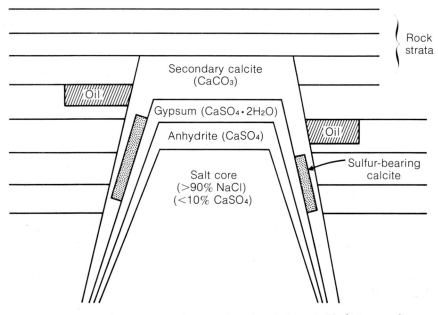

Fig. 6.10. Diagram of salt dome and cap rock as found along Gulf of Mexico, showing relative locations of salt core, cap rock zones, and associated oil fields. (From S. I. Kuznetsov, M. V. Ivanov, and N. N. Lyalikova, *Introduction to Geological Microbiology,* McGraw-Hill, New York, 1963. Used with permission of McGraw-Hill.)

of approximately 22.21. Values for sulfur compounds of volcanic and other igneous origin are similar but do deviate somewhat. In a general way the reduced compounds (sulfide, elemental sulfur) are all slightly depleted in ^{34}S, and the oxidized compounds (sulfate) are somewhat enriched in this isotope. These differences are frequently expressed in terms of the relative change in the heavier isotope, $\delta^{34}S$, in parts per thousand ($^{0}/_{00}$) compared to the isotope ratios in a meteoric standard. A positive value for the $\delta^{34}S$ signifies a greater amount of ^{34}S and a negative value indicates a greater amount of ^{32}S than in the standard. The greatest deviations are noted in sedimentary sulfur compounds, which vary up to 50 $^{0}/_{00}$ on either side of the standard value, apparently as a result of biological actions. Laboratory experiments indicate that the differences noted are brought about by a preferential utilization of ^{32}S in various biological processes, especially oxidation-reduction reactions. The fractionation of sulfur isotopes in some of these reactions is illustrated in Table 6.2. In the reduction of sulfate by *Desulfovibrio* $^{32}SO_4^{2-}$ is preferentially reduced over $^{34}SO_4^{2-}$, result-

Table 6.2.   Fractionation of Sulfur Isotopes in Biological Reactions*

Reaction	Typical Organism	Observed $\delta\ ^{34}S$ in Product†
$SO_4{}^{2-} \longrightarrow H_2S$	*Desulfovibrio*	-10 to -46
$SO_3{}^{2-} \longrightarrow H_2S$	*Saccharomyces*	-34 to -40
$H_2S \longrightarrow SO_4{}^{2-}$	*Thiobacillus* sp.	-10 to -18
$H_2S \longrightarrow S^0$	*Chromatium* sp.	-3 to -10
$H_2S \longrightarrow S_4O_6{}^{2-}$	*Thiobacillus* sp.	$+3.2$ to $+19$
$S^0 \longrightarrow SO_4{}^{2-}$	*Thiobacillus* sp.	-0.1 to $+1.4$
Cysteine $\longrightarrow H_2S$	*Proteus* sp.	-4 to -5
Sulfate assimilation	*E. coli*	-1 to -2.8

*Data from I. R. Kaplan and S. C. Rittenberg, Microbiological Fractionation of Sulphur Isotopes, *J. Gen. Microbiol.* **34**, 195 (1964).

$$\dagger\delta\ ^{34}S\ ^0/_{00} = \frac{(^{34}S/^{32}S)\ \text{sample} - (^{34}S/^{32}S)\ \text{starting material}}{(^{34}S/^{32}S)\ \text{starting material}} \times 1000$$

ing in sulfide enriched in ^{32}S (i.e., depleted in ^{34}S) compared to the starting material. The $\delta\ ^{34}S$ values for this process are inversely related to the rate of reduction, and at slow rates in the presence of a large reservoir of sulfate may reach -46 $^0/_{00}$. Other reactions showing significant isotope fractionation include the reduction of sulfite to hydrogen sulfide by yeast, the oxidation of sulfide to sulfate by *Thiobacillus,* the oxidation of sulfide to elemental sulfur by *Chromatium,* and the formation of hydrogen sulfide from cysteine by *Proteus.* Very little fractionation of the sulfur isotopes occurs in the assimilation of sulfate by bacteria and algae, and essentially none in the oxidation of elemental sulfur to sulfate either by thiobacilli or photosynthetic bacteria. The positive values observed in the polythionates produced in sulfide oxidation present an interesting anomaly, probably indicating nonenzymatic reactions or secondary chemical changes. Equilibrium chemical oxidation of sulfide to sulfate will give strongly positive $\delta^{34}S$ values.

A survey of the ratios of sulfur isotopes in various portions of salt domes reveals distinct fluctuations between the layers as illustrated in Figure 6.11. The sulfates of the salt core and the anhydrite layer show $^{32}S/^{34}S$ ratios around 21.8–21.9, approximately the same as for the sulfate of present day sea water. In the calcite zone, samples of the sulfates contain much more ^{34}S, shown by decreased $^{32}S/^{34}S$ values ranging as low as 20.85, whereas the elemental sulfur is correspondingly enriched in the lighter isotope, giving a $^{32}S/^{34}S$ ratio around 22.19. This is explained on the basis of a bacterial reduction of calcium sulfate yielding sulfide with an increased ^{32}S content. A subsequent chemical oxidation of this sulfide to elemental sulfur would preserve the isotopic ratio. Such a preferential reduction of the lighter isotope

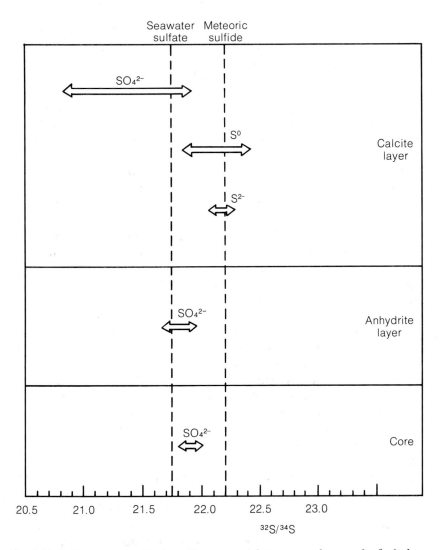

Fig. 6.11. Sulfur isotope ratios in sulfur compounds in core and cap rock of salt domes. The enrichment of [34]S in the sulfide and elemental sulfur, with corresponding depletion of [34]S in sulfate of the cap rock indicates bacterial processes were responsible. (Based on data of H. W. Feely and J. L. Kulp, Origin of Gulf Coast Salt Dome Sulfur Deposits, *Bull. Amer. Assoc. Geol.* **41,** 1802 (1957).)

would leave the residual sulfate richer in ^{34}S. The organic carbon needed by the sulfate reducing bacteria would be provided by the petroleum associated with the salt dome.

If this is a correct interpretation of the process, then the calcite must be secondary also. Obviously the calcium ion of calcite is derivable from the gypsum being reduced, but the source of the carbonate group presents a problem. The only available source of carbon is the petroleum. A connection between these two is suggested by the similarity of stable carbon isotope ratios in each.

A biological fractionation of stable carbon isotopes similar to that for sulfur occurs during photosynthetic carbon dioxide assimilation by plants or bacteria caused by a relatively greater incorporation of ^{12}C. This yields organic material richer in ^{12}C and therefore lower in ^{13}C content by approximately $-21^0/oo$ compared to the starting material. The $^{12}C/^{13}C$ ratios observed in modern plants is around 90.8–91.2, compared to 89.27 for atmospheric CO_2. This would serve to explain the variations found between different geological materials; presumably, biogenic materials such as coal and oil would be expected to contain more ^{12}C, and in fact they exhibit $^{12}C/^{13}C$ ratios in the neighborhood of 91.00, compared to 89.11 for magmatic carbon (diamonds, graphite). Selective biological utilization of ^{12}C in photosynthesis should also lead to a relative enrichment of ^{13}C in the primary calcite formed concurrently (see Equation 6.18). Marine limestones have $^{12}C/^{13}C$ ratios around 88.55–88.9. However, samples of the calcite of the salt domes show $^{12}C/^{13}C$ ratios similar to those of biogenic materials, including the petroleum from these salt domes, and is depleted in ^{13}C by -35 to $-55^0/oo$ compared to a biogenic marine calcite.

The isotopic approach has proven useful also in analyzing modern sulfur depositing systems. The elemental sulfur produced biogenically in Lake Ain-ez-Zauni, described previously, is found to be depleted in ^{34}S by $-32^0/oo$ compared to the sulfate in the same lake. This is consistent with the action of the sulfate reducing bacteria postulated to be involved. In the acid thermal springs (90°C) of Yellowstone Park, Wyoming, both sulfate and elemental sulfur are produced by the oxidation of plutonic hydrogen sulfide, and the question arises as to whether chemical or biological oxidations are responsible. In most locations it is found that the elemental sulfur and the sulfate do not differ significantly from the hydrogen sulfide with respect to $^{32}S/^{34}S$ ratio. A purely chemical equilibrium oxidation of sulfide to sulfate should give an increase in ^{34}S, and a bacterial oxidation of sulfide to sulfur should lead to an enrichment in ^{32}S. The observed lack of isotope fractionation is interpreted as an indication of a chemical oxidation of sulfide to elemental sulfur, followed by bacterial oxidation of the sulfur to sulfate, the latter reaction not involving a shift in isotopic ratios (see Table 6.2). In one location, Mammoth Hot Springs, where the temperature is less (approximately 70°C), the sulfur and sulfate did show some increase in ^{32}S (δ^{34}S -3 to -5) and this could be indicative of a bacterial oxidation of sulfide.

Formation of iron and manganese minerals. The encrustations of iron and manganese oxides which develop on certain bacteria (*Leptothrix, Gallionella*) suggest a possible biogenic origin for at least some deposits of these metals. This concept is strengthened by the finding of structures interpretable as fossilized bacterial filaments in iron oxide ores. A connection between bacteria and generation of iron and manganese minerals, while circumstantial, seems reasonable, as it did to Winogradsky and Beijerinck, and warrants more investigation. Two examples of types of possible bacterial involvement in production of these minerals are illustrated in Fig. 6.12. Sulfide minerals such as pyrite (FeS_2) could arise via the action of sulfate-reducing bacteria, whereas oxidations of filamentous "iron bacteria" could lead to the oxides, such as goethite ($Fe_2O_3 \cdot H_2O$) and hematite (Fe_2O_3). Manganese is chemically similar to iron in many respects and is often found in association with iron minerals, but differs in that it is seldom found in the form of sulfides. The principal mineralogical occurrences of manganese are as its oxides and carbonates, including pyrolusite (MnO_2) and rhodochrosite ($MnCO_3$).

Anaerobic conditions and the presence of organic residues in sediments favor development of *Desulfovibrio* and lead to formation of hydrogen sulfide from sulfates. A chemical reaction then takes place between H_2S and iron salts, yielding a black insoluble precipitate of amorphous iron sulfide called hydrotroilite ($FeS \cdot H_2O$), the occurrence of which accounts for the intense black color of many marine and fresh water muds. Further geochemical changes not involving bacteria take place as the sediments become more deeply buried and are responsible for transforming hydrotroilite into the more stable iron disulfides, pyrite and marcasite. It is not known whether other naturally occurring metallic sulfide minerals are produced biogenically, but it remains a distinct possibility. Laboratory experiments

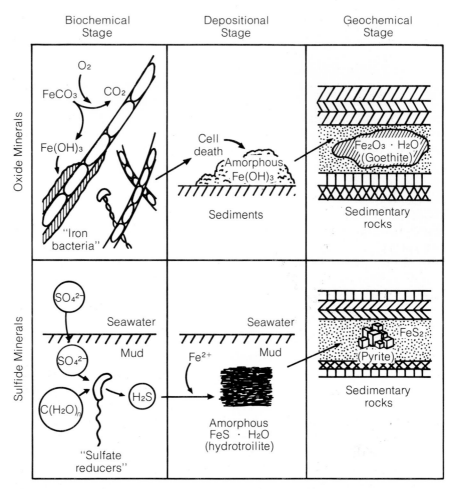

Fig. 6.12. Bacterial processes postulated as leading to formation of iron mineral deposits. Iron oxide minerals may arise from alteration of the ferric hydroxide precipitated by *Leptothrix, Gallionella,* and other iron-oxidizing bacteria. Sulfide minerals may result from reaction of iron with H_2S generated by *Desulfovibrio* in marine sediments.

with *Desulfovibrio* in artificial sea water have demonstrated the formation of a number of sulfide minerals, such as covellite (CuS) from chrysocolla ($CuSiO_3 \cdot 2H_2O$) and galena (PbS) from $PbCO_3$. The major impediment to the natural operation of such processes on a geologically significant scale is the requirement for sufficient quantities of the metal salts. Heavy metal compounds, except for iron, are generally present in very small amounts in natural waters.

Various observations of the precipitation of $MnCO_3$ and $FeCO_3$ in laboratory cultures of anaerobic bacteria, including *Desulfovibrio*, have been reported, but definitive evidence is lacking concerning bacterial intervention in the formation of geological deposits of rhodochrosite ($MnCO_3$) or siderite ($FeCO_3$). It is suggestive that siderite often is associated with shallow water sediments and sedimentary rocks derived from them, along with pyrite and organic matter in the form of coal. Siderite and rhodochrosite are stable only under reducing conditions, and are subject to bacterial and chemical oxidations in aerobic regions.

Iron-rich waters with neutral to alkaline pH values usually develop a characteristic flora with large numbers of the ribbon-forming *Gallionella* and filamentous sheathed bacteria such as *Leptothrix* (Sections 3.3 and 3.7). These organisms catalyze the oxidation of ferrous to ferric iron, and this becomes deposited on the sheath or ribbon as the hydroxide. Siderite included in cultures of *Leptothrix ochracea* is converted to amorphous brownish ferric hydroxide which encrusts the filaments:

$$4FeCO_3 + O_2 + 6H_2O \longrightarrow 4Fe(OH)_3 \downarrow + 4CO_2 \qquad (6.22)$$
siderite

The brownish gelatinous flocs and coatings developing in ferruginous waters consist of large masses of iron-impregnated sheaths, ribbons, and slime produced by these organisms. Some of these so-called "iron bacteria" deposit both iron and manganese hydroxides. The filamentous and ribbon-forming bacteria are not alone in the ability to oxidize ferrous and manganous compounds. A variety of other slightly studied unicellular forms, some not yet cultivated in the laboratory, have been observed in iron-bearing waters. These include rods and coccoid forms, many with large capsules which become impregnated with iron. Among these bacteria are *Siderocapsa, Sideromonas, Ferribacterium, Siderococcus, Naumanniella,* and *Metallogenium*. After the death of the cells, the iron hydroxide residue may become buried, and later converted chemically to typical iron oxide minerals, goethite and hematite.

Bacterial oxidations appear to be involved in the contemporary formation of nodules of iron ore found in lakes and bogs. These are concretions of iron oxides, ranging up to 10–15 cm in diameter, occurring in the shallow littoral zone in depths less than 5 m of lakes and swamps in the taiga-podzolic soil regions. Iron occurring in the upper horizons of acidic soils of the podzolic type tends to become reduced as an indirect consequence of microbial utilization of oxygen in respiration. Ferrous salts are more soluble and are leached away. Leaching is also promoted by the binding of iron by humic substances to form soluble complexes. The ferrous iron may be simply precipitated again as the downward percolation carries it into more alkaline lower layers, but if lateral movements along slopes transport it into the streams feeding ponds and peat bogs, it becomes subjected to bacterial oxidations to produce lake or bog iron ore.

6.12 | Bacterial Attack on Minerals

Bacterial actions play an important part in the solubilization and degradation of minerals, as well as in their formation. Three mechanisms appear to be involved in these processes: acidification, chelation, and oxidation-reduction reactions. Acidifications may result either from the formation of an acidic metabolite, or from a preferential utilization of alkaline substrates. An example of the latter situation is seen when bacteria assimilate ammonium ions, leaving the acidic anion (sulfate, chloride) in solution. Organic acids, such as lactic, acetic, butyric, succinic, and gluconic, are produced as end products of carbohydrate metabolism by various bacteria, and in addition, certain organisms form inorganic acids, such as nitrous and nitric (nitrifying bacteria), and sulfuric (thiobacilli).

A number of organic compounds, including certain bacterial metabolites, have the ability to act as chelating agents, that is to bind strongly to metal ions. This property may be only incidental in the case of many of these substances, but some compounds may function as metal transporting factors in cellular metabolism. Among the chelating compounds formed by bacteria are amino acids (glycine, histidine, glutamic acid), phenols (pyrocatechol, salicylic acid), phosphates (adenosinetriphosphate), organic acids (citric, 2-ketogluconic), and polyols (mannitol). Many of these compounds show strong affinity for geologically (and metabolically) important divalent metals, such as iron, copper, zinc, nickel, manganese, calcium, and magnesium.

An example of the chelation process is the reaction of iron and pyrocatechol to form the following complex:

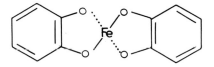

Humic substances with structures related to pyrocatechol accomplish much of the solubilization and transport of iron in soils mentioned in connection with bog iron formation.

A number of bacteria are known to form and release into the environment phenolic acids which chelate iron in a similar fashion, and such compounds may function as mobilizers of ferric iron for use by the organism. Among these are 2,3-dihydroxybenzoic acid, 2,3-dihydroxybenzoyl glycine, 2,3-dihydroxybenzoyl serine, salicylic acid, and 6-methylsalicylic acid. The importance of such compounds for the organism is shown by the fact that a mutant of *E. coli* unable to synthesize 2,3-dihydroxybenzoyl serine requires this compound or citrate for growth in media of low iron content. Recently, another compound, named enterochelin, which is a cyclic trimer of 2,3-dihydroxybenzoyl serine (Fig. 6.13), has been found as an excretion product of *E. coli* and is presumed to be a major factor for iron transport in the coliform bacteria. In other bacteria the transport into the cell may be accomplished by the so-called sideramine group of compounds. These are complex hydroxamic acid derivatives with a high affinity for iron, first recognized as growth factors for certain deficient organisms, but clearly of more general occurrence and significance. Among these are the so-called "terregens factor" (a growth factor for *Arthrobacter terregens*), coprogen (a growth factor for the coprophilic fungus, *Philobolus,* and also produced by the bacterium *Sarcina lutea*), the mycobactins (growth factors for *Mycobacterium paratuberculosis*), and the various ferrioxamines produced by *Streptomyces* species. The structure of one of these, ferrioxamine B, is shown in Fig. 6.13.

There is considerable evidence suggesting that bacteria take part in the "weathering" of rocks and in alterations of minerals in the soil. Bacteria are present on rocks undergoing weathering and are especially abundant on those colonized by lichens. Cultures of bacteria and streptomycetes isolated from such weathered rocks have proven capable of dissolving 50–80% of added silicate rocks. *Pseudomonas* cultures destroy the crystalline structure of silicate rocks such as

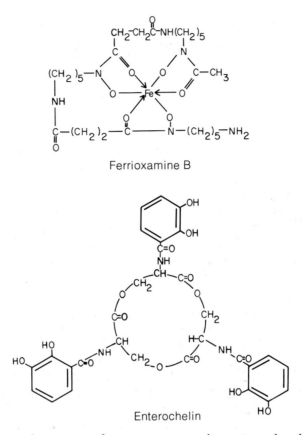

Fig. 6.13. Chemical structures of iron-transporting substances produced by bacteria. The cyclic dihydroxybenzoyl serine trimer, enterochelin, is formed by *Escherichia coli* and other enteric bacteria. The iron is bound as a hexadentate ligand through the phenolic hydroxyls, but is not shown here. Ferrioxamine B and similar substances are produced by *Streptomyces* species.

olivine, $(Mg, Fe)_2SiO_4$, or wollastonite, $Ca(SiO_3)$, possibly by the action of the 2-ketogluconic acid produced from glucose. Various bacteria reportedly improve soil fertility by releasing potassium from aluminosilicate complexes, and some attempts have been made to use deliberate inoculations with such "silicate bacteria" for soil improvement. Phosphate occurs mostly in silicate rocks as the complex fluorophosphate, apatite, $3Ca_3(PO_4)_2$, CaF_2; crushed phosphate rock is added to the soil as a phosphorus source in agricultural practice Inclusion of elemental sulfur along with the phosphate rock is found to result in more biologically available phosphorus. The explanation for this is that *Thiobacillus* species oxidize the sulfur to sulfuric acid, which in turn liberates the phosphate from the apatite.

Thiobacilli and leaching of metallic ores. As previously indicated, species of *Thiobacillus* oxidize hydrogen sulfide and elemental sulfur to sulfuric acid. These oxidations account for the acid production in sulfurous mineral waters, and the acid corrosion of concrete sewer pipes and other concrete structures exposed to sulfide fumes. Elemental sulfur occurring in exposed rock formations is also attacked *in situ* by thiobacilli, as revealed by acid production and the recovery of living organisms from these rocks.

One group of these bacteria is able to use as energy substrates not only sulfur compounds but also ferrous ions (see also Section 5.9). These organisms, exemplified by *T. ferrooxidans*, obtain useful energy by oxidation of ferrous iron to the ferric state:

$$\text{FeSO}_4 + \tfrac{1}{2}\text{O}_2 + \text{H}_2\text{SO}_4 \xrightarrow{\textit{T. ferrooxidans}} \text{Fe}_2(\text{SO}_4)_3 + \text{H}_2\text{O} \qquad (6.23)$$

ferrous   ferric
sulfate   sulfate

These organisms differ from the filamentous iron-oxidizers mentioned in Section 6.10 in catalyzing the oxidation of iron under acid conditions which retard chemical oxidation.

Some iron minerals such as pyrite and marcasite also are amenable to oxidation by these bacteria. The attack on pyrite proceeds with the utilization of both the iron and sulfur moieties, and begins by conversion of the iron sulfide to ferrous sulfate and sulfuric acid:

$$\text{FeS}_2 + 3\tfrac{1}{2}\text{O}_2 + \text{H}_2\text{O} \xrightarrow{\textit{T. ferrooxidans}} \text{FeSO}_4 + \text{H}_2\text{SO}_4 \qquad (6.24)$$

pyrite

The ferrous sulfate then can be further oxidized as shown in equation (6.23). The ferric sulfate produced is a powerful chemical oxidant and will oxidize more pyrite to the sulfate stage, while itself being reduced to ferrous sulfate. This allows the cyclic process shown in Fig. 6.14, in which *T. ferrooxidans* generates the ferric sulfate used for the chemical oxidation of sulfide mineral, and the latter reaction regenerates the ferrous sulfate needed by the bacterium. Pyrite is a common constituent in the unwanted rock residues or "tailings" from coal mines, and bacterial oxidation of such materials is the source

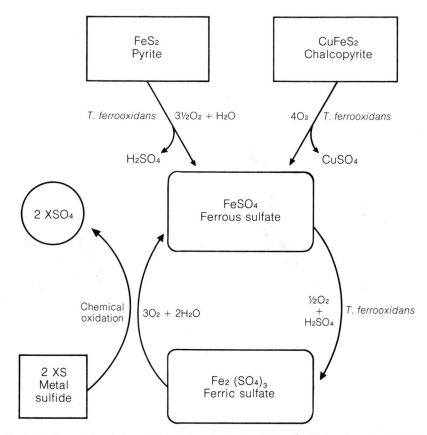

Fig. 6.14. Bacterial and chemical reactions in the leaching of sulfide minerals. Oxidation of sulfide ores by *Thiobacillus ferrooxidans* produces ferric sulfate, which acts as a chemical oxidant for oxidizing and solubilizing metallic sulfides not amenable to direct bacterial oxidation.

of acid mine drainage waters which often pollute streams in coal mining districts.

Several other metallic sulfides are susceptible to oxidation by these bacteria: chalcopyrite ($CuFeS_2$), bornite (Cu_5FeS_4), covellite (CuS), sphalerite (ZnS), and molybdenite (MoS_2). When the ore contains both iron and sulfur, as in chalcopyrite, each of these components will be oxidized, and the direct bacterial attack will be supplemented by the chemical oxidation by ferric sulfate. When the sulfide mineral contains a metal not oxidized by *T. ferrooxidans*, and no iron, only the sulfur portion will be oxidized, and solubilization is accomplished by the sulfuric acid formed. This type of reaction takes place with covellite, sphalerite, and molybdenite. In many ore-bearing rocks a variety of sulfide minerals including pyrite occur, and this permits leaching of materials not directly attacked by the bacteria, such as galena (PbS) and uranite (UO_2).

It has proven economically desirable to take advantage of the bacterial leaching process to extract valuable metals such as copper and uranium from low grade ore. A heap of the ore-bearing rock will develop a flora of iron-oxidizing thiobacilli and their metabolic reactions result in a leaching out of the desired metal. Water is sprayed over the pile, and carries the bacterial oxidation products into a catchment basin dug at the base of the rock pile. In the case of copper leaching operations the copper is then recovered from the copper sulfate accumulating in the catchment pond by an exchange reaction using metallic iron. The major reactions involved in the leaching of a copper ore such as chalcopyrite are:

$$CuFeS_2 + 4O_2 \xrightarrow{T.\ ferrooxidans} CuSO_4 + FeSO_4 \qquad (6.25)$$
chalcopyrite

$$FeSO_4 + \tfrac{1}{2}O_2 + H_2SO_4 \xrightarrow{T.\ ferrooxidans} Fe_2(SO_4)_3 + H_2O \qquad (6.26)$$

$$CuFeS_2 + Fe_2(SO_4)_3 + H_2O + 3O_2 \xrightarrow{non\text{-}biological} CuSO_4 + 5FeSO_4$$
$$+ 2H_2SO_4 \qquad (6.27)$$

$$CuSO_4 + Fe^0 \xrightarrow{non\text{-}biological} FeSO_4 + Cu^0 \downarrow \qquad (6.28)$$

References

Abelson, P. H., Chemical Events on the Primitive Earth, *Proc. Nat. Acad. Sci. U. S. A.* **55,** 1365 (1966).

Abelson, P. H. (ed.), *Researches in Geochemistry,* Wiley, New York, 1959.

Alexander, M., Biochemical Ecology of Soil Microorganisms, *Ann. Rev. Microbiol.* **18,** 217 (1964).

Davis, J. B., *Petroleum Microbiology,* Elsevier, Amsterdam, 1967.

Degens, E. T., *Geochemistry of Sediments,* Prentice-Hall, Englewood Cliffs, N. J., 1965.

Eglinton, G., and Sr. M. T. J. Murphy (eds.), *Organic Geochemistry*, Springer-Verlag, New York, 1969.

Feely, H. W., and J. L. Kulp, Origin of Gulf Coast Salt Dome Sulfur Deposits, *Bull. Amer. Assoc. Petrol. Geol.* **41**, 1802 (1957).

Gascoigne, J. A., and M. M. Gascoigne, *Biological Degradation of Cellulose*, Butterworths, London, 1960.

Greenwood, C. T., and E. A. Milne, Starch Degrading and Synthesizing Enzymes: A Discussion of Their Properties and Action Pattern, *Adv. Carbohyd. Chem*, **23**, 281 (1968).

Hardy, R. W. F., and R. C. Burns, Biological Nitrogen Fixation, *Ann, Rev. Biochem.* **37**, 331 (1968).

Ivanov, M. V., *Microbiological Processes in the Formation of Sulfur Deposits,* U. S. Department of Agriculture and National Science Foundation, Washington, D.C., 1968.

Jones, G. E., and R. L. Starkey, Fractionation of Stable Isotopes of Sulfur by Microorganisms and Their Role in Deposition of Native Sulfur, *Appl. Microbiol.* **5**, 111 (1957).

Jurášek, L., J. R. Colvin, and D. R. Whitaker, Microbiological Aspects of the Formation and Degradation of Cellulosic Fibers, *Adv. Appl. Microbiol.* **9**, 131 (1967).

Kaplan, I. R., and S. C. Rittenberg, Microbiological Fractionation of Sulfur Isotopes, *J. Gen. Microbiol.* **34**, 195 (1964).

Kemp, A. L. W., and H. G. Thode, The Mechanism of the Bacterial Reduction of Sulphate and of Sulphite from Isotope Fractionation Studies, *Geochim. Cosmochim. Acta,* **32,** 71 (1968).

Kuznetsov, S. I., M. V. Ivanov, and N. N. Lyalikova, *Introduction to Geological Microbiology*, McGraw-Hill, New York, 1963.

Mason, B., *Principles of Geochemistry*, Wiley, New York, 1966.

Mehta, N. C., P. Dubach, and H. Deuel, Carbohydrates in the Soil, *Adv. Carbohyd. Chem.* **16,** 335 (1961).

Norkrans, B., Cellulose and Cellulolysis, *Adv. Appl. Microbiol.* **9**, 91 (1967).

Oglesby, R. T., R. F. Christman, and C. H. Driver, The Biotransformation of Lignin to Humus—Facts and Postulates, *Adv. Appl. Microbiol.* **9**, 171 (1967).

Postgate, J. R., The Sulphur Cycle, in G. Nickless (ed.), *Inorganic Sulphur Chemistry*, Elsevier, Amsterdam, 1968.

Rogers, H. J., The Dissimilation of High Molecular Weight Substances in I. C. Gunsalus and R. Y. Stanier (eds.), *The Bacteria* Vol. II., *Metabolism,* Academic Press, New York, 1961.

Schoen, R., and R. O. Rye, Sulfur Isotope Distribution in Solfataras, Yellowstone National Park, *Science* **170**, 1082 (1970).

Siegel, S. M., Biochemistry of the Plant Cell Wall, in M. Florkin and E.

H. Stotz (eds.), *Comprehensive Biochemistry* **26**, Part A, Elsevier, Amsterdam, 1968.

Silverman, M. P., and H. L. Ehrlich, Microbial Formation and Degradation of Minerals, *Adv. Appl. Microbiol.* **6**, 153 (1964).

Siu, R. G. H., *Microbial Decomposition of Cellulose, with Special Reference to Cotton Textiles,* Reinhold Publishing Co., New York, 1951.

Swain, F. M., *Non-Marine Organic Geochemistry,* Cambridge University Press, Cambridge, England, 1970.

Van der Linden, A. C., and G. J. E. Thijsse, The Mechanism of Microbial Oxidations of Petroleum Hydrocarbons, *Adv. Enzymol.* **27**, 469 (1965).

Zajic, J. E., *Microbial Biogeochemistry,* Academic Press, New York, 1969.

seven | Epilogue

"Today, after a half century, there is no longer room for doubt that the fitness of organic beings for their life in the world has been won in whole or in part by an almost infinite series of adaptations of life to its environment . . ."

L. J. Henderson (1913)

The emergence of bacteriology as a distinct branch of biology resulted from events occurring in the last quarter of the nineteenth century in western European laboratories. The work of Pasteur and Koch consisted mainly in developing techniques that resulted in, among other things, the implication of bacteria as "causative agents" of specific diseases. This concept was ultimately put into terms called "Koch's postulates" namely: (a) that the bacterium be present in all cases of the disease; (b) that this organism be obtained in the laboratory as a pure culture; (c) that inoculation of this culture into a susceptible host result in the same disease; and (d) that the bacterium again be re-isolated from this host. Stated in this way, the "germ theory" assumes an absolute character that it does not, in fact, have. No mention is made, for example, of the apparently healthy host found harboring disease organisms, or of the modifying factors, such as host resistance and general physiological conditions, that mediate the course of a given disease, or of the fact that the inoculation of a pure culture into the blood stream or peritoneal cavity is not exactly the same as a natural infection. It must be stressed that the property of virulence (Section 4.18) is ecological, and that the appearance of a specific disease in a specific host depends upon a multitude of defined and undefined external and internal factors in addition to the presence of the bacterium itself.

The emergence of techniques for isolating and handling pure cultures of potentially pathogenic bacteria, and the empirical methods leading to prophylactic immunization, were intellectual triumphs of the first magnitude. To a lesser degree, the roles of bacteria and other microorganisms in fermentations, in soil fertility, in diseases of plants, and in the mineralization of organic matter, were phenomena that, although only sketched out in rough form, constituted the bulk of information subsumed under the name "bacteriology." In a word, nineteenth century bacteriology developed operational techniques that had value particularly in the solution of important practical problems of human and animal health, and in economically useful processes. An outstanding feature of the early history of bacteriology is the impression, derived from such work, of "one kind of organism—one process." The idea of a causative organism or etiological agent long has dominated the thinking of twentieth century bacteriologists, and modern textbooks on the subject continue to promulgate this doctrine.

In nature, bacteria are intimately related to all animate and inanimate components of their environment. These interrelationships, rather than dissection and analysis of inanimate parts of organisms, constitute the natural history of bacteria, when they are accorded recognition as entities in their own right. With this in mind, it becomes obvious that new approaches to the study of bacteria are required, and those of molecular biology will complement, but not substitute, for them.

Most bacteriological endeavors have developed the science vertically rather than horizontally, but, as stated above, pure culture studies on *Escherichia coli* and biochemical investigations of subcellular components do not yield results equivalent to an understanding of general bacteriology. Furthermore, the historical development of bacteriology was such that early in its course much more was learned about what bacteria did than about their relationships to other life forms. Once devised, techniques for dealing with bacteria have remained almost static, and petri dishes, agar and gelatin media, fermentation tests, and staining reagents are as much a familiar part of late twentieth century bacteriology laboratories as they were one hundred years earlier. The assumptions and errors, carried knowingly or unknowingly, in bacteriological methodology have been transmitted from generation to generation of students. We have written about the indeterminacy of knowledge (Section 1.2) concerning certain facets of individual bacterial activities, and the looseness of definitions of bacterial growth and death. The fact that much of what we know about bacteria is derived only from laboratory methods imposed on them has not bothered bacteriologists too much. Perhaps it is a case of being content with "Nature as we imagine her to be," and not "Nature as she is." Since the methods employed are pragmatic, those interested in "practical results" have not concerned themselves with this possibility. However, physicists have long recognized that the characteristics of their inanimate systems may be subtly altered by perturbations introduced by the probes used to investigate them.

The natural history of the early nineteenth century ultimately fragmented into the separate (and separated) disciplines of botany, zoology, paleontology, and so forth; and it is evident that bacteriology has ramified into and fused with other subdisciplines to form the areas of bacterial cytology, physiology, genetics, taxonomy, and various specialized medical and applied aspects. The amount of information accumulated in each of these is so enormous and so rapidly proliferating that specialists in one area cannot communicate easily (if at all!) with those in another area. A bacteriologist is now no longer one who studies bacteria as complete and intact living organisms, but rather one who specializes in certain features of certain bacteria, usually to the exclusion of all others.

While bacteriological investigations led to the specialisms and specialists mentioned above, a small minority of investigators recognized that the roles of bacteria in nature must be studied in ways other than those giving rise to more of the same kind of "facts." Impetus has been given to this pursuit by the belated recognition that many of the pressing problems of the last quarter of the twentieth century are biological problems involving bacteria. Environmental pollution looms large in this regard, and the essential activities of bacteria in streams, rivers, lakes, and oceans in the mineralization of organic pollutants cannot be overestimated. Furthermore, the formations of obnoxious byproducts, such as acid mine waters and hydrogen sulfide, are aspects of the same problem, namely, that these bacterial processes are judged useful or harmful insofar as they affect man in some way. To the extent that the activities are in his interest, man will encourage them, and if they are not, he will try to prevent them.It is our contention that to do either on a global scale requires information about the natural history of bacteria that has yet to be obtained, by methods yet to be developed.

In a strictly localized situation, such as a food processing plant, control of undesirable bacterial activities may be achieved by the use of bactericidal agents, but in an unbounded situation, such as the ocean, this is impossible. In addition, a detailed knowledge of the biological interrelations which bring about undesirable changes is not really necessary in a local frame, since in theory all activities may be halted. Spoilage problems in a food processing plant are controlled by applying standardized procedures of sanitization and sterilization which lead to the elimination of *all* bacterial processes. However, undesirable bacterial processes occurring in a large body of water could hardly be managed in this way, and the strategy would be to prevent conditions which permit the initiation of these processes in the first place. This, of course, presumes that conditions are definable and preventable—assumptions that are not always accurate. In some cases, once conditions leading to undesirable changes occur, they cannot be readily eliminated; pollution of arctic regions with sewage would be one such example. There are no operational solutions to problems of this kind except the prevention of the initial pollution, since once it occurs, bacterial control activities, being minimal, are of little value.

Even more urgently in need of solution are bacteriological problems involved in degrading the organic matter of sewage in the form of human body excretions and garbage in the form of vegetable and animal materials, since dumping at sea, incineration with consequent air pollution, and simply piling up such waste in heaps are dubious and limited methods. It is a matter of high priority that the bacterial activities involved be fully understood, and that the possibilities of accelerating bacterial processes be studied. Not only are existing methods too slow to cope with the mounting increase in the rate of sewage production, but also new and presently "nonbiodegradable" materials are appearing which must be processed in a different way.

A complete description of the naturalistic behavior of bacteria is the aim of ecobiological studies, since it seems clear that environmental conditions determine what bacterial types are best suited for survival and proliferation. Unfortunately, the role of any particular kind of bacterium must be inferred in an environment undergoing rapid transformation by its resident organisms, but it would be most advantageous to know what the conditions were prior to the appearance of the bacterium whose role is to be specified.

This, then, is the crucial arena for biological endeavors, and a serious and forceful attempt must be made to fill the embarrassing lacunae of our knowledge in these many and difficult bacteriological problems. In retrospect, these considerations on the natural history of bacteria suggest the words used by Bates in his writings on the nature of the subject when he said, "I think natural history . . . represents a growing point of science, an area in which ignorance is more impressive than our knowledge, yet an area in which our knowledge is hopefully increasing."

References

Bates, M., *The Nature of Natural History,* Rev. ed., Scribner's Sons, New York, 1961.

Gaden, E. L. Jr., *Global Impacts of Applied Microbiology II,* Interscience, New York, 1969.

Stanier, R. Y., Adaptation, Evolutionary and Physiological: or Darwinism Among the Microorganisms, in R. Davis and E. F. Gale (eds.), *Adaptation in Micro-Organisms,* 3rd Symp. Soc. Gen. Microbiol., Cambridge University Press., Cambridge, England, 1953.

van Niel, C. B., Natural Selection in the Microbial World, *J. Gen. Microbiol.* **13,** 201 (1955).

Index